DATE DUE

Statistics for the Environment 2

Statistics for the Environment 2

Water Related Issues

Proceedings of SPRUCE II: the second SPRUCE Conference held in Rothamsted Experimental Station, Harpenden, UK, 13–16 September 1993.

SPRUCE is an International initiative concerned with *Statistics in Public Resources, Utilities and in Care of the Environment.*

Sponsored by:

COMMISSION
OF THE
EUROPEAN COMMUNITIES

DIRECTORATE-GENERAL
FOR SCIENCE, RESEARCH
AND DEVELOPMENT
DG12

Statistics for The Environment 2

Water Related Issues

Edited by

Vic Barnett
Rothamsted Experimental Station, UK

and

K. Feridun Turkman
Centre of Statistics, University of Lisbon, Portugal

JOHN WILEY & SONS
Chichester · New York · Brisbane · Toronto · Singapore

Other Wiley Editorial Offices

John Wiley & Sons, Inc., 605 third Avenue,
New York, NY 10158-0012, USA

Jacaranda Wiley Ltd, 33 Park Road, Milton,
Queensland 4064, Australia

John Wiley & Sons (Canada) Ltd, 22 Worcester Road,
Rexdale, Ontario M9W 1L1, Canada

John Wiley & Sons (SEA) Pte Ltd, 37 Jalan Pemimpin #05-04,
Block B, Union Industrial Building, Singapore 2057

Library of Congress Cataloging-in-Publication Data

Statistics for the environment : water-related issues / edited by Vic
Barnett and K. Feridun Turkman.
        p.   cm.
    "Proceedings of SPRUCE II, the second SPRUCE Conference held in
Rothamsted Experimental Station, Harpenden, UK, 13–16 September
1993"—Prelim.
    "Sponsored by Commission of the European Communities"—Prelim.
    Includes bibliographical references and index.
    ISBN 0 471 95048 —3
    1. Environmental sciences—Statistical methods—Congresses.
2. Hydrology—Statistical methods—Congresses.   3. Rain and
rainfall—Statistical methods—Congresses.   4. Sea level–
–Statistical methods—Congresses.   5. Water—Pollution—Statistical
methods—Congresses.   I. Barnett, Vic.   II. Turkman, K. Feridun.
III. Commission of the European Communities.   IV. SPRUCE Conference
(2nd : 1993 : Rothamsted Experimental Station)
GE45.S73S833   1994
553.7'072—dc20   94-10082
                                                              CIP

British Library Cataloguing in Publication Data

A catalogue record for this book is available from the British Library

ISBN 0471 95048 3

Typeset in 10/12pt Photina by Thomson Press (India) Ltd
Printed and bound in Great Britain by Biddles Ltd, Guildford and Kings Lynn

To Audrey

Contents

PART III WATER QUALITY, SUPPLY AND MANAGEMENT

Preface

Environmental matters will continue to feature prominently in the attention of the scientists, sociologists and politicians of all countries over the immediate future. Water is fundamental to every living being, and as such is at the forefront of environmental concerns. This is manifest in the problems of pollution, nourishment, management, conservation, flood control, climate, energy supply, and so on.

It is natural therefore that the SPRUCE initiative, concerned as it is with statistics in Public Resources, Utilities and in Care of the Environment, should have organised its second international conference on the specific theme of statistical models and methods applied to environmental problems concerned with water. The multidisciplinary conference, held in Rothamsted Experimental Station in the UK over the period 13–16 September 1993, covered such diverse topics as energy generation, pollution and water quality, coastal protection, and so on. The dual emphasis was on soundly constructed statistical models and methods and on their use in specific practical applications. It was particularly appropriate that this conference was held in Rothamsted during the 150 th Anniversary celebrations of that prestigious research institute, which is very much concerned (*inter alia*) with relevant water-related environmental issues in agriculture.

The executive committee for the conference consisted of V. Barnett (Chairman of SPRUCE), K. F. Turkman (Vice-Chairman) R. W. Payne (Local organiser), C. W. Anderson, P. Calow, P. J. Diggle, A. O'Hagan, J. Riley and J. A. Tawn.

The invited speakers reflected major emphases in international work on various aspects of the theme, and included T. M. Addiscott, K. Beven, S. G. Coles, D. R. Cox, V. Isham, V. Klemeš, D. Mollison, A. O'Hagan, P. J. Robinson, R. L. Smith, J. A. Tawn, W. Urfer, and J. K. Vrijling. Additionally, more than 40 contributed papers were featured.

This volume (a sequel to *Statistics for the Environment*, 1993, edited by V. Barnett and K. F. Turkman) presents an integrated and comprehensive coverage of the material of the conference (mainly all invited papers), thematically structured into four sections entitled

Part I: Rainfall and Climate
Part II: Sea Levels and Wave Energy
Part III: Water Quality, Supply and Management
Part IV: Hydrological Modelling

A consolidated and cross-referenced bibliography is provided.

As for the Lisbon conference, this second international SPRUCE II Conference was held in association with the International Statistical Institute, and its importance was signalled by major support from the EC. The section DG XII (G-4) provided funds to invite 18 scientists, involved in water-related environmental interests principally from the member states of the EC, to participate in the conference, and has assisted with the costs of publishing these proceedings. This generous support is very much appreciated.

SPRUCE will continue to contribute to the field of environmental statistics, both in broad statistical issues and in relation to more specific fields of interest: such as our current theme of water. Plans are already in hand for the third SPRUCE International conference (SPRUCE III) to be held in Merida, Mexico in December 1995; it will be concerned with all aspects of the problem of *Pollution*.

We hope that this volume, covering the rich fields of statistical models, methods and applications on the theme of water-related issues, will prove to be interesting and valuable.

V. Barnett and **K. Feridun Turkman**
October, 1993

PART I
Rainfall and Climate

1

Stochastic Models of Precipitation

D. R. Cox and Valerie Isham

Nuffield College, Oxford and University College London, UK

1.1 INTRODUCTION

The mathematical representation of rainfall calls for a range of methods depending on the time and spatial scales involved. Thus the study of annual rainfall levels at a particular site is likely to require a quite different approach from that for daily data, and again from that for, say, 10 min rainfall. Also, in many contexts it is essential to consider more than one site, i.e. to regard rainfall as a spatial–temporal field.

Further there are at least three broad types of mathematical model of rainfall, namely

(i) empirical statistical models in which the distribution of, say, daily rainfall is represented via an empirical equation linking it to experience in previous years and to other relevant explanatory variables, e.g. seasonal factors (Stern and Coe, 1984);

(ii) the models of dynamic meteorology in which large systems of simultaneous nonlinear partial differential equations are solved numerically, the equations being chosen to represent fairly realistically the physical processes involved (Mason, 1986);

(iii) intermediate stochastic models in which a modest number of parameters are used to represent the rainfall process, the parameters being intended to relate to underlying physical phenomena such as rain cells.

All three have a valuable role, and the choice between them depends on the purpose of the analysis (short-term prediction, longer-term prediction and planning, data reduction, and scientific understanding).

Statistics for the Environment 2: Water Related Issues. Edited by Vic Barnett and K. Feridun Turkman
© 1994 John Wiley & Sons Ltd.

We shall concentrate on the last of these types of model. Such models are particularly appropriate for the analysis of data collected on a short time scale, e.g. hourly, both for summarising the complicated time series data that result via a few physically interpretable parameters and also as a basis for 'synthetic' hydrology (e.g. for providing simulated input for rainfall run-off models in the design of storm-drainage systems).

The physical basis for the models that we shall consider is the following. There are two broad types of rainfall: convective and nonconvective. As its name implies, the former is due to the effects of convection currents and a typical example is the short-lived and extremely intense rainfall of a summer thunderstorm. In contrast, nonconvective rainfall (often called cyclonic in mid-latitudes) is associated with large-scale uplift, and results in the more prolonged, but generally less intense, rainfall so familiar to those living in the UK. Frontal precipitation, usually treated as nonconvective, may contain a mixture of these two types.

Nonconvective storm systems can be conceived on a range of scales. For Western Europe, these are typically as follows. The precipitation field is a region of perhaps $10^4\,\mathrm{km}^2$ or more, lasting 12 h or so, and consisting of a series of low-pressure fronts. These frontal rainfall areas or rainbands, are regions of high-intensity rainfall lasting for between about 90 min and 4 h, and have a spatial size of the order of 10^3–$10^4\,\mathrm{km}^2$. Within the rainbands are the rain cells, which are the smallest precipitation elements observable by radar, and are represented by small intense radar echoes of the order of 10–$50\,\mathrm{km}^2$. These cells last on average for about 40 min and tend to cluster within the rainbands.

Essentially the same type of multilevel representation can be used also for convective rain, although then the cells are typically less frequent, more intense and of shorter duration.

The property of self-similarity, with its connection with the notion of fractals, could only hold over time scales for which the probability of zero rainfall is small. To explore self-similarity, a hierarchy of scales of variation would be called for. Here we consider models in which cells are clustered within storms but there is no other structure.

The analysis of rainfall, even when restricted to the type of model studied here, has a large literature, and we shall not attempt a systematic review, concentrating largely on work that has arisen out of our collaboration with I. Rodriguez-Iturbe and including some new results on multisite models.

1.2   SINGLE-SITE MODELS

1.2.1   Preliminaries

The models reviewed in this chapter are all constructed in continuous time. Rainfall data are collected in discrete time, and can be aggregated on a variety of time scales, with different effects being evident at different scales. It will often be important that a model can be used reliably to predict the main features of rainfall at scales other than those used in its fitting to empirical data. Building the model in continuous time and then obtaining its properties for rainfall cumulated over discrete intervals seems the best way to achieve this aim.

First we consider models for a single site (Rodriguez-Iturbe *et al.*, 1987, 1988) in which storms arrive randomly (i.e. in a Poisson process), each storm consisting of a random number of cells each depositing rain for a period. The arrival of cells forms a stochastic series of points in time subject to clustering. Two main models have been used in the literature to represent such a clustered point process: the Neyman–Scott process and the Bartlett–Lewis process.

The former is defined by a Poisson distribution of cluster centres, by the choice of a random number C to give the number of cells attached to any centre, and by a distribution defining cell position relative to the cluster centre, the positions of the cells' origins being independently drawn from that distribution. By contrast, in the Bartlett–Lewis process intervals *between* successive cells in the cluster are independently distributed.

It is not entirely clear when these two processes can be distinguished empirically, but in any case it is most unlikely that any general conclusion can be drawn from one model that cannot be obtained at least approximately from the other. We have used the Bartlett–Lewis form partly because it seems slightly more plausible on physical grounds and partly because it is more tractable mathematically for our purposes (at least when, as we assume, the intercell intervals and cell durations are all exponentially distributed and it therefore possesses a strong Markov structure).

1.2.2 Simplest model

The simplest such Bartlett–Lewis-type model with any pretence at realism has five parameters describing respectively

  (i) the rate λ of arrival of the Poisson process of storms;
 (ii) the mean number μ_C of rain cells per storm;
(iii) the mean $1/\beta'$ of the exponentially distributed interval between successive cells within a storm;
 (iv) the mean $1/\eta$ of the exponentially distributed duration of a rain cell;
  (v) the mean intensity or depth μ_X of the precipitation per cell.

It is assumed that different cells, and indeed different storms, may overlap. Thus five is the minimum number of parameters needed to describe such a system, even when the distribution of the depth per cell is assumed to have a single parameter and the depth within any cell is assumed constant over its duration. Clearly there is much scope for additional complication.

As mentioned above, the tractability of this model relies on the Markov structure in time built into its specification. Although it does not contribute to this property, it is convenient to assume that the cell depth X also has an exponential distribution. However if, for example, a heavier-tailed distribution were preferred, it would be easy to incorporate this.

Note that an equivalent, and in some ways more useful, specification of the model is to assume that the storm has a lifetime that is exponentially distributed (with parameter γ) and that cell origins occur in a Poisson process of rate β, following the storm origin and truncated by the storm lifetime. Then, necessarily, the number C of cells per storm (including one starting at the storm origin) has a geometric distribution with mean $\mu_c = 1 + \beta/\gamma$, and $\beta' = \beta + \gamma$.

A slightly more complex version of this model in which one additional parameter representing a random multiplier effect applying to different storms is introduced, appears to give a reasonably good description of the various data sets to which it has been fitted (see e.g. Rodriguez-Iturbe *et al.*, 1988). Specifically, we first reparameterise the model, replacing β by $\kappa\eta$ and γ by $\phi\eta$, and then assume that η varies randomly from storm to storm with a gamma distribution having shape parameter α and scale parameter ν. Within a storm, all cells have the same value of η. The effect of this is that all the storms have the same structure but are allowed to run on different time scales.

1.2.3   Model properties

We shall not here discuss the mathematics of analysing such a process. Because the process in its simplest form is built from components with a strong Markov structure, a number of properties of the rainfall process can be deduced mathematically in reasonably simple form. Other properties can be found by simulation; the process is of highly variable form, as indeed are the empirical data, so that it is a considerable advantage to have analytical solutions, quite apart from the qualitative insight obtained from them.

The following are the main properties that can be obtained in simple explicit form:

  (i) the proportion of dry time points;
 (ii) the moments of the marginal distribution of rainfall-depth and, in particular, the mean, variance and skewness;
(iii) the autocovariance function of rainfall depth;
(iv) the distribution of the length of dry periods.

In order that the models are not tied to a particular time interval, the initial formulation is in continuous time, although the data are aggregates over short intervals such as 1 h or 15 min. It is thus important that the above properties can all be determined analytically for aggregated data. Indeed, in the fitting procedure we use properties at various levels of time aggregation.

Other properties (e.g. of extremes, and of the joint distribution of the precipitation in successive time periods) can be evaluated by simulation once the relevant parameters are given numerical values.

1.2.4   Model fitting

The fitting of even a five-parameter version of such a model and the assessment of the adequacy of its fit raise major statistical issues. Even if a likelihood function could be calculated, it would not be an appropriate basis for fitting at least the simplest form of the model, because the idealisation involved leads to sample paths with some (short-term) deterministic features.

The method of fitting used in Rodriguez-Iturbe *et al.* (1988) was to equate selected theoretical values to their historical values, month by month, and to use remaining theoretical and simulated values to check the adequacy of fit. An improved method is to use a larger set of theoretical values combined via a simulation-based estimate of their covariance matrix and generalised least squares. Methods of assessing the standard errors of parameter estimates are also needed.

A quite different approach would have been to isolate individual storms for separate study. Note, however, that in these models not only may different storms overlap but, more seriously, there may be one or more dry periods within a single storm, so that the isolation of separate storms is not straightforward.

Adequacy of fit is judged via those properties not used in the fitting procedure and also, rather importantly, by whether the estimates derived for the underlying parameters are physically plausible. Lack of fit may also be detected via discrepancies between the use of simulated and real data when these are used as input to models of other hydrological processes.

Broadly similar issues arise in fitting relatively complicated stochastic models to empirical data in a wide variety of applied areas.

1.3   SPATIAL–TEMPORAL ASPECTS

1.3.1   Preliminary comments

The rainfall field is essentially spatial–temporal. The work outlined in Section 1.2 concentrates on a single site. If the spatial aspects alone are of concern, a possibility is to integrate over a substantial time period, perhaps the passage of a storm (Rodriguez-Iturbe *et al.*, 1986). Of more interest, however, is to study the spatial and temporal aspects together. These are being examined empirically in the wide-ranging HYREX (Hydrological Radar Experiment) investigations begun recently under NERC support. Here radar rainfall measurements are being obtained every 5 min for about thirty 2×2 km squares in the basin of the river Brue in SW England, and about 50 rain gauges have been allocated to the basin. There are several objectives, including the calibration of radar data against rain gauge data, the study of the water balance in the basin and the study of the local spatial–temporal structure of the rainfall field. Appropriate siting of the rain gauges is thus a compromise; some principles have been set out for guidance and are the basis for the allocation used, although inevitably certain practical constraints have to be met.

1.3.2   Types of model

Models that may be fitted to spatial–temporal rainfall data are of two broad types. The first regard the sites as a discrete set of points at each of which the marginal process is of the single site kind sketched above, appropriate correlation being induced between sites. Some possible models for this situation will be discussed in Section 1.3.4. In many ways a more interesting, but more challenging, approach is via a 'genuine' i.e. *continuous* spatial model; a very simple version of this was outlined by Cox and Isham (1988). However, in either case it is desirable that the further development and fitting of such models be preceded by careful empirical analysis to determine the types of spatial dependence arising in practice. The HYREX project will provide an ideal opportunity for such analysis.

1.3.3   Model development in continuous space

The models that have been discussed in Section 1.2 describe the temporal evolution of rainfall at a single spatial location. It is assumed that only rain cells affecting the specific location are represented in the model and that they are only of interest during the period for which they affect the cell. Thus, implicitly, the (effective) duration of a cell at the location is a representation of an unknown (and, for this purpose, irrelevant) combination of its actual lifetime, its shape and the velocity of its motion. The properties studied are necessarily all properties of the temporal process, that is, of the joint distribution of cumulative rainfall over a set of time periods. In particular, in fitting such models, it is essential that the temporal autocorrelation structure of the model for rainfall aggregated over discrete time periods should match well that of the empirical data. However, the process at a single location is really only the marginal univariate process corresponding to a very much more complicated spatial–temporal random field. As described in Section 1.1, the structure of this field occurs at a sequence of spatial scales. It is of particular interest and importance to model this field and to determine, for example, its spatial and spatial–temporal autocorrelations as well as its purely temporal properties.

In recent years various authors have developed exploratory models that serve as our starting point for the study of this problem. In the single-site model described above a rain storm consists of a temporal cluster of cells all located at a fixed point in space. In the model investigated by Waymire *et al.* (1984) it is assumed that rain events occur in a Poisson process in time. Each rain event consists of a non-homogeneous spatial (two-dimensional) Poisson process of storms, where, in principle, the rate of this Poisson process of storm centres may be a function of spatial position, thereby allowing for possible orographic effects. This rate function can also vary to take account of seasonal effects. In practice, Waymire *et al.* (1984) restrict their attention to the homogeneous case. Associated with each storm centre there is a cluster of rain cells, which has a Neyman–Scott structure (in three-dimensional space–time) independently for each cluster. For each cluster, there is a random number of cell origins that are each displaced from the storm centre, where this displacement has an exponential distribution in time and a normal distribution in space, and allowance is made for the movement of the rain event and of the rain cells within the event. The cell intensity at the cell origin is a fixed quantity, and the intensity decays exponentially in time and quadratic–exponentially with spatial distance away from the origin. Thus, as compared with our single-site model, this model incorporates an extra layer of clustering, of storms within rain events, over and above the clustering of cells within storms. The cells, however, all have an identical spatial–temporal structure. The process of empirical interest is the total rainfall intensity at a particular space–time point, and Waymire *et al.* (1984) determine its second-order properties.

Note, however, that in this model any rain cell has infinite spatial extent and lasts for ever following its origin. Thus there will be rain everywhere in the plane, at all times, and in order to match the models to data it will be necessary to set a rather arbitrary cut-off level for the total rain intensity, below which it is deemed to be dry. Also, observed data will consist of values of the rainfall intensity integrated over discrete space–time regions, and the properties of a process aggregated in this way are not discussed.

The properties of some variants of this model, which include assuming independent random rainfall intensities at the cell origins and alternative decay functions for the spatial and temporal spread of the rain cell, are systematically examined by Rodriguez-Iturbe *et al.* (1986). Almost throughout, however, the clustered structure of the rain cells is dropped in favour of the simpler assumption of a non-clustered homogeneous Poisson process of cell origins. In their paper the authors concentrate on determining the properties of the spatial process comprising the total rainfall generated by a single rain event, that result from the detailed spatial–temporal structure assumed in the model.

This line of work is developed further by Rodriguez-Iturbe and Eagleson (1987), who still concentrate on the properties of a single rain event, but where now the second-order spatial–temporal properties are investigated, both of the rainfall intensity process itself in continuous time and of the rainfall depth cumulated over discrete time intervals.

Phelan and Goodall (1990) extend the Waymire *et al.* (1984) model to allow a more general cell structure, in which the temporal and spatial decay of the cell intensity with distance from the cell origin can vary between cells. They present a most interesting and helpful assessment of the appropriateness of this model for the analysis of radar data from tropical rainfall fields. Note, however, that in fitting the model (using a nonlinear least-squares approach) they rely on the manual identification of rain cells by inspection of the radar data.

In Cox and Isham (1988) a spatial–temporal model with some rather different features is proposed and explored. The aim of this work was to extend the earlier model (Rodriguez-Iturbe *et al.*, 1987, 1988) for the temporal process of rainfall at a single site to a fully spatial–temporal process in such a way that the marginal temporal process at a single spatial location of the earlier model is preserved. The model concentrates on representing the basic features of the rainfall process in terms of a few interpretable parameters. It is constructed in continuous space and time, but ultimately it is its properties cumulated over arbitrary discrete spatial regions and time intervals that are needed in order to assess its adequacy as a suitable model for rainfall. In contrast with the papers already described, the emphasis is on the final observed rainfall process, rather than with a single rain event, and rain cells are assumed to be finite in both temporal duration and spatial extent.

In this model storms are assumed to occur in a homogeneous three- (two space and one time) dimensional Poisson process. Following each storm origin, and at the same spatial location, cell origins occur in a temporal Bartlett–Lewis-type cluster. Thus, as in the single-site process, one may imagine the cell origins following the storm origin in a Poisson process that is terminated after an exponentially distributed storm lifetime. The cells themselves are idealised as cylinders having random radii and heights (the rainfall intensities of the cells), lasting for random lifetimes. During their lifetimes, all the cells in a single storm move with the same random velocity. Distinct storms are assumed to be mutually independent, as are distinct cells within a storm, apart from the structure imposed by the process of cell origins and their common velocity. Inevitably, the more independence assumptions that are incorporated into the model the more tractable it is regarding the determination of its properties. In this model cells have a finite extent in both space and time and a constant depth (intensity) throughout both. This latter restriction could be removed without difficulty by allowing a 'jitter' component as for the single-site model. A storm consists of a cluster of cells of varying sizes, strung out in a straight line in space and moving with a common velocity through it. This line can be thought of as a (highly idealised) representation of a rainband or rain front.

Various properties of this model have been investigated, e.g. the marginal distribution of rainfall intensity at a point in space and the second-order properties of the spatial–temporal process. It is usually not possible to do more than to obtain general expressions for properties as expectations of particular functions with respect to the joint distribution of cell variables (radius, intensity, velocity), and difficult to know what explicit form to assume for this. As is to be expected, many of the expressions simplify substantially when each storm consists of a single cell, so that there is no clustering of cells, since the amount of dependence in the system is thereby reduced enormously. However, this assumption is almost certainly undesirable, since we know that for the single-site process a clustered process of cells is necessary to provide a reasonable fit to data.

One property of interest for this spatial–temporal model is the duration of a cell at a fixed spatial position. In the single-site model this variable was taken to have an exponential distribution, while in the elaboration of the model the parameter of this exponential distribution was itself assumed to be random, having a gamma distribution. However, in the present model this duration will be affected not only by the lifetime of the cell but also by its velocity and radius. In general, a cell that passes over the fixed position will do so for a time determined by a random chord of the relevant disc, possibly foreshortened depending upon the time of birth and death of the cell. Cox and Isham (1988) investigated possible choices for the distribution of the cell radius, but in the special non-clustered case only.

In the study of fluid turbulence there is a well-known hypothesis first formulated by G. I. Taylor (1938). The frozen field hypothesis postulates the equivalence of the spatial autocorrelation at a fixed time point and the temporal autocorrelation at a fixed position in space, if the ratio of the spatial displacement (argument) of the former to the time lag of the latter can be suitably interpreted as an 'average' velocity of the field. For example, the hypothesis will hold exactly if the spatial–temporal field can be represented as a fixed (i.e. frozen) spatial field which moves in time with a constant velocity. It has been observed empirically (Zawadski, 1973) that this hypothesis is approximately valid for spatial–temporal rainfall fields as long as the spatial and temporal displacements are not too large. In particular, Zawadski found good agreement at time lags under 40 min, which is the typical duration of a rain cell.

The frozen field hypothesis was found by Waymire *et al.* (1984) to be approximately valid for the spatial–temporal model they developed, for times less than the mean cell duration, and it is extremely important that spatial–temporal models should reflect such empirically verifiable properties. It is therefore very satisfactory to see that the hypothesis also holds for our (Cox and Isham, 1988) model, at least for the special non-clustered case, as long as

(i) the time lag is small relative to the mean lifetime of a rain cell, so that the temporal autocorrelation is unaffected by the death of cells; and

(ii) the velocity of the cells does not vary much from one to another.

1.3.4   Multisite models

From our investigations of the spatial–temporal model (Cox and Isham, 1988), it is clear that it will be difficult to make much progress in such models without extensive analysis of suitable data to guide choice of appropriate dependences between variables and forms for their distributions. However, since many of the available data sets consist of rainfall sequences at a small number of spatially separated sites, we have turned our attention first to the consideration of multivariate or, as we prefer, multisite models for such sequences. As before, we look for models that preserve the structure of our single-site models for their marginal processes at each site.

Let us assume that there are n sites at which the rainfall depth is to be modelled. Suppose that there is a master process of storms that is a Poisson process (in time) of rate λ and counting measure N, and that these storms evolve independently of each other. Consider one particular storm and assume that

$$P(\text{storm affects site } i) = p_i, \quad \text{where } \lambda_i = \lambda p_i,$$

$$P(\text{storm affects sites } i \text{ and } j) = p_{ij}, \quad \text{where } \lambda_{ij} = \lambda p_{ij}.$$

Since, as far as between-site characteristics are concerned, we shall concentrate upon second-order properties between pairs of sites, there is no need here to define joint probabilities that a storm affects more than two sites. An obvious special case is of independence over sites, i.e. $p_{ij} = p_i p_j$, though this is unlikely to be realistic unless sites are very widely separated.

At each individual site we require the model to have the Bartlett–Lewis structure with either five parameters (fixed η)

storm rate λ_i,   cell origin rate β_i,   mean cell depth $\mu_i = \mu_{X^{(i)}}$,
cell termination rate η_i,   storm termination rate γ_i,

or six parameters (random η),

as above, but with η having a gamma distribution with index α_i and scale parameter ν_i.

Throughout the rest of this chapter the cell depth $X^{(i)}$ at site i is assumed to have an exponential distribution, so that $E([X^{(i)}]^2) = 2[E(X^{(i)})]^2$. It is straightforward to remove this assumption, with the introduction of a second moment parameter for $X^{(i)}$.

To estimate the parameters of the model, the marginal properties of each site can be fitted separately, as in the single-site case, to give the five or six site parameters. If there are parameters in common between sites, e.g. β, then this can be taken into account in the fitting. The lag-0 cross-correlation or, to allow for systematic directional effects, the maximum cross-correlation, being probably the best determined of the between-site properties, can be used to determine λ_{ij}.

We now have to decide what sort of assumptions should be made about the dependences between the evolutions at distinct sites of storms having a common storm origin. At one extreme we could suppose that, given that a storm affects a particular site, its evolution is independent of that at each other site affected by the same storm. In this case the parameters of the model are the full set of five or six for each site together with the λ_{ij}. With this model, the only intersite correlation comes from the master process of storm origins, and is unlikely to provide a high enough cross-correlation to be realistic. In particular, having obtained estimates of the λ_i from the single site analysis, and since necessarily we must have $\max_j \lambda_{ij} \leqslant \lambda_i$ for consistency, the maximum fitted cross-correlation is unlikely to be big enough to match that observed, except perhaps for distant sites, a case not of great interest.

At the other extreme, we could assume identical rather than independent storms, i.e. that, given that a storm affects a particular site, its evolution is identical at all affected sites. We may choose to modify this slightly to allow the possibility that the cell depths at site i are all scaled by a constant a_i. In this case the cell termination rate η (fixed or random), cell origin rate β, cell depth μ_X (without loss of generality, $\mu_X = 1$) and storm termination rate γ are common over all the n sites; the other parameters being the λ_i and λ_{ij} and cell-depth scaling constants a_i.

Properties of this model are straightforward to determine. In effect, we have a master process $Y(t)$ of the single-site type, for which the properties have already been described. Any cell is observed at both sites i and j with probability p_{ij}, and if so its depth is scaled by quantities a_i and a_j respectively. So, for example, if $Y_i(t)$ denotes the process at site i then the cross-covariance function for sites i and j is given by

$$c_{ij}(s) \equiv \mathrm{cov}\,(Y_i(t),\,Y_j(t+s)) = p_{ij}a_ia_jc_Y(s), \qquad (1.3.1)$$

which holds for $i = j$ if p_{ij} is replaced by p_i, where $c_Y(s)$ is the autocovariance function for the master process, as given in Rodriguez-Iturbe *et al.* (1987, equation (3.4)) for the fixed-η case or Rodriguez-Iturbe *et al.* (1988, equation (2.2)) for random η.

If we now consider the discrete-time processes $\{Y_{i,k}^{(h)}, k = \ldots, -1, 0, 1, \ldots\}$ obtained by aggregating the rainfall at each site over successive time intervals of length h, it follows that cross-correlation of lag-0 is

$$\mathrm{cov}\,(Y_{i,k}^{(h)},\,Y_{j,k}^{(h)}) \equiv c_{ij}^{(h)}(0) = p_{ij}a_ia_j \int_0^h 2(h-u)c_Y(u)\,du, \qquad (1.3.2)$$

and hence that

$$\mathrm{corr}\,(Y_{i,k}^{(h)},\,Y_{j,k}^{(h)}) = \frac{p_{ij}}{(p_ip_j)^{1/2}}. \qquad (1.3.3)$$

Thus the lag-0 cross-correlation of the aggregated data (for the fixed- or random-η model) does not depend on the interval of aggregation h. This is inconsistent with the particular empirical data we have examined, which show an increasing cross-correlation with increasing period of aggregation. Note, though, that short-term 'jitter' or error of observation could be an explanation of this increase.

We now try to construct an intermediate model that is both tractable and intuitively realistic. Let us suppose that the Poisson process of cell origins $N_u(t)$ (of rate β), corresponding to a particular storm origin at u is the same at all affected sites, but that the storm truncation mechanism acts independently at each site (as before, storm lifetimes are exponentially distributed with parameters γ_i, so that the mean number of cells in a storm experienced at site i is $1 + \beta/\gamma_i$). We also assume that, for a particular cell at a particular site, the depth and duration are independent variables, that depths of cells with a common origin but at different sites are independent, but that durations of such cells are dependent. Specifically, we assume that the duration L_i at site i is given by $L_i = l_i L$, where the l_i are scalar constants and L is an exponential variable with parameter η, the value of η being the same for all cells belonging to a single storm (over all sites) but possibly varying randomly from one storm to another. It is convenient to write $\eta/l_i = \eta_i$.

For a fixed-η version of this model, the only constraint imposed on the single-site parameters for which allowance must be made in the fitting is that $\beta_i = \beta$. Otherwise, the dependence between durations of cells with a common origin is only evident in its effect on the forms of the between-site properties such as cross-correlations. The random-η version is slightly more complicated, and will be discussed further below.

For this model, the second-order properties are straightforward to determine. In particular, for $i \neq j$, the cross-covariance function in the fixed-η case is

$$
\begin{aligned}
c_{ij}(s) = \lambda_{ij} \int_{-\infty}^{t} du\, \mathrm{E}\Bigg[\int_{u}^{t} dN_u^{(i)}(v)\, X_v^{(i)}(t-v) e^{-\gamma_i(v-u)} \\
\times \int_{u}^{t+s} dN_u^{(j)}(w)\, X_w^{(j)}(t+s-w) e^{-\gamma_j(w-u)} \Bigg],
\end{aligned}
\tag{1.3.4}
$$

where $X_v^{(i)}(t-v)$ is the depth at site i at time t, of a cell having its origin at time v. By expressing $\mathrm{E}[dN_u^{(i)}(v)]$ as $[\delta(u-v) + \beta]\,dv$, and splitting this integral up into five terms, containing the contributions from

(i)   the first cell at each site (these cells have a common origin at u i.e. $v = w = u$),
(ii) the first cell at i and a cell other than the first at j ($w > v = u$),
(iii) as (ii) with i and j reversed ($v > w = u$),
(iv) the same cell at each site, but not the first ($v = w > u$), and
(v)   different cells at each site, neither being the first ($v \neq w; v, w \neq u$),

it is a matter of routine integration to show that the cross-covariance has the form

$$
c_{ij}(s) = J_{ij} e^{-\eta_j s} + K_{ij} e^{-\gamma_j s} + L_{ij} e^{-s/(1/\eta_j - 1/\eta_i)}
\tag{1.3.5}
$$

where

$$
J_{ij} = \frac{M_{ij}}{\eta_j} - \frac{N_{ij}}{(\eta_j - \gamma_j)(\eta_j + \gamma_i)(\eta_i + \eta_j)}
$$

$$K_{ij} = \frac{N_{ij}}{(\eta_j - \gamma_j)(\eta_i + \gamma_j)(\gamma_i + \gamma_j)},$$

$$L_{ij} = \left(\frac{1}{\eta_j} - \frac{1}{\eta_i}\right)^+ M_{ij},$$

$$M_{ij} = \lambda_{ij} E(X^{(i)} X^{(j)}) \left(1 + \frac{\beta}{\gamma_i + \gamma_j}\right),$$

$$N_{ij} = \lambda_{ij} \beta E(X^{(i)}) E(X^{(j)}) (\beta + \gamma_i + \gamma_j),$$

and $x^+ = x$ if $x > 0$, $x^+ = 0$ otherwise. Note that the factor $1 + \beta/(\gamma_i + \gamma_j)$ is simply the mean number of cells of a storm which are common to the two sites. For $i = j$, the expression is given in Rodriguez-Iturbe *et al.* (1987, equation (4.12)).

For aggregated data, the corresponding cross-covariance $c_{ij}^{(h)}(k)$ at lag k, for $k > 0$, is given by

$$c_{ij}^{(h)}(k) = J_{ij} e^{-\eta_i k h} (e^{\eta_i h} + e^{-\eta_i h} - 2)/\eta_j^2 + K_{ij} e^{-\gamma_j k h} (e^{\gamma_j h} + e^{-\gamma_j h} - 2)/\gamma_j^2$$
$$+ L_{ij} e^{-\delta_{ij} k h} (e^{\delta_{ij} h} + e^{-\delta_{ij} h} - 2)/\delta_{ij}^2, \tag{1.3.6}$$

where $\delta_{ij} = \eta_i \eta_j /(\eta_i - \eta_j)$. For lag $k = 0$, the cross-covariance has a similar but symmetric form, namely

$$c_{ij}^{(h)}(0) = J_{ij}(e^{-\eta_j h} - 1 + \eta_j h)/\eta_j^2 + J_{ji}(e^{-\eta_i h} - 1 + \eta_i h)/\eta_i^2$$
$$+ K_{ij}(e^{-\gamma_j h} - 1 + \gamma_j h)/\gamma_j^2 + K_{ji}(e^{-\gamma_i h} - 1 + \gamma_i h)/\gamma_i^2$$
$$+ L_{ij}(e^{-\delta_{ij} h} - 1 + \delta_{ij} h)/\delta_{ij}^2 + L_{ji}(e^{-\delta_{ji} h} - 1 + \delta_{ji} h)/\delta_{ji}^2. \tag{1.3.7}$$

With this model, no particular assumptions about the joint distribution of depths for the same cell at distinct sites have been made. One plausible choice is that they scale in a similar way to the durations, i.e. $X^{(i)} = a_i X$, so that $E(X^{(i)} X^{(j)}) = a_i a_j E(X^2)$.

For the random-η version of this model, $\eta_i = \eta/l_i$ and η is a random variable having a gamma distribution with index α and scale parameter v. Thus η_i also has a gamma distribution with index α but scale parameter $v_i = v l_i$. In fitting this model to the single-site data, the six parameters at site i (λ_i, $\kappa_i = \beta_i/\eta_i$, $\phi_i = \gamma_i/\eta_i$, μ_i, α_i and v_i) must satisfy the constraints $\alpha_i = \alpha$ and also $v_i/\kappa_i = v\eta/\beta = $ constant, θ, say. In other words, we must fit the four parameters λ_i, κ_i, ϕ_i and μ_i for each site, together with the two extra parameters α and θ.

Now, if we rewrite the cross-covariances for the aggregated data derived above, substituting for the parameters η_i, γ_i and β in terms of ϕ_i, l_i, $\kappa = \beta/\eta$ and η, and then take expectations over η, we can obtain the cross-covariances for the random η model. The argument whereby we need only the expectation of the covariance is essentially that, for the non-aggregated case, we can write

$$c_{ij}(s) = \int\int E[Y^{(i)}_{t-u}(t)Y^{(j)}_{t+\tau-v}(t+\tau)]\,\text{cov}\,(dN(t-u), dN(t+\tau-v)) \qquad (1.3.8)$$

$$= \lambda_{ij}\int E[Y^{(i)}_{t-u}(t)Y^{(j)}_{t-u}(t+\tau)]\,du, \qquad (1.3.9)$$

where $Y^{(i)}_{t-u}(t)$ denotes the total rainfall at site i and at time t, from a storm with origin at $t-u$. The expectation in this equation is over all components of randomness for the storm, and, in particular, we may envisage the expectation taken in two steps: in the first conditioning on η gives the fixed η cross-covariance, while the second, over η, yields the unconditional result.

Thus the cross-covariances for the aggregated data in the random-η model are obtained as functions of the parameters λ_{ij}, μ_i, l_i, ϕ_i, κ, α and v. Estimates of the λ_{ij} can now be obtained by the substitution of estimates of the remaining parameters into the sample cross-covariances. Since $l_i = \kappa_i/\kappa$ and $v = \theta\kappa$, it appears that we are short of an estimate of the unknown κ and have one more parameter than estimate. In fact this is not so, because in defining $L_i = l_i L$ and $E(L) = 1/\eta$ a redundant parameter has been introduced. (Without loss of generality, we could have taken $l_1 = 1$, say.) Expressions for the cross-covariances in term of the estimated parameters are given in Appendix 1.1.

1.3.5   Further extensions of the multisite model

In our present model, for a particular storm experienced at sites i and j, the cell durations are scaled, i.e. $L_i = l_i L$, where the l_i are constants, while the cell origins are identical and the storm lifetimes are independent. Two possible extensions that are at least potentially plausible are

(i) to scale the intervals between the cell origins, i.e. if Z_i is an interval between successive cell origins in the storm at site i then $Z_i = l_i Z$, where Z is exponentially distributed with parameter β so that Z_i is also exponentially distributed but with parameter β/l_i;

(ii) to scale the storm durations, i.e. if S_i is the storm duration at site i then $S_i = l_i S$, where S is exponentially distributed with parameter γ so that S_i is also exponentially distributed but with parameter γ/l_i.

Note that if we incorporate both of these changes into the model then $\beta_i/\gamma_i = \beta/\gamma$, and the number of cells in a storm is the same at all affected sites. Effectively, the storm is identical (apart from the cell depths) at each site, but simply experienced on different time scales. If we incorporate (i) only then, since $\kappa_i = \beta_i/\eta_i = \beta/\eta$, the effect is that both α and κ must be constant over all sites, in contrast to α and v/κ being constant as in the model investigated above. With (ii) only, ϕ must be constant as well as α and v/κ.

The cross-covariance properties corresponding to each of these options can be deduced in a manner similar to that used for the basic version. However, before pursuing such algebra further, experience with a substantial amount of data analysis is needed, not only to determine under what circumstances it is reasonable to keep a parameter of the model fixed over sites, but also to discover what assumptions are appropriate on the stochastic dependence of storm and cell characteristics as experienced at different sites. Even if, say, cell durations at two sites can be assumed to be exponentially distributed with the same parameter η, these durations could be independent, identical or dependent in some other way. Non-trivial analysis of appropriate data will be needed to answer such questions.

ACKNOWLEDGEMENTS

VSI is grateful to the SERC for a Visiting Fellowship held at the Isaac Newton Institute for Mathematical Sciences, Cambridge, during the tenure of which some of the work on this paper was carried out, and to that Institute for their hospitality.

APPENDIX 1.1

The cross-covariances for the multisite random η model described in Section 1.3.4 are given below for $i \neq j$. For $i = j$, the autocovariances are given in Rodriguez-Iturbe *et al.* (1988, equations (2.3) and (2.4)).

For lag k, $k > 0$,

$$
\begin{aligned}
c_{ij}^{(h)}(k) = {}& J_{ij}'(\theta\kappa_j)^\alpha \frac{\Gamma(\alpha - 3)}{\Gamma(\alpha)} \{ [h(k-1) + \theta\kappa_j]^{3-\alpha} + [h(k+1) + \theta\kappa_j]^{3-\alpha} \\
& - 2[hk + \theta\kappa_j]^{3-\alpha} \} + K_{ij}'(\theta\kappa_j)^\alpha \frac{\Gamma(\alpha - 3)}{\Gamma(\alpha)} \{ [h(k-1)\phi_j + \theta\kappa_j]^{3-\alpha} \\
& + [h(k+1)\phi_j + \theta\kappa_j]^{3-\alpha} - 2[hk\phi_j + \theta\kappa_j]^{3-\alpha} \} \\
& + L_{ij}'[\theta(\kappa_j - \kappa_i)]^\alpha \frac{\Gamma(\alpha - 3)}{\Gamma(\alpha)} \{ [h(k-1) + \theta(\kappa_j - \kappa_i)]^{3-\alpha} \\
& + [h(k+1) + \theta(\kappa_j - \kappa_i)]^{3-\alpha} - 2[hk + \theta(\kappa_j - \kappa_i)]^{3-\alpha} \},
\end{aligned}
$$

where

$$
J_{ij}' = M_{ij}' - \frac{N_{ij}'}{(1/\kappa_j + \phi_i/\kappa_i)(1/\kappa_i + 1/\kappa_j)(1 - \phi_j)},
$$

$$
K_{ij}' = \frac{N_{ij}'}{\phi_j^2(1/\kappa_i + \phi_j/\kappa_j)(\phi_i/\kappa_i + \phi_j/\kappa_j)(1 - \phi_j)},
$$

$$L'_{ij} = M'_{ij} H(\kappa_j - \kappa_i),$$

$$M'_{ij} = \lambda_{ij} E(X^{(i)} X^{(j)}) \frac{1 + \phi_i/\kappa_i + \phi_j/\kappa_j}{\phi_i/\kappa_i + \phi_j/\kappa_j},$$

$$N'_{ij} = \lambda_{ij} E(X^{(i)}) E(X^{(j)}) (1 + \phi_i/\kappa_i + \phi_j/\kappa_j),$$

and $H(x) = 1$ if $x > 0$, $H(x) = 0$ otherwise.

For lag-0,

$$c_{ij}^{(h)}(0) = J'_{ij} \frac{\Gamma(\alpha - 3)}{\Gamma(\alpha)} [(\theta\kappa_j)^\alpha (h + \theta\kappa_j)^{3-\alpha} - (\theta\kappa_j)^3 - (\theta\kappa_j)^2 h(3 - \alpha)]$$

$$+ J'_{ji} \frac{\Gamma(\alpha - 3)}{\Gamma(\alpha)} [(\theta\kappa_i)^\alpha (h + \theta\kappa_i)^{3-\alpha} - (\theta\kappa_i)^3 - (\theta\kappa_i)^2 h(3 - \alpha)]$$

$$+ K'_{ij} \frac{\Gamma(\alpha - 3)}{\Gamma(\alpha)} [(\theta\kappa_j)^\alpha (h\phi_j + \theta\kappa_j)^{3-\alpha} - (\theta\kappa_j)^3 - (\theta\kappa_j)^2 h\phi_j(3 - \alpha)]$$

$$+ K'_{ji} \frac{\Gamma(\alpha - 3)}{\Gamma(\alpha)} [(\theta\kappa_i)^\alpha (h\phi_i + \theta\kappa_i)^{3-\alpha} - (\theta\kappa_i)^3 - (\theta\kappa_i)^2 h\phi_i(3 - \alpha)]$$

$$+ L'_{ij} \frac{\Gamma(\alpha - 3)}{\Gamma(\alpha)} \{ [\theta(\kappa_j - \kappa_i)]^\alpha [h + \theta(\kappa_j - \kappa_i)]^{3-\alpha}$$

$$- [\theta(\kappa_j - \kappa_i)]^3 - [\theta(\kappa_j - \kappa_i)]^2 h(3 - \alpha) \}$$

$$+ L'_{ji} \frac{\Gamma(\alpha - 3)}{\Gamma(\alpha)} \{ [\theta(\kappa_i - \kappa_j)]^\alpha [h + \theta(\kappa_i - \kappa_j)]^{3-\alpha}$$

$$- [\theta(\kappa_i - \kappa_j)]^3 - [\theta(\kappa_i - \kappa_j)]^2 h(3 - \alpha) \}.$$

2

Spatial Modelling of Rainfall Data

R. L. Smith

University of North Carolina, Chapel Hill, USA

2.1 INTRODUCTION

This chapter continues the discussion of rainfall modelling by Cox and Isham (1994) and Robinson (1994) (Chapters 1 and 3 of this volume). Cox and Isham present a general overview of stochastic models for rainfall, drawing particular attention to the contrast between those constructed primarily by empirical modelling of the data and those in which physical characteristics such as rain cells are explicitly modelled. Robinson discusses numerous climatological issues associated with rainfall, and presents some simple statistical techniques designed to shed light on some of those issues. We continue Robinson's analysis by presenting some specific models illustrated with daily rainfall data from North Carolina. From the point of view of the classification outlined by Cox and Isham, the models are of the empirical type, consisting principally of extensions of the models introduced by Stern and Coe (1984) to consider more general features, especially joint distributions of rainfall series at two nearby sites. However, it is also hoped that the analyses presented will be the first step in introducing a more broadly based approach to the spatial analysis of rainfall data, which may well involve more specific modelling of physical features as has been proposed by Cox and Isham.

The data are daily rainfall data extracted from the large network of North Carolina meteorological stations. For most of the present analysis we have confined ourselves to five stations: Asheville, North Fork and Highlands, all of which are in the mountainous western part of the state, and Smithfield and Whiteville, which are on the flat coastal plain in the eastern part of the state. As discussed by Robinson, one would expect the patterns of rainfall to be influenced

Statistics for the Environment 2: Water Related Issues. Edited by Vic Barnett and K. Feridun Turkman
© 1994 John Wiley & Sons Ltd.

by topography, so there is considerable interest in the contrasts between these two regions.

Rainfall data differ from other forms of climatological data in that there are many zero values corresponding to no rainfall on a particular day. It might be thought possible to apply standard methods based on temporal and spatial correlations without attention to this feature. We shall argue that such an approach ignores some of the most interesting and informative features of the data. For example, a significant proportion of the recorded data values are 'trace', signifying that there is rainfall on a particular day but too little to obtain a precise measurement. Setting all such values to zero would be tantamount to treating 'trace' days as equivalent to dry days. A more reasonable approach would seem to be one in which the problem is broken into two parts: first analyse the pattern of wet and dry days (treating 'trace' days as wet), and then perform a separate analysis of rainfall amounts given that they are positive. Our task is therefore to develop appropriate models for temporal and spatial dependence based on data of this structure.

An outline of the chapter is as follows. In Section 2.2 we review previous models for rainfall data, concentrating on the class of models introduced by Stern and Coe (1984) for time series at a single site. In Section 2.3 we present some initial analyses of the data, aimed primarily at identifying an appropriate marginal distribution taking the 'trace' values into account. In Section 2.4 we develop models for the time series of observations at a single site. The main difference from Stern and Coe (1984) is the explicit treatment of 'trace' values as censored data points. Section 2.5 continues this analysis to consider the joint distribution of rainfall events at two sites, while Section 2.6 is directly concerned with dependence measures of the kind introduced by Robinson (1994). Finally, Section 2.7 reviews what has been achieved and presents suggestions for future work.

2.2   THE STERN–COE MODEL AND SOME EXTENSIONS

Stern and Coe (1984) presented a general approach to the analysis of daily rainfall data from a single station. Although similar ideas have been presented by other authors, both previously and subsequently, this paper does appear to be the most comprehensive overview of the issues raised by such data, especially from a statistician's perspective. A recent paper by Woolhiser (1992) has reviewed subsequent developments and has discussed current research issues raised by this approach.

The main features of the model proposal by Stern and Coe are the following. The analysis is broken up into separate models for rainfall occurrences and rainfall amounts. By rainfall occurrences, we mean the binary process indicating whether or not it is raining on a particular day. This process is modelled as a higher-order Markov chain whose transition probabilities vary continuously according to season. For example, if we let X_n be 1 if day n is wet and 0 if day n is dry then a typical model might be a second-order Markov chain

$$\Pr\{X_n = 1 \mid X_k, k < n\} = \Pr\{X_n = 1 \mid X_{n-1} = i, X_{n-2} = h\}$$
$$= p_{hi}(n), \tag{2.1}$$

where $p_{hi}(\cdot)$, for each combination of $h = 0, 1$, $i = 0, 1$, varies seasonally according to

$$\log\left[\frac{p_{hi}(n)}{1 - p_{hi}(n)}\right] = a_{hi0} + \sum_{k=1}^{m}\left(a_{hik}\sin\frac{2\pi n}{366} + b_{hik}\cos\frac{2\pi n}{366}\right). \tag{2.2}$$

Here m is the number of Fourier terms included, which must be determined by an analysis of deviance or some other form of model selection procedure. Of course, the model is defined in a similar way for other orders of Markov dependence, and choosing the appropriate order is also part of the model selection process.

Suppose the measured rainfall amounts are denoted by $\{Y_n, n = 1, 2, \ldots\}$. Stern and Coe defined $Y'_n = Y_n - \delta$, where δ is some minimum detection limit, and then modelled $\{Y'_n\}$, conditionally on being positive, as independent gamma-distributed random variables. The gamma density for Y_n on day n may be expressed in the form

$$f_n(y) = \frac{1}{\Gamma(\kappa)}\left[\frac{\kappa}{\mu(n)}\right]^{\kappa}y^{\kappa-1}\exp\left[-\frac{\kappa y}{\mu(n)}\right], \tag{2.3}$$

in which the shape parameter κ is kept constant and the mean function $\mu(n)$ assumed to be seasonal with a logarithmic link function: $\log\mu(n)$ has a Fourier series representation of the same form as (2.2).

Stern and Coe discussed a variety of issues related to parameter estimation, model selection and testing fit of these models, making considerable use of ideas from the theory of generalised linear models and the GLIM software package. They also discussed the application of the models to a variety of problems of meteorological and climatological interest, for example the length of a dry spell. It is clear that some direct extensions are possible within their framework—for example, dependence among the Ys could be incorporated by making $\mu(n)$ in (2.3) a function of previous values of X_n and/or Y_n—but other features, such as a detailed consideration of alternative distributions for Y_n, are not so easily handled with their methods. The gamma distribution seems to be widely but not universally accepted. Woolhiser (1992) refers to several papers in which a mixed exponential distribution has been used, for example Woolhiser and Roldán (1982).

Woolhiser (1992) reviewed a number of recent extensions designed to address climatological questions of current interest. For example, Wilks (1989) modelled the dependence of daily rainfall on monthly totals, the latter being divided into three categories corresponding to the lower 30%, middle 40% and upper 30% of all months. The apparent motivation of this was to be able to utilise long-range weather forecasts to predict some characteristics of daily rainfall. The long-range forecasts typically contain some prediction of monthly rainfall totals, but not of daily rainfall. However, it is also the case that general circulation models, used in studies of climate change, produce projections of monthly rainfall totals. Therefore the possibility exists that methodology similar to that employed by Wilks could be used to study the influence of the greenhouse effect on daily rainfall characteristics. This is related to the ideas introduced by Robinson (1994), except that Robinson's analysis was related to the Pacific–North American index (PNA) rather than monthly rainfall totals. There is clearly much scope to consider different 'indicators' of climate change in conjunction with this type of statistical methodology, and this is one aspect that we hope to pursue in the future.

A second generalisation reviewed by Woolhiser (1992) is the incorporation of the El Niño–Southern Oscillation index (ENSO). This is another variable that could possibly be tied in with climate change studies. A third generalisation could be the inclusion of elevation as an additional covariate. Apart from possibly improving the modelling of the data, this could help in projecting rainfall at higher elevations than those available in the data base (meteorological stations tend to be situated on the valley floor rather than at the top of a mountain). In North Carolina this feature is very unlikely to be relevant on the coastal plain, where nearly all the stations are below 60 m in elevation, but it may well be worth incorporating into the mountain data. We shall not attempt that here, however.

2.3   PRELIMINARY ANALYSIS OF THE DATA

Our initial analysis of daily rainfall data is focused especially on how best to deal with the 'trace' values, and their implications for the marginal modelling of rainfall distributions. As discussed by Robinson (1994), all data are recorded in inches to the nearest 0.01 inch, and a trace value is recorded whenever there is rain but not enough for the operator to determine a numerical value.

Consider Table 2.1. This represents all transition counts among the three states 'zero', 'trace' and 'positive', for Asheville, for the period over which data are

Table 2.1   Transition counts for Asheville summer data

| | | **Day *n*** | | | |
		Zero	Trace	Positive	Total
Day	Zero	576	134	241	951
n − 1	Trace	122	41	127	290
	Positive	250	119	414	783

available, which is September 1964 to December 1986, but restricted to the summer months (June, July, August). Thus, for example, out of 951 days on which no rainfall was recorded, the next day has no rainfall on 576 occasions, trace rainfall on 134 occasions and positive rainfall on 241 occasions. One could use this table to compute the transition probabilities of a three-state Markov chain, in which the states 'zero', 'trace' and 'positive' are coded respectively as 1, 2 and 3. For example, p_{12}, the probability that tomorrow is a trace day given that today is a dry day, may be estimated as $134/951 = 0.141$.

In subsequent analysis we shall use higher-order Markov chains to model the pattern of wet and dry days. For the moment, however, we consider just first-order Markov chains, and study the question of whether the 'trace' state can be combined with either of the other two. Consider first a null hypothesis H_0 that the reduced process in which 'trace' and 'zero' are combined into a single state retains the Markov property. A general condition for a subset of states E in a Markov chain to be collapsible to a single state while retaining the Markov property is that $p_{ij} = p_{i'j}$ whenever $i \in E$, $i' \in E$ and $j \notin E$. In the present case this reduces to testing the hypothesis $H_0 : p_{13} = p_{23}$ against the alternative $H_1 : p_{13} \neq p_{23}$. For the data in Table 2.1, p_{13} and p_{23} are estimated to be $241/951$ and $127/290$ under H_1, and the common value $368/1241$ under H_0. Therefore the likelihood ratio test statistic is

$$241 \log \frac{241}{951} + 710 \log \frac{710}{951} + 127 \log \frac{127}{290} + 163 \log \frac{163}{290} - 368 \log \frac{368}{1241}$$
$$- 873 \log \frac{873}{1241} = 17.310.$$

which is clearly significant against the standard $\frac{1}{2}\chi_1^2$ null distribution. Thus we conclude that the 'trace' and 'zero' states cannot be combined into one. Now let us consider combining the 'trace' and 'positive' states. In this case the null hypothesis is $H_0 : p_{21} = p_{31}$, the alternative $H_1 : p_{21} \neq p_{31}$, and the likelihood-ratio test statistic is

$$122 \log \frac{122}{290} + 168 \log \frac{168}{290} + 250 \log \frac{250}{783} + 533 \log \frac{533}{783} - 372 \log \frac{372}{1073}$$
$$- 701 \log \frac{701}{1073} = 4.722,$$

which is still significant, but not nearly so much. Thus in this case a formal test would strictly speaking reject both null hypotheses, though much less strongly in the second case.

This conclusion sets a pattern that is largely reflected in the other analyses conducted. In the case of Asheville, with the winter data (December, January, February) the test statistics for the same two null hypotheses are respectively 1.192 and 9.252, which would support combining the 'trace' and 'zero' states, but this result appears exceptional. Of the 20 tests conducted (5 stations and 4 seasons), only 2 give stronger support to combining 'trace' and 'zero' than they do to combining 'trace' and 'positive', and in the majority of cases the test statistics for the latter hypothesis are not significant. Indeed, 7 of 8 cases involving the two coastal stations support this conclusion, the test statistic for combining 'trace' and 'positive' being less than 1.0 and therefore not significant. In contrast, in 7 of the 8 cases the alternative test statistic for combining 'trace' and 'zero' is significant.

On the basis of these results, we therefore conclude that, in most cases, it is reasonable to combine the 'trace' and 'positive' states into a single state for analysing the rainfall occurrence process. The results do not support treating 'trace' as a measurement of 0.

We now turn to the distribution of rainfall amounts. As noted in Section 2.2, most previous authors, including Stern and Coe (1984), have used a continuous distribution such as the gamma distribution to model rainfall amounts Y' above a small value δ representing the smallest possible measurement. In the present case rainfall amounts are recorded to the nearest 0.01 inch, so it seems reasonable to take $\delta = 0.005$.

In considering possible rainfall distributions, we have tried testing a suite of distributions appropriate for positive variables. These are

  (i) exponential,
 (ii) gamma,
(iii) Weibull,
 (iv) lognormal,
  (v) generalised Pareto,
 (vi) transformed generalised Pareto (i.e. $(Y')^\lambda$ has a generalised Pareto distribution for some $\lambda > 0$),
(vii) transformed gamma.

The generalised Pareto distribution has distribution function

$$F(y; \xi, \sigma) = 1 - \left(1 + \xi \frac{y}{\sigma}\right)_+^{-1/\xi}, \quad y > 0,$$

where $\sigma > 0$ and ξ may be positive or negative, the limit $\xi \to 0$ corresponding to the exponential distribution with mean σ. This distribution is widely used in extreme value theory for modelling excesses over a threshold (see Coles, 1994— Chapter 4 of this volume), and was included here because of the obvious analogy between rainfall processes having a zero or positive value, and threshold exceedances in a continuous-state process.

Figure 2.1 shows so-called Q–Q plots of observed versus expected values of the order statistics, for the winter data at Asheville, under six of the above seven models. The lognormal distribution was omitted because, as things turn out, the fit is very bad indeed. Also shown on the plots are the minimised negative log likelihood after the parameters have been fitted by maximum likelihood. All these plots are based on the positive data values with trace values omitted.

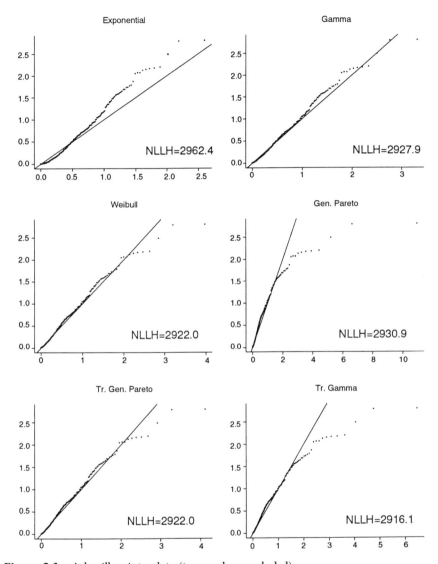

Figure 2.1 Asheville winter data (trace values excluded).

It can be seen that the Weibull, transformed generalised Pareto and transformed gamma distributions have the best likelihood fits, though the Q–Q plot for transformed gamma, which has the best likelihood fit of all, suggests a poor fit in the upper tail. Combining the likelihood values with subjective judgement of the Q–Q plots, it would appear that the gamma, Weibull and transformed generalised Pareto are the best fits, so providing some support to the traditional use of the gamma distribution. Similar results were obtained for other cases tried.

However, suppose we include the trace values in this analysis? We can consider the trace values to be censored data points lying between 0 and 0.005, while the rest of the positive data can be considered as grouped data, the grouping intervals being of length 0.01 and centred at the recorded data values. For example, a recorded value of 0.55 inches will be treated as a grouped value lying between 0.545 and 0.555. Unlike the analysis of Stern and Coe (1984), this version of the problem is not so easily treated within a statistical package such as GLIM, but

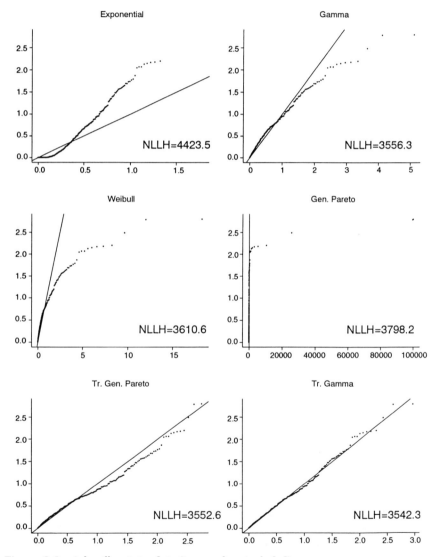

Figure 2.2 Asheville winter data (trace values included).

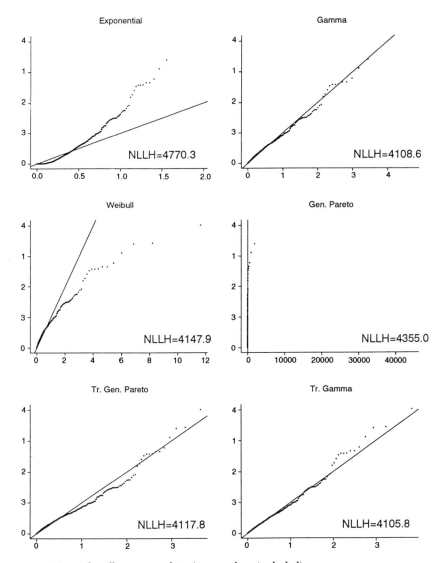

Figure 2.3 Asheville summer data (trace values included).

nevertheless is straightforward to handle directly as a numerical maximum likelihood problem. The results, and associated Q–Q plots, are shown in Figure 2.2. Again the Asheville winter data were used for this illustration.

In this case it can be seen from the Q–Q plots that the gamma distribution does not perform so well. In fact a clear winner, in this particular case, is the transformed gamma distribution, which does best based both on likelihood values and the plot. An explanation for the deterioration in the performance of the

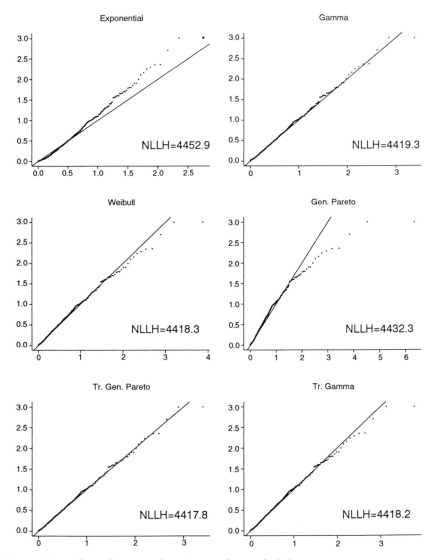

Figure 2.4   Whiteville winter data (trace values included).

gamma distribution is that, when the parameters are modified to take account of all the (assumed censored) trace values, the distribution no longer fits well in the upper tail. Since we should like to have a model that includes the trace values, this is of concern. The exponential and Weibull distributions clearly do not fit at all well, and the generalised Pareto performs terribly, as does the lognormal (not shown). An explanation for the poor performance of the lognormal distribution lies in the large number of trace or very small values, which skews the distribution of the logged data greatly to the left. It would appear that something similar is making things work badly for the generalised Pareto distribution, though it does seem, here and in subsequent cases, that adding a transformation parameter to the generalised Pareto distribution greatly improves the fit. The primary motivation for introducing the generalised Pareto, however, was to explore the analogy between rainfall event and threshold exceedances in a continuous-state process, but because this analogy does not seem to be very helpful, it is not pursued in the subsequent analysis.

A similar analysis was applied to several other data sets. For the Asheville summar data (Figure 2.3), the plots and likelihood values again support the transformed gamma distribution, though in this case the ordinary gamma distribution is not far behind. As an example of a coastal station, Figure 2.4 shows the winter data from Whiteville. In this case, four of the six distributions seem to be providing acceptable fits.

It was decided also to add the mixed exponential distribution to this classification. The mixed exponential distribution has a distribution function of the form

$$F(y) = 1 - \phi e^{-y/\mu_1} - (1 - \phi)e^{-y/\mu_2}$$

and has been widely used in studies of this nature, in particular by Woolhiser and Roldán (1982), who found it superior to the gamma distribution on the basis of numerous maximum-likelihood fits. When this was applied to the present data, an interesting conclusion emerged. If the trace values were omitted from the analysis then the mixed exponential distribution indeed emerged as the best, in most cases where it was tried. However, for the censored data analysis in which trace values were included, it did much worse (based on maximised likelihoods) than the gamma or transformed gamma distributions.

Our main conclusion from this analysis is that there is no reason to exclude the trace values: they can be treated as censored data points, and parametric models fitted by numerical maximum likelihood. As for the actual distribution adopted, there is still plenty of room for debate, but our analysis supports the conclusion that either the gamma or transformed gamma distribution performs reasonably in each of the cases examined.

2.4   TIME SERIES MODELS FOR A SINGLE SITE

In the light of the preliminary discussion of Section 2.2 and the exploratory analysis of Section 2.3, we are now able to formulate suitable models for the time series of daily rainfall totals at a single site. Let X_n be 1 if the site is wet on day n, 0 otherwise, counting all trace values as wet. Let Y_n denote the amount of rainfall on day n. Following the grouped data analysis of Section 2.3, we assume that the recorded value is 'trace' if $0 < Y_n < 0.005$, and otherwise is rounded to the nearest 0.01 inch. We consider models in which $\{X_n\}$ is treated as a kth-order Markov chain, for some $k \geqslant 1$, and $\{Y_n\}$ have a gamma or transformed gamma distribution conditionally on certain covariates. In our general formulation we shall allow these covariates to include past values of the process, so a Markov type of structure is permissible for $\{Y_n\}$ as well for $\{X_n\}$.

It is a challenging question how best to deal with seasonality. In the Stern–Coe approach all parameters are represented by a Fourier series after transformation using a suitable link function. This approach is very general, but suffers from a lack of parsimony as soon as the models get at all complicated. As discussed by Robinson (1994), in North Carolina we anticipate qualitative differences between the pattern of rainfall events, with cyclonic precipitation predominating in winter and convective precipitation in summer. This suggests that it may be better to fit entirely separate models for the two seasons rather than to try for one general model encompassing both. In fact, preliminary analyses using the Stern–Coe approach suggested that at least two Fourier terms would be needed in any representation of their type. Since two Fourier terms means five coefficients to estimate (a constant plus two sine and two cosine terms), and this must be multiplied by the number of parameters in the model, it can be seen that the number of coefficients to estimate quickly becomes prohibitive. Climatologists traditionally prepare different analyses for the different seasons of the year, and this is the approach that has actually been followed here.

Let us first consider the order of Markov chain for $\{X_n\}$. Table 2.2 shows the fitted negative log-likelihood value (NLLH) for a Markov chain of order 0 (i.e. independent observations), 1, 2 and 3, for each of our five stations, for the summer (June, July, August) and winter (December, January, February) seasons. These models have respectively 1, 2, 4 and 8 parameters, so the changes in degrees of freedom from one model to the next are respectively 1, 2 and 4. The 95% points of the $\frac{1}{2}\chi^2_v$ distributions are 1.92, 3.00 and 4.74 for $v = 1, 2$ and 4. Moving from order 0 to order 1, for all 10 cases, the change in NLLH is well in excess of 1.92, and therefore highly significant, as expected. Moving from order 1 to order 2, the NLLH is significant in all cases except the Asheville, Smithfield and Whiteville summer data sets, though in the case of Whiteville it only just fails to be

Table 2.2 Comparison of negative log likelihoods for Markov chains of various orders

Data set	Years of data	Order 0 NLLH	Order 1 NLLH	Order 2 NLLH	Order 3 NLLH
Asheville summer	1965–1986	1398.880	1333.641	1332.727	1330.690
Asheville winter	1964–1986	1397.376	1345.364	1332.440	1330.897
Highlands summer	1948–1986	2440.554	2291.580	2284.322	2280.880
Highlands winter	1948–1986	2216.438	2115.610	2103.255	2095.377
North Fork summer	1948–1986	2388.998	2259.201	2248.821	2246.242
North Fork winter	1948–1986	2240.830	2165.867	2143.730	2142.062
Smithfield summer	1948–1986	2312.449	2229.413	2229.366	2227.864
Smithfield winter	1948–1986	2277.962	2193.105	2177.995	2176.684
Whiteville summer	1954–1986	1997.198	1908.547	1905.581	1903.235
Whiteville winter	1954–1986	1850.576	1806.959	1794.697	1790.887

significant at the 5% level. The only case for which the order-3 model is significant against order 2 is Highlands winter. From these comparisons we conclude that a second-order Markov chain is the appropriate choice in the majority of cases, but the results also suggest that temporal dependence in summer is of a shorter range than that in winter, consistent with the notion of short-duration convective storms being the predominant mode of precipitation during the summer months.

Consider now the process of rainfall amounts $\{Y_n\}$. The basic model is that, conditionally on $Y_n > 0$, Y_n^λ has a gamma distribution, with density given by (2.3). Here λ and the gamma shape parameter κ are constants, and $\log \mu(n)$ is allowed to depend linearly on certain covariates. For a model of order k in X and order l in Y, the covariates are the 2^k indicator variables formed by all possible combinations of $X_{n-k}, \ldots, X_{n-1}$, and $Y_{n-l}, \ldots, Y_{n-1}$. For the purpose of defining the covariates, Y_{n-j} for $j \geq 1$ is defined to be 0 on days with no rain and 0.005 on days with trace rainfall.

Some examples of the resulting model fits are given in Tables 2.3–2.3. Table 2.3 shows the summer results for Smithfield. From a comparison of the NLLH values reported, it can be seen that there is essentially no need to improve on the simple model in which $k = l = 0$. Also shown in the table are the estimated values of κ and

Table 2.3 Comparison of Markov models for rainfall amounts: Smithfield summer data

k	l	NLLH	Parameters	κ	λ
0	0	5944.033	3	0.49	0.96
1	0	5943.174	4	0.50	0.95
2	0	5941.203	6	0.50	0.95
3	0	5939.640	10	0.49	0.97
1	1	5942.903	5	0.50	0.95
2	1	5940.917	7	0.50	0.95
2	2	5940.714	8	0.50	0.95
3	1	5939.221	11	0.49	0.96
3	2	5939.012	12	0.49	0.96

Table 2.4 Comparison of Markov models for rainfall amounts: Highlands winter data

k	l	NLLH	Parameters	κ	λ
0	0	6485.521	3	0.29	1.40
1	0	6471.157	4	0.27	1.50
2	0	6470.525	6	0.27	1.50
3	0	6469.900	10	0.27	1.50
1	1	6470.910	5	0.27	1.51
2	1	6470.091	7	0.27	1.50
2	2	6470.089	8	0.27	1.50
3	1	6469.544	11	0.27	1.50
3	2	6469.492	12	0.27	1.51

Table 2.5   Comparison of Markov models for rainfall amounts: North Fork summer data

k	l	NLLH	Parameters	κ	λ
0	0	6927.653	3	0.71	0.91
1	0	6926.998	4	0.71	0.91
2	0	6923.909	6	0.71	0.92
1	1	6924.577	5	0.70	0.93
2	1	6920.963	7	0.70	0.93

λ for each of the models fitted. The values of κ are all significantly different from 1, but those of λ (which each have standard error around 0.08) are not. Thus in this case we could not improve on the Stern–Coe model of independent gamma variates.

In general we should expect things to be more complicated for the mountain stations and for the winter data, and, as an example of that, Table 2.4 gives the winter results for Highlands. In this case there is a sharp drop in NLLH as a result of including X_{n-1} among the covariates, but no further significant reduction if additional covariates are included. The standard errors of the estimates of κ and λ are approximately 0.03 and 0.13 respectively, so in this case the estimates of $\kappa = 0.27$ and $\lambda = 1.50$ are both significantly different from 0. Thus in this case the preferred model is a transformed gamma model with first-order Markov dependence in $\{X_n\}$.

Most of the other results that have been examined are consistent with the results of Table 2.3 and 2.4. The preferred models have orders either $k = l = 0$ or $k = 1, l = 0$. As an example of one which goes against this pattern, Table 2.5 shows some of the summer results for North Fork. In this case the preferred model apparently has $k = 2, l = 1$.

To summarise, in this section we have examined Markov models for both $\{X_n\}$ and $\{Y_n\}$. A second-order Markov chain seems adequate for $\{X_n\}$, though for the summar data sometimes a first-order chain fits as well. The models for $\{Y_n\}$ again use either a gamma or transformed gamma distribution, either independent or incorporating a first-order Markov dependence on X_{n-1}, though, as we saw in Table 2.5, sometimes a more complicated model appears to be needed.

2.5   BIVARIATE MODELS

We now turn our attention to the construction of models for the bivariate time series of daily rainfall totals at each of two sites. Our ultimate objective is to find measures of dependence between the two sites, which can then be used as part of a general study of spatial dependence that will then help to answer some of the questions posed in Robinson (1994).

The general idea will be to follow the same principles as in the previous section, modelling the rainfall occurrence process separately from the process of rainfall amounts, using Markov dependence structures for each, but now employing bivariate instead of univariate distributions. One technical difficulty about this is the need for general families of bivariate distributions with given marginals. If we consider a bivariate random variable (Z_1, Z_2) in which the marginal distribution functions are expressed in the form

$$\Pr\{Z_j > z\} = 1 - G_j(z) = \exp[-H_j(z)], \quad j = 1, 2,$$

G_j being the distribution function and H_j the cumulative hazard function, then one bivariate family is given by

$$\Pr\{Z_1 > z_1, Z_2 > z_2\} = \exp\{-[H_1(z_1)^{1/\alpha} + H_2(z_2)^{1/\alpha}]^{\alpha}\}, \tag{2.4}$$

with $0 \leqslant \alpha \leqslant 1$, the limits $\alpha \to 1$ and $\alpha \to 0$ corresponding to independence and total dependence ($Z_1 = Z_2$ with probability 1) respectively.

The dependence function (2.4) is widely used in reliability and extreme value theory, where it is often called the logistic model (Tawn, 1988). The fact that this is a valid bivariate distribution function for all G_1 and G_2 follows immediately by pointwise transformation of the margins. It will be used here primarily as a means of characterising the dependence between the marginal distributions of rainfall, though a comparison of different forms of dependence function has not been attempted.

In the case of binary data—where, say, Z_1 and Z_2 take on only the values 0 and 1—we may define $p_j = \Pr\{Z_j = 1\}$ ($j = 1, 2$), and the formula becomes

$$\begin{aligned}
p_{12} &= \Pr\{Z_1 = 1, Z_2 = 1\} \\
&= \exp\{-[(-\log p_1)^{1/\alpha} + (-\log p_2)^{1/\alpha}]^{\alpha}\},
\end{aligned} \tag{2.5}$$

which is simply an alternative parametrisation of the joint probability. In this case $\alpha = 0$ is the maximally dependent case $p_{12} = \min(p_1, p_2)$, $\alpha = 1$ the independent case $p_{12} = p_1 p_2$, and the range may be extended to $\alpha > 1$ (if $p_1 + p_2 \leqslant 1$ then the minimum possible value of p_{12} is 0, corresponding to $\alpha = \infty$, while for $p_1 + p_2 > 1$ the minimum is $p_1 + p_2 - 1$).

In the following we shall write $\rho = 1 - \alpha$ to emphasise the analogy with a correlation coefficient.

Following the same kinds of procedures as in the previous section, we shall first construct a Markov chain model for the joint process of rainfall events. Denoting the observation for day n as $(X_{n,1}, X_{n,2})$, where $X_{n,1}$ is 1 if site 1 is wet on day n and $X_{n,2}$ is 1 if site 2 is wet on day n, we consider models of the following structure:

(a) $p_1 = \Pr\{X_{n,1} = 1 | \text{past}\}$ is a function of $X_{n-k,1}, \ldots, X_{n-1,1}$ and $X_{n-l,2}, \ldots, X_{n-1,2}$;

(b) $p_2 = \Pr\{X_{n,2} = 1 | \text{past}\}$ is a function of $X_{n-l,1}, \ldots, X_{n-1,1}$ and $X_{n-k,2}, \ldots, X_{n-1,2}$;

(c) $p_{12} = \Pr\{X_{n,1} = 1, X_{n,2} = 1 | \text{past}\}$ is defined by (5) with $\alpha = 1 - \rho$.

As an example, we consider the Highlands and North Fork summer data sets.

Table 2.6 Joint Markov models for bivariate rainfall occurrences: Highlands–North Fork summer data

k	l	NLLH	Parameters	k	l	NLLH	Parameters
0	0	4634.299	3	2	0	4452.881	9
1	0	4472.867	5	2	1	4200.946	11
1	1	4207.264	7	2	2	4199.718	15

Table 2.7 Joint Markov models for bivariate rainfall occurrences: Highlands–North Fork winter data

k	l	NLLH	Parameters	k	l	NLLH	Parameters
0	0	4105.818	3	2	0	3981.968	9
1	0	4007.211	5	2	1	3640.233	11
1	1	3651.891	7	2	2	3629.362	15

A sampling of possible models and their NLLH values is given in Table 2.6.

In this case the best model would appear to be that corresponding to $k = 2$, $l = 1$, but the most striking feature of the analysis is the result for $k = l = 1$, which shows the dramatic improvement in fit as soon as p_1 and p_2 are allowed to depend on the past values of both processes.

The corresponding winter results are shown in Table 2.7. In this case the optimum model would appear to be $k = l = 2$, but again it is the result for $k = l = 1$, when compared with those for $l = 0$, that is the most remarkable feature. Similar results were obtained for the two coastal stations, Smithfield and Whiteville. Asheville was not included in this analysis because its recording time is midnight, as against early morning for the others, so the bivariate distribution would be less easy to relate to actual rain storms.

We now consider a corresponding analysis for rainfall amounts. Let $(Y_{n,1}, Y_{n,2})$ denote the amounts at the two sites on day n. Conditionally on both values being positive, we assume that $Y_{n,1}$ has a gamma distribution of the form (2.3) with parameters κ_1 and $\mu_1(n)$, and $Y_{n,2}$ has a gamma distribution with parameters κ_2 and $\mu_2(n)$. To simplify matters slightly, we do not consider a transformation parameter in this case. The parameters κ_1 and κ_2 are treated as constants, but $\log \mu_j(n)$ ($j = 1, 2$) is allowed to depend linearly on certain past values. If these past values are $X_{n-k_1,j}, \ldots, X_{n-1,j}, X_{n-1,3-j}, \ldots, X_{n-1,3-j}, Y_{n-k_2,j}, \ldots, Y_{n-1,j}$ and $Y_{n-l_2,3-j}, \ldots, Y_{n-1,3-j}$ then we describe the model as being of order (k_1, l_1, k_2, l_2). Here X is being treated as a factor variable, e.g. if $k_1 = 2$ then there are, for each $j = 1, 2$, four coefficients corresponding to four possible values of the pair $X_{n-2,j}, X_{n-1,j}$. On the other hand, Y is treated as a continuous variable that enters linearly into $\log \mu_j(n)$.

For the joint distribution of $(Y_{n,1}, Y_{n,2})$, we again assume the form (2.4) with

Table 2.8 Joint Markov models for bivariate rainfall amounts: Highlands–North Fork winter data

k_1	l_1	k_2	l_2	NLLH	Parameters	k_1	l_1	k_2	l_2	NLLH	Parameters
0	0	0	0	11 895.182	5	2	1	1	1	11 784.855	17
1	0	0	0	11 885.838	7	2	2	0	0	11 843.686	17
1	0	1	0	11 882.013	9	2	2	1	0	11 838.695	19
1	1	1	0	11 844.033	11	2	2	1	1	11 780.679	21
1	1	1	1	11 843.108	13	2	1	2	1	11 784.191	19
2	1	0	0	11 846.924	13	2	2	2	1	11 779.629	23
2	1	1	0	11 842.346	15	2	2	2	2	11 777.746	25

$\alpha = 1 - \rho$ an unknown constant. This is not a particularly elegant form of joint distribution, because it requires evaluating the incomplete gamma function, but that is something that can be implemented numerically without difficulty, and it has the advantage of being a very general approach. As in the univariate analysis, the Y values are actually treated as grouped observations, with 'trace' values being grouped in the interval $(0, 0.005)$. For those days on which only one of the sites has rain, we simply evaluate the log-likelihood contribution for that site, ignoring the dependence. Those days on which neither site has rain, of course, contribute nothing to the log likelihood.

Table 2.8 shows an example of the results of this analysis. It is striking to see how much the negative log likelihood falls as additional dependence terms are introduced. Although there is room for debate as to the best model, the $(2, 1, 1, 1)$ model stands out as a still moderately simple model that has much lower NLLH than the ones before it in the table.

For the Highlands–North Fork summer data, a similar analysis again produced the $(2, 1, 1, 1)$ model as the best. In the case of Smithfield and Whiteville, the analysis resulted in the model $(1, 1, 1, 1)$ for the winter data and $(0, 0, 0, 0)$ for the summer data. Thus it is only in the very last case—the summer data on the coastal plains—that the results are simple enough to support independent gamma variables as the model of choice.

For the final part of this section we now consider the dependence parameter ρ. In fact, there are two ρ parameters for each station pair and season: one corresponding to the rainfall occurrence analysis as in Table 2.6 and 2.7, and the other to the rainfall amounts analysis as in Table 2.8. We shall write ρ_1 for the parameter associated with rainfall occurrences and ρ_2 for the parameter associated with rainfall amounts. The results, using in each case what we have judged to be the best model within the class of Markov models considered, are in Table 2.9.

All of these estimates have a standard error of around 0.02. Thus it can be seen that, although in all stations the dependence between rainfall occurrences is stronger than that between rainfall amounts, the latter are still highly significant. The spatial dependence is stronger in winter than in summer, and stronger for the coastal plain stations than for the mountain stations. The latter point is reinforced

Table 2.9 Estimated dependence parameters under best Markov model

Stations/season	p_1	p_2
Highlands–North Fork winter	0.47	0.35
Highlands–North Fork summer	0.24	0.09
Smithfield–Whiteville winter	0.62	0.36
Smithfield–Whiteville summer	0.41	0.12

by the fact that the distance between the two coastal plain stations Smithfield and Whiteville (70 nautical miles) is actually greater than that between the mountain stations Highlands and North Fork (55 nautical miles).

2.6 ANALOGUES OF VARIOGRAM ANALYSIS

One of the most widely adopted current methods in the analysis of spatial data is based on the variogram, both as an exploratory device and as a means of fitting models for spatial correlation. Consider a process $Z(t)$, where t is a spatial variable. For example, in geostatistics $Z(t)$ may be the density of an ore measured at a location t. The variogram of Z is defined by

$$V(t_1, t_2) = \text{Var}\{Z(t_1) - Z(t_2)\}.$$

If the process is *intrinsically stationary* then $V(t_1, t_2)$ is a function of the vector difference $t_1 - t_2$, and if it is also *isotropic* then it depends only on the scalar distance $\| t_1 - t_2 \|$. In the stationary isotropic case it is a simple matter to plot the empirical variogram against $\| t_1 - t_2 \|$. There are also a variety of parametric models for the variogram as a function of distance, and sometimes these are fitted directly from the empirical variogram via a least-squares or weighted least-squares approach. Other methods of fitting, such as maximum likelihood or restricted maximum likelihood, are in principle more powerful, but are not universally accepted at the moment. A detailed discussion of these and other issues in variogram analysis was given by Cressie (1991).

It is possible that we could apply variogram analysis directly to the measured rainfall amounts, including the zeros. This would, however, go against much of the discussion in this chapter, where we have treated the rainfall occurrences separately from the rainfal amounts. In this section we therefore confine attention to rainfall occurrences, i.e. $Z(t)$ is defined to be 1 if it is raining at site t and 0 otherwise, and consider various measures of spatial dependence that can be used in a way analogous to variogram analysis.

The obvious idea is to take some measure ρ of the dependence between two stations, and to plot that as a function of distance. Throughout we let p_1 and p_2 denote the marginal probabilities of rainfall at each of the two stations, and p_{12} the joint probability. Among the measures we might consider are the following.

(i) The conditional probability of rainfall at one station given the other—this actually leads to two values, p_{12}/p_1 and p_{12}/p_2, since conditional probabilities are not symmetric. This procedure is directly analogous to that discussed by Robinson (1994). Since there are two probabilities to calculate for any pair of stations, they will both be included in any subsequent plot. Apart from the lack of symmetry, a disadvantage of this as a measure of dependence is that the range of values is restricted and this can interfere with the specification of a mathematical model.

(ii) The covariance $p_{12} - p_1 p_2$ is an obvious choice, and the one most closely related to the variogram itself, and it does have the advantage of being a symmetric measure. A disadvantage is that the range of values is still restricted: if we know p_1 and p_2, then we know that the covariance lies between $-p_1 p_2$ and $\min(p_1, p_2) - p_1 p_2$, and this can again be inconvenient if we are trying to specify a mathematical formula for covariance as a function of the distance between the stations.

(iii) In biostatistics a common device for measuring the dependence between two binary random variables Z_1 and Z_2 is the odds ratio

$$r = \frac{E\{Z_1 Z_2\} E\{(1 - Z_1)(1 - Z_2)\}}{E\{Z_1 (1 - Z_2)\} E\{(1 - Z_1) Z_2\}}.$$

Therefore we could consider plotting the odds ratio or its logarithm as a function of distance. In fact we have used $\rho = (r - 1)/(r + 1)$ to create a quantity analogous to the correlation coefficient. It has the advantage of being symmetric, and restricted only to the range $(-1, 1)$. (I am grateful to David Cox for personal comment, as a result of which this suggestion was added to the analysis.)

(iv) A fourth possibility is to let ρ be the same as α defined by (2.5), which has the advantage of allowing the analysis to be directly tied in with the models for dependence in the previous section. For the analysis considered here, this is estimated directly, for given p_1, p_2 and p_{12}, by solving (2.5) to find $\alpha = \rho$. In contrast with the preceding section, this ignores the time series dependence in the data, but on the other hand it is a more direct method of measuring dependence. This measure of dependence only applies to positively correlated random variables, but this is realistic for rainfall data, and the permissible range of ρ is $[0, 1]$ whatever the values of p_1 and p_2.

For each of these definitions of ρ, we can try plotting ρ against distance and so consider various functional forms for the relationship. For this purpose, we used some data summaries prepared by Peter Robinson, which contained joint occurrence information for the year 1981 for a large number of daily rainfall stations in North Carolina. Again, we have separated the mountain and coastal plain areas of the state, ignoring a third region, known as the Piedmont, which lies between the two. Considering just the mountain area for the moment, Figure 2.5 shows a plot of the summer data, and Figure 2.6 one of the winter data. The four figures labelled (i)–(iv) correspond to the four definitions of ρ given above.

As in Robinson (1994), an attempt was made to fit various functional forms to these plots. The curves added to the plots in Figures 2.5 and 2.6 correspond to Robinson's proposal

$$\rho \equiv \rho(d) = a - b\log d, \qquad (2.6)$$

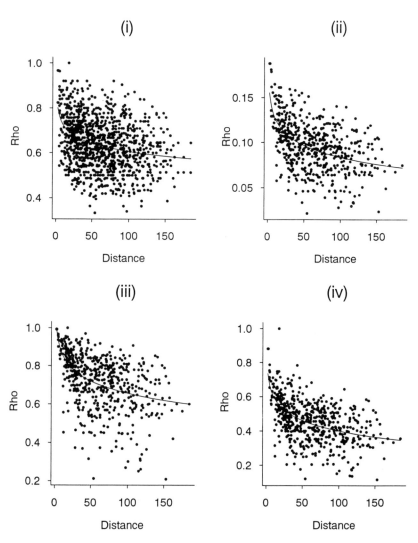

Figure 2.5 Dependence plots for mountains: summer data.

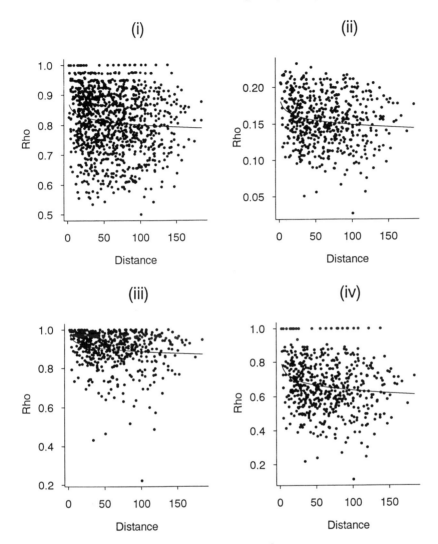

Figure 2.6 Dependence plot for mountains: winter data.

where d is distance, fitted by ordinary least squares. This is open to the obvious objection that the range of (2.6) goes outside the interval (0, 1), but in fact other forms of functional fit do not obviously improve on (2.6). A more rigorous analysis would require a much more detailed consideration of what kinds of functional forms of $\rho(d)$ are consistent with a valid stochastic model for the process, but the purpose of the present analysis is primarily exploratory, so we have not pursued these considerations.

The resulting parameters are summarised in Table 2.10.

Table 2.10   Fitted values of a and b in (6)

Region	Season	(i) a	(i) b	(ii) a	(ii) b	(iii) a	(iii) b	(iv) a	(iv) b
Mountains	Summer	0.84	0.052	0.17	0.019	1.07	0.090	0.81	0.088
		(0.017	0.004	0.005	0.0013	0.027	0.007	0.023	0.006)
Mountains	Winter	0.88	0.018	0.19	0.0082	0.99	0.023	0.82	0.040
		(0.016	0.004	0.007	0.0018	0.020	0.005	0.033	0.008)
Coastal Plain	Summer	1.06	0.089	0.26	0.035	1.36	0.14	1.14	0.15
		(0.021	0.005	0.008	0.0019	0.038	0.0091	0.034	0.0082)
Coastal Plain	Winter	0.96	0.025	0.23	0.012	1.01	0.015	0.93	0.040
		(0.019	0.004	0.008	0.0020	0.017	0.004	0.040	0.010)

The figures in parentheses are nominal 'standard errors' based on the least-squares fitting procedure, though, in view of the lack of independence, they cannot be regarded as anything more than nominal measures of accuracy, whose main purpose is to compare the different procedures. From the table, we see that, in every case, the slop b is smaller for the winter data than for the corresponding summer data, implying that spatial dependence falls off with distance much more rapidly in the summer than in the winter. Also, in most cases, if we compare corresponding mountains and coastal plain data, the parameter a is larger for the coastal plain but the parameter b is larger in the mountains, implying that the spatial correlation is stronger on the coastal plain for short distances, but also drops off more rapidly with distance. However, within the actual range of distances in the sample, all the spatial dependences are stronger for the coastal plain than for the mountains. These results are consistent with those found in Section 2.5 and in Robinson (1994).

Comparing the different measures of dependence (i)–(iv), no clear-cut conclusion emerges. All four measures show a marked scatter about the fitted curve and all have similar ratios of 'standard errors' to parameter estimates in Table 2.10. Figures 2.6 (i), (iii) and (iv) have a strange appearance because a number of the estimated ρ values are 1 in the winter: there are some pairs of stations for which there were no days on which there was rain at station 1 but not at station 2. However, we should expect this phenomenon to disappear if we used a data record of longer than one year.

Another way to try to improve on (2.6) is to allow for an anisotropic model — specifically, the form of anisotropy known as geometric anisotropy (Cressie 1991, page 64) was tried. In this, d is taken to be not the Euclidean distance between two points, but rather the distance after taking a linear transformation of the plane, the linear transformation being chosen, just as the constants a and b in (2.6), to minimise the sum of squares of residuals in the plot. However, the results suggested that the degree of anisotropy in the data is slight and probably not significant.

2.7 DISCUSSION

This chapter has developed models for both the temporal and spatial dependences in a rainfall process, in which the rainfall occurrences and rainfall amounts were modelled separately. The results support the notion that dependences among the occurrences are stronger than dependences among the amounts, but in most cases the latter cannot be neglected. The analysis of Section 2.6 has indicated the possibilities of using these ideas to make more general statements about spatial dependence, but there are clearly limitations in this, that indicate the need for models for the joint distribution of rainfall at more than two sites.

In principle, of course, one could continue the analysis of Section 2.5 to obtain joint models for three sites, four sites, etc., but it is clear that the number of parameters will quickly become prohibitive. We believe that this points towards the development of more physically-based models for rainfall, as discussed by Cox and Isham (1994). From that point of view, the main virtue of the models in the present chapter will be to help identify important features of the data which such a model must reflect. Current work is in progress in this direction.

A philosophical point is that much of the theory of spatial and time series analysis is still wedded to the multivariate normal distribution, for which, of course, specification of all bivariate distributions is sufficient to specify the whole process. Our discussion of trace values, and of the need to separate the analyses of occurrences and amounts, has shown clearly that concepts based on the normal distribution are not very helpful in this analysis. It follows that consideration of bivariate distributions cannot be expected to be sufficient to derive all the quantities of interest.

Perhaps the main virtue of the present analysis is that, by reducing the characteristics of daily rainfall to a comparatively small number of parameters, it allows us to study the dependence of those parameters on various measures of the larger climate, such as the circulation indices discussed by Robinson (1994). In this way, it is possible to tie in the analysis with concept of climate change including, but not limited to, the greenhouse effect. This is another aspect that is presently being investigated.

ACKNOWLEDGEMENTS

This work was partially supported by the National Science Foundation Mathematics–Geostatistics Initiative, though a joint grant to Peter Robinson and myself. I am extremely grateful to Peter Robinson for many conversations about this work and about climatology generally. I also thank Geoffrey Pegram and David Cox for suggestions made at the spruce II meeting, which resulted in additions to Sections 2.2 and 2.6 respectively.

3

Precipitation Regime Changes over Small Watersheds

P. J. Robinson
University of North Carolina, Chapel Hill, USA

3.1 INTRODUCTION

The current concern about climate changes and their possible impact on human activities has heightened awareness that climate is not static. Examinations of past climates have indicated fluctuations and variations on a wide variety of time and space scales, but for almost all analyses an underlying static baseline has been assumed. Without such a baseline, most analyses become more complex, but they must be undertaken if rational response strategies to possible climate changes are to be developed. In most cases this requires a re-examination of many of the analysis techniques commonly used in climatology, and a similar examination of the statistical techniques. This chapter considers some aspects of this problem, emphasizing climatological problems in the analysis of precipitation in a situation where the underlying mean climate is not necessarily constant. Some of the complementary statistical problems are considered in Chapter 2 of this volume (Smith, 1994).

Precipitation can be analysed on a variety of time and space scales, some having implications for the fundamental understanding of climatic processes, others having practical applications in hydrology. For the present purposes, consideration is restricted to small watersheds having drainage areas of the order of a few thousand square kilometres, and to precipitation recorded on a daily or hourly basis. At these scales there is a major practical application in assessment of the potential for flash floods. Although these floods are commonly associated with

Statistics for the Environment 2: Water Related Issues. Edited by Vic Barnett and K. Feridun Turkman
© 1994 John Wiley & Sons Ltd.

heavy rainfall, a direct extreme value or threshold exceedance analysis is not appropriate, since floods are a function of soil moisture as well as precipitation, so that a large storm on a dry soil may cause no problem, while a much smaller one, following several days of prolonged rain, may have severe consequences. This has long been known to hydrologists, who have developed an arsenal of analysis tools to deal with the problem, including area–depth–duration curves (see e.g. Buishand, 1993) and stochastic time series models (see e.g. Stern and Coe, 1984). These tools, however, imply a stationary climate and, to a great extent, a random distribution of daily precipitation amounts. Hence they must be refined or replaced if useful information of climate change analyses are to be produced.

The climatological basis for an analysis of precipitation in a form that allows consideration of the non-stationary and non-random character is outlined in this chapter. It treats precipitation as a series of individual events that taken together constitute the precipitation regime of an area. The objective is not to analyse or forecast individual storm events, but to develop a typical or composite sequence that retains a realistic set of statistical and climatological characteristics of the events. This sequence must be developed in a way that allows modification, again with an emphasis on realism based on observed characteristics, as the underlying climate changes.

The conceptual approach adopted is to classify the individual precipitation events according to their mode of formation, determine the event characteristics as a function of this formation process, and ascertain the frequency of events in each class in the present climate and in any changed situation. This has the potential to allow an analysis similar to that undertaken by Cox and Isham (1988, 1994). However, examination of the precipitation data for a variety of regions indicates many of the climatological problems that must be considered before the results of any such analysis can be used routinely. Thus the emphasis here is on the climatological aspects of the problem of specifying precipitation, with a quantitative description of precipitation events leading to some indications of statistical and climatological areas where fuller investigation are needed to allow the development of scenarios of future climate.

3.2   PRECIPITATION EVENTS

A precipitation event is defined, for a particular observing point, as a period during which precipitation is either continuous or contains only dry intervals of less than a pre-determined length (Robinson and Henderson, 1992). The length of the dry interval will partly be determined by the temporal resolution of the available data and partly by the purposes of the analysis itself. Each event has a specific duration giving a specific precipitation amount. The precipitation regime is characterised by the frequency distribution of event durations and amounts, and includes the distribution of the separation time between individual events. Since climatology is usually concerned with typical conditions and their variability, this regime is best expressed on a seasonal basis, with values determined from a long-term, multi-year analysis. Although this definition was developed specifically for conditions at a single ground-based precipitation observing station, it is possible to extend it to simultaneous observations at neighbouring stations and thus to incorporate the spatial dimension of the precipitation regime and its constituent events.

There are a variety of mechanisms in the atmosphere that lead to the production of precipitation events. Thus there are also a variety of possible classification schemes. A very convenient and simple one, amenable to analyses appropriate for assessing hydrological changes associated with climate change, divides the processes into two broad categories: cyclonic and convective. In practice, these are more akin to the extremes of a continuum rather than being mutually exclusive, but they give precipitation with markedly different characteristics.

Cyclonic precipitation is associated with widespread, relatively gentle uplift of a mass of air, giving the feature variously known as a depression, a frontal cyclone or often, and rather loosely, a low-pressure front. Whatever the name, the feature is a region of fairly uniform horizontal stratus cloud, which commonly gives drizzle or light rain. Regions within the general area will have more intense uplift and heavier rain. Satellite and radar observations clearly indicate the resulting rainbands or individual rain cells. In addition, particularly at frontal surfaces, there will be a spatially well organized region of such enhanced uplift and heavier precipitation. These will frequently be in the form of a line of individual, well defined rain cells. Indeed, the cells at the front can easily exhibit many of the characteristics of convective precipitation.

Convection is a much more localized process, with relatively rapid vertical motions dominating, so that cumulus clouds, with marked vertical development, are created. These tend to give rather short sharp showers. As the convection becomes more vigorous, usually as the result of intense solar heating of the Earth's surface, greater vertical development occurs, and thunderstorms with greater precipitation amounts, result. In extreme, and somewhat special, conditions the convection will create tornados.

The precipitation at any point outside the tropics can be characterised as a sequence of wet periods interspersed with dry intervals, each wet period having the character of either a cyclonic or a convective event. The wet periods, depending on the time scale and resolution used, have been associated with, or labelled as, storms, depressions, frontal passages, rain cells, or rainbands. In this work they are simply called events, and the sequence of events is regarded as the precipitation regime.

The mix of events that creates the precipitation regime will vary with geographical location, with local topography, with season, and with the dominant mode of the atmospheric circulation. A major objective in the analysis of precipitation regimes is to identify the nature and magnitude of these controls. The basic effects of the controls are clear. On a long-term basis the regime will be a response both to the common or preferred circulation pattern for the season, which largely controls depression movement and cyclonic activity, and to location, which greatly influences the intensity of surface heating, the major influence on convective activity. Thus, for example, most of western Europe is dominated by depression passage for most of the year, and surface heating is relatively minor, so that cyclonic precipitation dominates. In the summer, and in the Mediterranean basin, warmer conditions encourage a higher proportion of convective activity. For the south-eastern United States, the seasonal contrasts are much more marked. In winter cyclonic activity dominates, but in summer convective activity is of at least equal importance. Interior North America has less activity, and lower precipitation totals throughout the year. In summer convection is important, while in winter rather small depressions are most common.

Most information about future climates, notably the changes that might occur as a result of increases in the concentrations of the atmospheric greenhouse gases created by anthropogenic actions, come from global climate models (Houghton *et al.*, 1992). These models, commonly used to simulate both the present climate and that after an effective doubling of atmospheric carbon dioxide, cannot yet provide reliable precipitation regime information (Robinson *et al.*, 1993). Nevertheless, they are beginning to give indications of circulation patterns that suggest the frequency, location and movement of individual depressions, and to estimate the potential for convective activity (Manabe and Brocolli, 1990). Thus it is theoretically possible to develop scenarios of future precipitation regimes by combining results of analyses of past conditions as a function of circulation patterns with the ciculation patterns projected by the models of future climates (Robinson and Finkelstein, 1991). There have been some analyses linking precipitation to circulation, the most successful developing statistical regression relationships on a monthly basis (Klein and Bloom, 1987). A more physically realistic event-based approach has been developed by Robinson and Henderson (1992), and tentative links with circulation patterns have been established (Henderson and Robinson, 1994). The work reported here is an outgrowth and extension of those works.

3.3   PRECIPITATION DATA

The identification and analysis of precipitation events requires observational data with a high spatial and temporal resolution. One minute, one kilometre data, for example, would allow detailed analysis of the intensity, duration and spatial pattern of individual rain cells and the structure of individual cyclonic and convective events. Such analyses are highly desirable to enhance our understanding of the underlying physical processes, but for hydrological purposes hourly or daily data with a few kilometre resolution might be all that is needed (Bruneau, 1993). Indeed, for many areas of the world, that is all that is available at present. Thus a major constraint on the type of analysis that can be undertaken comes from the data that are available.

For long-term climatological analyses of the type discussed here, the major data source is the network of surface rain-gauge stations. Other observational techniques are becoming available, but are not yet widespread or routine. Satellite observations have the potential to provide the desirable high spatial and temporal resolution data, but the provision of truly quantitative measures is still a major research effort and not yet a routine operational process (Arkin and Ardanuy, 1989). Radar observations, also having the potential for detailed spatial and temporal observations, are commonly non-quantitative. Here recent advances in technology suggest that radar is likely to be a significant source of data in the not too distant future. Certainly it has already been used in selected areas to aid in understanding the precipitation process and in the analysis of hydrological effects (Haggett, 1988; Croft and Shulman, 1989).

Most national meteorological services operate surface-based rain-gauge networks. These commonly provide a long period of record, so that climatological analyses, rather than single-storm analyses, can be undertaken, while the availability of a network of stations allows interstation comparison and some spatial analysis. Nevertheless, the data must be treated with caution, since they are prone to a variety of problems. The actual catch of any gauge will depend on its size, shape and height above the ground. While most nations have adopted standard gauges, so that interstation comparisons within nations are possible, there are no international standards. Indeed, there is no consensus as to the most desirable characteristics of rain gauges (Hughes *et al.*, 1993). Since the catch of any gauge commonly decreases as windspeed increases, changes in the configuration of the area surrounding the gauge changes the exposure and the catch. Slow secular effects, such as nearby tree growth, can therefore influence long-term precipitation trends. For event analyses, convective storms usually have higher wind speeds than do cyclonic ones, and so may be more severely affected by such changes. When recording gauges are used, especially at remote sites, there must always be concern for the detection and adequate recording of instrument failures. Added to the uncertainties created by the measurements themselves are the potential errors arising from the whole observational and archiving process. A good quality control system should rectify most major errors, but the spatially and temporally discontinuous nature of precipitation ensures that detection of small, subtle errors is difficult. The US National Weather Service records precipitation to one hundredth of an inch. In this work the data are rounded to the nearest tenth of an inch as a simple way of minimising these minor, undetected errors. Further, the non-standard units are retained, primarily since this work emphasises practical applications. In the United States SI units are rarely used in anything except scientific investigations.

Both of the two primary rain-gauge networks in the United States—one measuring daily precipitation, the other giving hourly values—are used here. While use of hourly data is most desirable, there is a low station density for this network. At any one time there may be in excess of 150 stations operating in the south-eastern United States. However, stations having long-term continuous records were required for the event analysis. Hence, after allowance for missing and unreliable data, a set of 26 high-quality stations, representing an area of approximately 800 000 km², was used for the temporal analysis. These data, the Hourly Precipitation Data, were obtained from the National Climatic Data Center (Hatch, 1983). This very-low-density network was inadequate for any spatial analysis. Consequently the daily data network, denser and more reliable, was used. For the south-east, and throughout much of the United States, the network density is of the order of 1 station per 1000 km². These data, the Climatological Data were also obtained from the National Climatic Data Center.

3.4   TEMPORAL VARIABILITY OF REGIMES

A prerequisite for the development of an event climatology was specification of the separation interval between storms. Using the hourly precipitation data, this interval must be in multiples of an hour. After consideration of the need to balance the desirability of keeping that interval as long as possible, in order to ensure that the various rainbands within a cyclonic system were treated together as a single storm, against the requirements that the interval be as short as practical, to ensure that individual convective events were correctly isolated, a separation interval of 2 h was chosen (Robinson and Henderson, 1992). This was found to be appropriate for the south-eastern United States, but may not be appropriate for other areas, such as Western Europe, where the intensity of the processes may be very different.

Although the binary classification of event types implies that there are two distinct regimes, the actual data indicated a continuum in both precipitation amount and event duration. This was apparent for stations in the south-eastern United States, the Great Lakes region and Colorado. Two stations were selected for presentation and comparison here; Grand Junction, Colorado, and Asheville, North Carolina. Both stations, typical of their states, are in mountain valley locations, but the former is in a continental interior, the latter close to the coast. Thus differences arising from geographical location and circulation patterns, as well as seasonal differences, could be anticipated. One fundamental difference was in the number of events. Largely because there were far fewer events in Colorado, the interval between events was much longer (Figure 3.1). For both stations and seasons the relatively large number of events separated by less than 6 h may partly be attributed to the method used to define events. Nevertheless, it also indicated that in Colorado the clustering of events was a common feature of the precipitation regime. Indeed, particuarly in summer, it appears that a series of events separated by less than two days occurred as a cluster, to be followed by a dry period lasting six or more days. In North Carolina this clustering was less apparent, and extended dry intervals were rather rare.

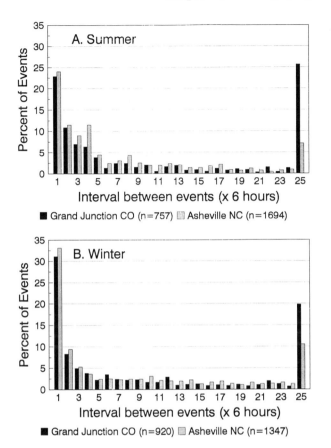

Figure 3.1   Percentage distribution of the length of time between events for Grand Junction, Colorado, and Asheville, North Carolina, in (A) summer and (B) winter. *n* indicates the number of events in the analysis.

Individual events in both states were typically of short duration. However, there were interstation difference that clearly showed the influence of geographical location (Figure 3.2). Grand Junction in summer had virtually no events lasting longer than 8 h, while Asheville had occasional ones lasting over 24 h. This reflects the almost exclusive convective activity in Colorado, contrasted with the mixture of cyclonic and convective in North Carolina. In winter, when both states are more likely to have a mixture, differences were less marked, although Colorado still had a tendency for shorter duration storms. There were a similar set of results for the amounts of precipitation (Figure 3.3). In both states in both seasons the majority of events produced 0.1 inch of precipitation or less. At Grand Junction greater amounts were more likely to occur in summer than in winter, although rarely did an event given more than 1.0 inch of rain. In Asheville there

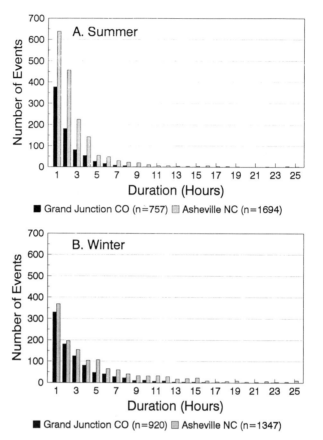

Figure 3.2   Distribution of event durations for Grand Junction, Colorado, and Asheville, North Carolina, in (A) summer and (B) winter. *n* indicates the number of events in the analysis.

was little difference in the distribution by season, and amounts in excess of 2.0 inch were not uncommon.

These results indicate that Colorado has fewer, lighter and shorter events than North Carolina, and that both regions have marked differences between seasons. Although it is not possible to separate unambiguously convective from cyclonic events when considering precipitation at individual points, these results suggest that a climatological interpretation of the precipitation regime in these terms is valid. Moreover, it is possible to suggest that the duration and interval results can be summarised, and potentially analysed in more detail, in terms of a Poisson process, while the amount data suggest an exponential or gamma distribution. Thus the present results provide some indications of the realistic distributions to be used with the stochastic time series models of point precipitation often used in hydrology, while exploration of the way in which the distribution parameters

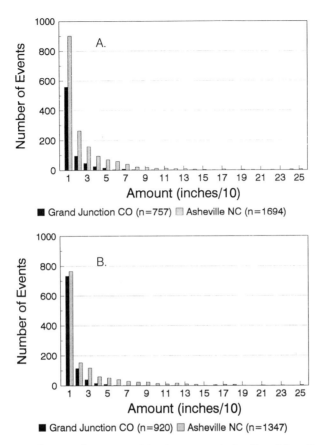

Figure 3.3 Distribution of event precipitation amounts for Grand Junction, Colorado, and Asheville, North Carolina, in (A) summer and (B) winter. *n* indicates the number of events in the analysis.

differ with season and location may provide further insights into the nature of the precipitation regimes.

Differences in event character as a function of circulation pattern can be explored using one of the many available circulation indices. Most indices are appropriate only for particular regions. For the south-eastern United States, and Asheville, NC in particular, the most appropriate index is the Pacific–North America (PNA) index (Wallace and Gutzler, 1981). With this, a negative value indicates zonal flow such that the main storm track in the south-east is directly from west to east. A high positive value indicates meridional flow and a tendency for storms to swing south-eastward from the Rocky Mountains towards Louisiana and the Gulf of Mexico, and then move north-eastward to cross the east coast somewhere south of New York (Figure 3.5). Although this index can be

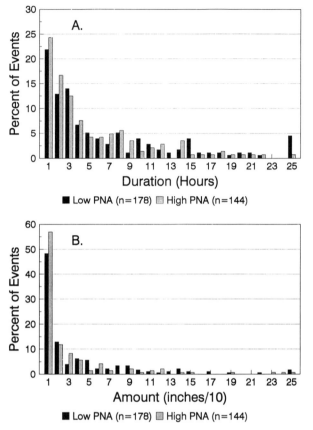

Figure 3.4   The distribution of (A) event duration and (B) event amounts for Asheville, North Carolina, in winter as a function of the Pacific–North America Index. *n* indicates the number of events in the analysis.

calculated for any season, only in winter is the overall flow regime consistently coherent enough to allow useful comparisons. Comparing composites of five winter with high PNA to five winters with low (highly negative) values (Figure 3.4), it is clear that shorter, lighter events were more common with a high PNA than with a negative one. This Asheville result was typical of most of the north-western portion of the region, with a reversal in response towards the south and east (Henderson and Robinson, 1994) (Figure 3.5). While these regional results display spatial coherence, they were derived from a rather sparse network of point observations, and thus must be treated with caution. Indeed, it is safest to assume that they indicate valid climatological interpretations at individual points rather than region-wide values. True spatial results must be developed using a much denser network.

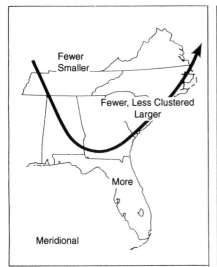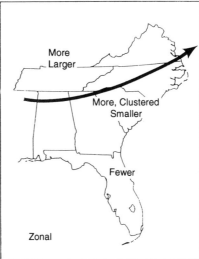

Figure 3.5 Generalised map of event characteristics associated with meridional and zonal circulation patterns in the southeast United States of America. (After K. G. Henderson and P. J. Robinson, Relationships between the Pacific/North American teleconnection patterns and precipitation events in the southeastern United States. *International Journal of Climatology* 1994. Reproduced by permission of Royal Meterological Society © 1994 Royal Meterological Society.)

3.5 SPATIAL VARIABILITY OF REGIMES

The spatial variability of events cannot be investigated with the same data setas the temporal analysis, because the hourly network is too sparse for any meaningful interpolation between observation stations. Consequently, the daily dataset, rather than the hourly one, had to be used, and a direct analysis of the same series of events as for the temporal study was not possible. Since in the areas considered here individual events are likely to have a general west-to-east movement, use of daily values tends to smooth out the location and increase the area of influence of a particular event. In addition, there is no convenient means of assessing whether on a particular day there is a single event covering or moving over the whole area or whether there is a series of isolated localised ones. It is theoretically possible to obtain this information by examining each daily synoptic weather map, but this forces a focus on individual storm systems rather than the climatologically typical or characteristic storm size of prime interest here. The result of using the daily dataset is therefore that there is no means of separating cyclonic and convective events for the spatial analysis. However, the temporal analysis had already indicated that there was, in practice when using rain-gauge data, a continuum of events, so this was not regarded as a severe problem.

The specification of the typical spatial extent of daily events was approached through the development of conditional probabilities of precipitation for pairs of stations. Probabilities were developed for all stations in the area taking daily observations at 7 a.m. local time. Precipitation was treated as a binary variable, each station each day being either wet or dry. The observational record contains a dummy value called 'trace', which indicates that precipitation has occurred but is below the measurement limit of the rain gauge. All days with a trace were treated as wet days. This approach takes no account of directionality in the probabilities, which could be very important when a series of rapidly moving storms cross a region. Rather, the aim was to investigate whether it was possible to isolate the storm size, rather than the storm shape, as a function of geographic location, season, topography and circulation.

Two states were selected for analysis; Colorado and North Carolina. Although there are marked differences in relief between these two, both have a mountainous west and a rather flat east, allowing some topographic contrasts and comparisons. Thus there were four regions, each having over 1000 station pairs available for analysis. In order to ensure a constant number and location of observing stations in this exploratory analysis, the data used were restricted to the 1980–1985 period. These years represented both dry and wet conditions in each region, and incorporated some variability of circulation patterns. Only winter (DJF) and summer (JJA) are considered here.

The conditional probability was defined simply as

$$P_{ij} = \frac{r_b}{r_b + r_i},$$

where P_{ij} is the probability that it is raining at station j given that it is raining at station i. r denotes the number of raindays, with subscripts i and b indicating rain at station i only and at both stations respectively. Since i and j are effectively interchangeable, there is no directionality in this analysis, but it is not the case that $P_{ij} = P_{ji}$.

The nature of the relationship between the conditional probability P_{ij} and station separation distance d_{ij} is not well established. Some tentative results based on correlations of daily amounts have been suggested (Sumner, 1983; Berndtsson, 1989; Hevesi et al. 1992), but there is no recognised model for the binary rain/no rain approach required here. A preliminary inspection of scattergrams suggested that some form of distance decay function was appropriate. Data from a wet, a dry and an intermediate season for each of the four areas were used to test linear, logarithmic and power law model formulations (Table 3.1). Judged simply on correlation coefficient results, with an r^2 of approximately 0.05 (0.1) significant at the 99% (99.9%) level, several models were reasonable. The logarithmic formulation was used, since not only did this reproduce the distance decay shape at short distances, but it also indicated a nearly constant probability at large distances. This latter probability was around 0.3 in Colorado and 0.5 in

Table 3.1 Frequency of occurrence, of a possible 24, of correlation coefficients > 0.05 and > 0.10 for various formulations of the distance–probability relationship

r^2	$a + b(x)$	$a + b \ln x$	$a + bx^{0.5}$	$a + bx^{-1}$	$a + bx^{-2}$	$a + bx^{-3}$	$a + be^{-x}$
> 0.05	18	18	18	7	17	14	2
> 0.10	9	9	9	2	7	4	0

North Carolina. Since individual stations have a probability of precipitation of approximately 0.25 and 0.30 in Colorado and North Carolina respectively, these conditional probabilities at large distances appear reasonable. Consequently, a model of the form

$$P_{ij} = a + b \ln d_{ij}$$

was used for the analysis of the full dataset. Although in the development of the conditional probabilities, $P_{ij} \neq P_{ji}$, in this model it is assumed that $P_{ij} = P_{ji}$, which implies that any observed difference is due to local randomness. Further, the model is not constrained to $0 < P_{ij} < 1$, although this is the case for all practical distances.

On the broadest scale, there is a distinct difference in the results for Colorado and North Carolina irrespective of topography and season (Figure 3.6). In general, a reflects the overall frequency of precipitation while b indicates the

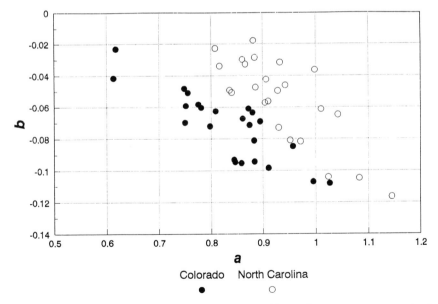

Colorado   North Carolina
  ●   ○

Figure 3.6 Scatter diagram of the a and b values of logarithmic-regression results for Colorado and North Carolina for all seasons and topographic areas.

relative frequency of small- and large-area storms. High *a* indicates a high overall probability and high (more negative) *b* indicates a steeper descent at short distances, suggestive of a predominance of localized precipitation events. Points close to the upper left in Figure 3.6 imply relatively low overall precipitation frequencies, with most events covering a wide area. Moving towards the lower right, overall frequencies increase but precipitation tends to be concentrated in small areas. Thus the results here indicate that North Carolina has more raindays covering a larger area when compared to Colorado. Since North Carolina is more susceptible to the frequent passage of mature depressions with widespread rainfall, while Colorado is more prone to localised convective activity, the general results accorded with the known synoptic climatology of the two areas.

Considering the topographic and seasonal results separately, there was a distinct difference between winter and summer in North Carolina (Figure 3.7B), with the former having more frequent events covering a greater area, while the scatter of points, indicative of year-to-year variability, was much smaller. The North Carolina mountains commonly had lower *a* and *b* values than did the plains regions. The Colorado conditions were mixed (Figure 3.7A), but the mountain results gave higher overall probabilities and a more peaked distribution. Differences between summer and winter, however, were minor, except that year-to-year variability in summer was less than in winter.

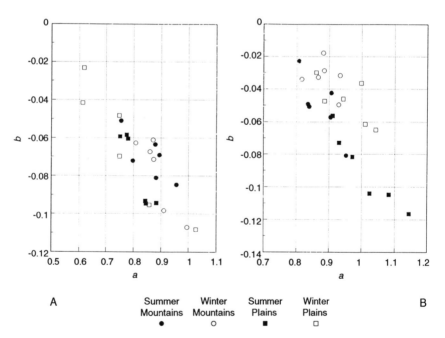

Figure 3.7 Logarithmic regression results for (A) Colorado and (B) North Carolina separated by season and topography.

For the six years used here there was relatively little relation between probability and PNA (Figure 3.8). Probability here was expressed as the chance of precipitation at a point 300 km away from a station, given that precipitation is occurring at that station. The value of 300 km was chosen since it approached the limit of the actual observed interstation distance and was sufficiently far from the observing station that the model curve was almost flat. The results were inconclusive, but suggest that the probability decreased as PNA increased in the North Carolina plains and possibly also for both Colorado regions. The range of PNA values used, however, was small, since the emphasis in selecting the years for investigation was on a series of years with stable observing network. More detailed analyses, for more extreme PNA conditions, are needed before firm conclusions can be drawn. The results, however, re-emphasised the difference between Colorado and North Carolina in overall probabilities, and also indicated the generally higher probability in the mountains for the former state.

These results strongly suggest that the present method of analysis is suitable for investigating the spatial distribution of daily precipitation. The influence of season, topography and geographical location is clearly demonstrated, while the links to circulation patterns need further investigation. Nevertheless, the analysis method is amenable to considerable refinement. In particular, investigations of the directionality of the conditional probabilities may lead to a better specification

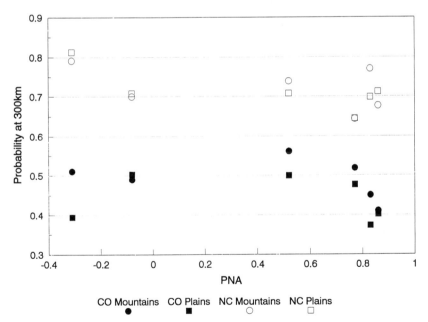

Figure 3.8 Relationship between the Pacific–North America Index (PNA) and the probability of precipitation 300 km from a station where precipitation is occurring as a function of location and topography.

of the direction of storm movement, or to understanding of the way in which topography influences the shape of an individual storm. These possibilities are considered by Smith (1994). Further, extension to include amount of precipitation, rather than the current binary approach, offers challenges and possibilities, potentially leading, through geostatistical methods, to a better specification and understanding of the volume of water falling on a drainage basin during a particular precipitation event.

3.6   DISCUSSION AND CONCLUSIONS

The results presented above represent a preliminary attempt to extend analyses of precipitation on the time and space scales already used for many hydrological investigations to acknowledge explicitly the influence of a changing climate. While statistically this implies non-stationarity, climatically it implies that the frequency and magnitude of the various atmospheric circulation patterns must change. It was postulated here that these circulation patterns are expressed through particular precipitation regimes. These regimes can be specified at individual points through development of distribution functions for the duration and the amount of precipitation in individual events. Further, there are also distribution functions for the spatial scale of these events. The individual events range through a continuum from short, localised convective storms to long-lived widespread periods associated with cyclonic activity. Thus the development of distribution functions is equivalent to the integration of the individual synoptic events, which can be interpreted in common daily meteorological terms, to more general, climatological statements concerning their frequency and typical range of characteristics.

Throughout this work the emphasis has been on a relatively simple analysis of the observational data to test whether the underlying concept is appropriate. The early results suggest that the approach has merit, despite the limitations imposed by the relatively sparse rain-gauge network. Distinct differences resulting from geographical location, season and topography are apparent. These differences are explicable in terms of the known climatology of the regions investigated (Barry and Perry, 1973). Summaries of the effects of these influences on the distribution functions must be developed to ensure that realistic values are incorporated into hydrological models. The present results also indicate that changes in regime are also apparent as a result of circulation changes, although more extensive investigations are needed before these can be quantified with any certainty.

This approach to precipitation analysis also points the way to more advanced statistical analyses. One obvious line continues the work of Cox and Isham (1994) in refining the approach emphasizing the passage of rain cells and rainbands. Another line, focused on the spatial aspects of the precipitation, is being pursued by Smith (1994). Climatologically, further analyses of the observational record, embracing a wider diversity of geographical locations and time periods, and whenever possible utilising a denser observational network, is desirable. In particular, closer investigation of the changes of regime as circulation patterns change are required, since these are likely to provide the key to the application of this approach to the specification of future climates as a result of climate change.

Although this approach is designed for use with the surface rain-gauge network, in the more distant future the ability to use radar or satellite information must be considered. Indeed, the present type of analysis may provide clues assisting in the analysis design for these temporally and spatially refined datasets. Meanwhile, the approach is beginning to suggest that we can say something about the potential changes of precipitation regimes on small watersheds as a result of climatic change.

ACKNOWLEDGEMENTS

Many helpful discussions with Richard L. Smith are gratefully acknowledged. This work is supported by a grant from the US National Science Foundation (jointly with Richard Smith).

4

A Temporal Study of Extreme Rainfall

S. G. Coles

University of Nottingham, UK

4.1 INTRODUCTION

Aggregation of rainfall occurs in both the spatial and temporal domains. In terms of designing structures that can withstand the extremal behaviour of the rainfall process, it is necessary to take account of both these aggregation effects. A reservoir, for example, must be capable of storing rainfall runoff derived from a large geographical region, and which may have accumulated over several days of persistent rain. For the purposes of design, and for scientific understanding, it is desirable to obtain a complete space–time description of the rainfall process, and specifically of the extremes of this process. Furthermore, it is widely accepted that since the aim is to estimate the probability of events that are rarer than those already observed, a reasonable paradigm for modelling extremal behaviour is to derive models that have some asymptotic motivation.

The goal of an asymptotically motivated space–time model is an ambitious one, and though we make some tentative suggestions here, no real attempt to obtain such a model is made. Instead, we shall analyse in some detail the temporal aspects of a series of rainfall data from a network of data sites in the south-west of England. This analysis is based on a modelling approach developed in Smith *et al.* (1994) and Coles *et al.* (1994). The idea is to use multivariate extreme value models within a Markov framework to describe the time series of events, with 'non-extreme' events suitably censored. It will be shown that this leads to an analysis that

Statistics for the Environment 2: Water Related Issues. Edited by Vic Barnett and K. Feridun Turkman
© 1994 John Wiley & Sons Ltd.

(i)   models seasonal effects;
(ii) allows for the possibility of trends;
(iii) accounts for, and gives a description of, temporal dependence;
(iv) enables the calculation of the probabilistic properties of various temporal aspects of the process, such as the mean duration of extreme conditions.

The extent of the spatial analysis presented here will be to apply the above analysis to a number of different sites within the region, leading to an informal assessment of the degree of spatial homogeneity in the various aspects of the temporal process. Potentially, this forms the basis for the development of a space–time model.

4.2   DATA DESCRIPTION

The data consist of daily accumulations of rainfall, recorded to the nearest tenth of a millimetre, at a network of 11 sites in the south-west of England. A map showing the grid locations of the 11 sites is shown in Figure 4.1. For the most part, analysis will be restricted to a 24 (or 25 for some sites) year period during

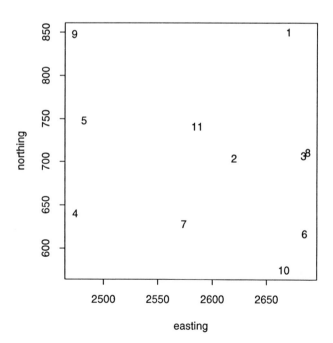

Figure 4.1   Grid reference map of locations of rainfall data sites.

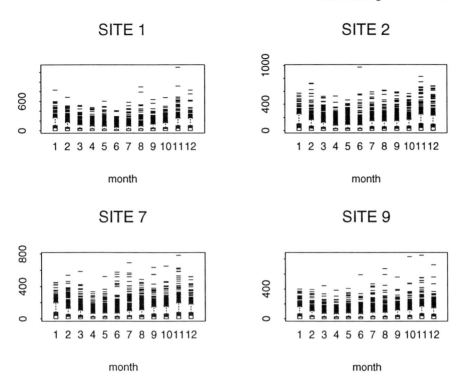

Figure 4.2 Boxplots of daily rainfall measurements at four data sites stratified by month.

which all recording sites were operational. However, considerably longer records are available at some of the sites, so we shall also consider an analysis based on the complete data records. For illustration we also give detailed results only for a subset of the sites, chosen to give a reasonable geographical coverage of the region.

Figure 4.2 gives monthly boxplots of the data for four of the sites. Our principal interest is in the extremes, which appear as horizontal marks in the boxplots (in other contexts these are considered to be outliers). A number of features are evident. First, there is evidence of seasonal variability in extremal behaviour, though it is by no means obvious that this variation is particularly smooth. There is similarity in the seasonal variation across sites, but it is not clear whether the similarities are strong enough to permit modelling the seasonal behaviour at different sites with the same model structure.

Figure 4.3 gives boxplots for the same sites but now with the data stratified by year. This time there appears to be no temporal structure to the variability in extremal behaviour at any of the sites. This suggests that models involving trends are unlikely to be necessary.

Finally, there is the issue of spatial and temporal dependence. Figure 4.4 shows concurrent time series plots for the same four data sites during a period in which

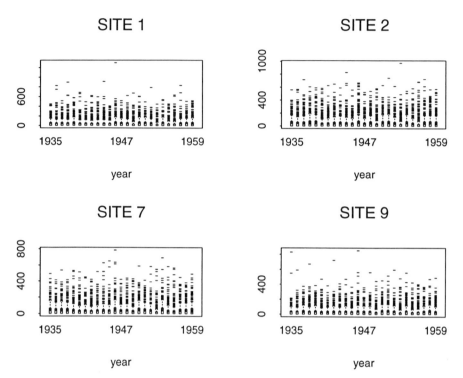

Figure 4.3   Boxplots of daily rainfall measurements at four data sites stratified by year.

several days of heavy rain occurred. The similarity of the time series reflects the fact that the meteorology of the region tends to cause extreme conditions to occur at all sites simultaneously, thus inducing spatial dependence at extreme levels. Looking through each of the series, it is clear that there is also some evidence of temporal dependence, but not markedly so at the most extreme levels. That is, there is only a very limited clustering of extreme events.

The modelling of the extremes in Section 4.4 will consider all of the temporal features described above—seasonality, trends, dependence. We shall also assess informally the stability of these aspects across sites. The issue of spatial dependence, which is not considered here, complicates this assessment.

4.3   MODEL DESCRIPTION

A thorough development of this model structure is given by Smith *et al.* (1994). A seasonal analysis of extremely low temperatures using this approach is described by Coles *et al.* (1994), and Smith (1994a) uses the stationary version to model a series of ozone measurements at extreme levels. We shall describe the model in stages.

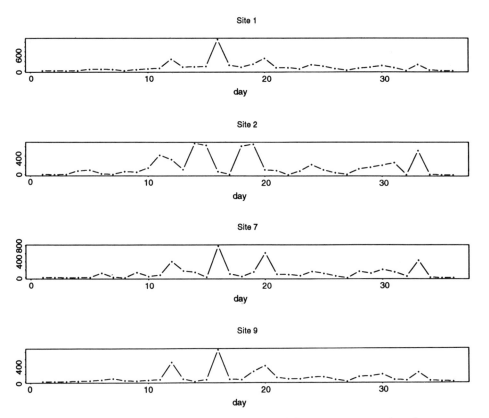

Figure 4.4   Concurrent time series plots of daily rainfall measurements at four data sites.

4.3.1   The independent model

Suppose $X_1, X_2, \ldots$ is an independent and identically distributed series. For a sufficiently high threshold u, and under suitable regularity conditions, a limiting approximation for the distribution of the excesses,

$$Y_i = X_i - u \,|\, X_i > u$$

is given by the generalised Pareto distribution (GPD). This has distribution function

$$F_u(y; \sigma^*, \xi^*) = 1 - (1 + \xi^* y/\sigma^*)_+^{-1/\xi^*}, \tag{4.1}$$

where σ^* and ξ^* are scale and shape parameters respectively and $x_+ = \max(x, 0)$. This result is due to Pickands (1971); Davison and Smith (1990) give an extensive account of the statistical applications of this model. This involves the determination of a suitable threshold u, and estimation of the parameters σ^*, ξ^* and λ_u, where λ_u is the exceedance rate of u. One approach is to maximise the two-parameter likelihood

$$\prod_{j=1}^{n_u} f(y_j), \tag{4.2}$$

where $f(\ )$ is the density of the GPD excess distribution, and the product is taken over all terms for which $x_j > u$. This gives the maximum-likelihood estimates (MLEs) for σ^* and ξ. The MLE of λ_u is simply obtained as the empirical excess rate, i.e. n_u/n, where n is the total number of observations. This is equivalent to maximising the three-parameter likelihood

$$\prod_{i=1}^{n} f^*(x_i), \tag{4.3}$$

where $f^*(x_i) = \lambda_u f(y_i)$ if $x_i > u$, but $f^*(x_i) = F(u) = 1 - \lambda_u$ if $x_i \leqslant u$. Thus the likelihood is censored at the threshold u. This approach has greater flexibility when we include non-stationary effects.

One further complication of this model is the parametrisation; the model should be a reasonable approximation at all sufficiently high levels, but the parameters σ^* and λ_u both depend on the particular choice of u. Again, this proves to be particularly inconvenient when modelling non-stationarity, where different thresholds may be appropriate in different periods. To obtain a more stable parametrisation, consider the classical result from extreme value theory (see e.g. Leadbetter *et al.*, 1983). In statistical terms, this states that a limiting approximation to the distribution of $M_N = \max_{i=1,...,N}\{X_i\}$ is the generalised extreme value (GEV) distribution with distribution function

$$G(z; \mu, \sigma, \xi) = \exp\left\{-\left[1 + \xi\left(\frac{z-\mu}{\sigma}\right)\right]_+^{-1/\xi}\right\}. \tag{4.4}$$

Taking N to be the number of observations in a year, and assuming this is sufficiently large for the asymptotics to be valid, (4.4) gives an approximation for the annual maximum distribution. Now this result is consistent with the GPD model for the excesses of u with the parameters related by

$$
\left.
\begin{aligned}
\sigma^* &= \sigma + \xi(\mu - u), \\
\xi^* &= \xi, \\
\lambda_u &= 1 - \exp\left\{-N^{-1}\left[1 + \xi\left(\frac{u-\mu}{\sigma}\right)\right]^{-1/\xi}\right\},
\end{aligned}
\right\}
\tag{4.5}
$$

Thus substituting these expressions in (4.3) gives a likelihood in terms of the parameters μ, σ and ξ, which, at levels sufficiently high for the asymptotics to give a reasonable approximation, are independent of the choice of u. Moreover they have a direct interpretation in terms of the annual maximum distribution.

4.3.2   Modelling dependence

We now wish to account for the possibility that there is temporal dependence in the series $X_1, X_2, \ldots$. We make two assumptions:

(i) the series is first-order Markov;
(ii) the distribution of successive pairs (X_i, X_{i+1}) is in the domain of attraction of a bivariate extreme value distribution.

The first of these assumptions determines a structural framework for modelling temporal dependence, while the second characterises its specific form in a way which is consistent with the asymptotic paradigm for modelling extremes. Specifically, the Markov structure allows the likelihood to be expressed simply as

$$f(x_1, \ldots, x_n) = \frac{\prod_{i=2}^{n} f_{1,2}(x_{i-1}, x_i)}{\prod_{i=2}^{n-1} f_1(x_i)}, \tag{4.6}$$

where f_1 and $f_{1,2}$ denote marginal and joint densities respectively. Now, we wish to use this likelihood to estimate extremal behaviour, but without having to make assumptions about distributional forms at non-extreme levels. Therefore our approach is to use the second assumption to determine an appropriate form for $f_{1,2}$ and f_1 at extreme levels, but otherwise to censor the likelihood.

   At extreme levels, as described above, the correct form for f_1 is the GPD density. An analogous bivariate density arises for $f_{1,2}$ by using the fact that (X_i, X_{i+1}) are in the domain of attraction of a bivariate extreme value distribution. There are a variety of ways of representing bivariate extreme value distributions; one convenient characterisation is that (Z_1, Z_2) have such a distribution if their distribution function has the form

$$G(z_1, z_2) = \exp[-V(\tilde{z}_1, \tilde{z}_2)],$$

where V is a homogeneous function of order -1, and

$$\tilde{z}_i = \left[\frac{1 + \xi_i(z_i - \mu_i)}{\sigma_i}\right]^{1/\xi_i}$$

(see e.g. Resnick, 1987, Chapter 5). Thus the marginal distributions are each GEV, and the transformation $Z \to \tilde{Z}$ gives marginals with standard form—the so-called standard Fréchet distribution. Dependence is determined by the function V. Since V has no finite parametrisation, it is common in practice to adopt a parametric form for V that covers a broad range of dependence types over the domain of the parameter space (Coles and Tawn, 1991). In this paper we restrict attention to the logistic model with

$$V(z_1, z_2) = (z_1^{-1/\alpha} + z_2^{-1/\alpha})^\alpha,$$

where $0 \leqslant \alpha \leqslant 1$, the upper and lower limits corresponding to independence and perfect dependence respectively. Tawn (1988) discusses this model in detail.

Ledford and Tawn (1994) have shown that when $X_1 > u_1$ and $X_2 > u_2$, a suitable approximation to the distribution of (X_1, X_2), based on the assumption of (X_1, X_2) being in the domain of attraction of a bivariate extreme value distribution, is given by

$$F(x_1, x_2) = \exp[-V(\tilde{x}_1, \tilde{x}_2)] \tag{4.7}$$

where

$$\tilde{x}_i = \left\{ -\log\left[1 - \lambda_i \left(1 + \xi_i \frac{x_i - u_i}{\sigma_i} \right)_+^{-1/\xi_i} \right] \right\}^{-1}. \tag{4.8}$$

This represents a slight modification of the form used in Smith *et al.* (1994) and Coles *et al.* (1994), but is a better approximation in the near-independence case.

Thus the corresponding density of (4.7) gives the required form of $f_{1,2}$ in the case where both arguments exceed their respective threshold. In other situations, analogous with the univariate case, we still use the model (4.7), but censored at the threshold. So, for example, when $X_1 < u_1$ but $X_2 > u_2$,

$$f_{1,2}(x_1, x_2) = \left. \frac{dF}{dx_2} \right|_{x_1 = u_1}.$$

4.3.3   The non-stationary model

The above model is easily adapted to accommodate non-stationarity. As it stands, the extremal behaviour is determined by the parameters μ, σ, ξ and α. By making these parameters functions of time ($\mu(t), \sigma(t), \xi(t)$ and $\alpha(t)$ say) through parametric models, both seasonality and trends can be absorbed into the model. For seasonal effects there are two possibilities: either a 'separate-seasons model' where each parameter is allowed to take a different value within each season, or a 'continuous-time model' in which each parameter is a periodic function of time involving a set of hyper–parameters. Continuous-time models are generally preferable, since they avoid the arbitrariness of blocking by season. Similarly, trends can be included by allowing some of the parameters to be, say, linearly dependent on the date of observation.

We see now the advantages of the GEV parametrisation described in Section 4.3.1. First, if non-stationarity is evident in the data then 'extreme' is a relative condition. An unusually large rainfall in April may not be particularly heavy by October's standards. Thus, in the context of our modelling approach, it makes sense to specify different thresholds at different times of the year. Also, using the GEV parametrisation means that the parameters $(\mu(t), \sigma(t), \xi(t))$ maintain their interpretation on the same scale at all times. That is, on any particular day t_0, $(\mu(t_0), \sigma(t_0), \xi(t_0))$ are the GEV parameters of the annual maximum distribution obtained from a stationary, independent series in which t_0 is a typical day.

4.4 RAINFALL MODELLING

4.4.1 Modelling at four sites

We now use the model described in Section 4.3 as a means for exploring the temporal aspects of the rainfall data described in Section 4.1, concentrating first on the four sites (numbers 1, 2, 7 and 9) whose data were shown graphically in that section. Initially, 'separate monthly' models were fitted to each of these sites, with thresholds selected to give 25 exceedances within each of the twelve months over the 25-year period. The corresponding estimates for one of these sites (site 9) are shown as asterisks on Figure 4.5. It is difficult to interpret the seasonal variation in extremal behaviour from these plots; a more direct summary is obtained by plotting quantiles of the GEV distributions corresponding to each month, obtained from the fitted model. The asterisks in Figure 4.6 show the 0.95 GEV quantiles for each month, and for each site. Recall that these are the estimated 0.95 quantiles of the annual maximum distribution relating to a year consisting of independent daily observations with the marginal characteristics of each respective month.

Figure 4.6 reveals a similar picture to the boxplots of Section 4.1: some evidence of seasonal structure which is consistent across sites, but not necessarily of simple form. However, it should be pointed out that, although not shown on these plots, the confidence interval on each of these monthly estimates is considerable. Furthermore, meteorological considerations suggest that some of the erratic behaviour from month to month suggested by this figure is not tenable.

Also shown on Figures 4.5 and 4.6 are curves (with approximate 95% confidence limits) of the corresponding parameter and quantile estimates obtained from a continuous-time model. A range of different models were fitted where each of the basic GEV parameters, and the dependence parameter α, had a simple Fourier form. So, for example,

$$\mu(t) = \beta_{\mu,1} + \sum_{i=1}^{n_\mu} \left(\beta_{\mu,2i-1} \sin \frac{2\pi t}{365} + \beta_{\mu,2i} \cos \frac{2\pi t}{365} \right), \tag{4.9}$$

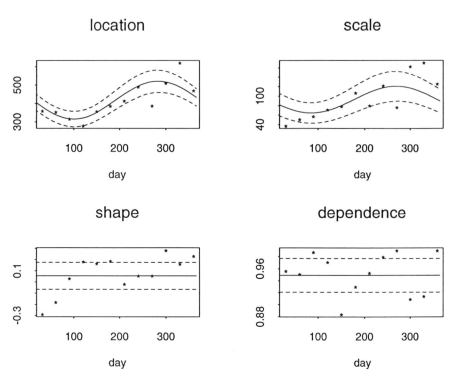

Figure 4.5 GEV (μ, σ, ξ) and dependence parameters fitted to rainfall data at site 9. Asterisks correspond to separate monthly fits; the continuous curve relates to the $(1, 1, 0, 0)$ harmonic model, with pointwise 95% confidence intervals shown as dashed curves.

where t is measured in days. The notation $(n_\mu, n_\sigma, n_\xi, n_\alpha)$ is used to specify each possible model, where, for example, n_μ is the number of harmonic pairs used to model the seasonal variation in μ. The curves drawn on Figures 4.5 and 4.6 are all made on the basis of the best-fitting $(1, 1, 0, 0)$ model; that is, a single sinusoid for μ and σ, but constant values for ξ and α. Likelihood comparisons showed that at three of the four sites this simple continuous-time model was not a significantly poorer fit than the separate monthly model. Though supported by meteorological considerations, it is perhaps surprising, in view of the separate monthly fits, that this relatively simple model structure is found to be an adequate simplification of the temporal behaviour. Furthermore, at each of these three sites higher-order models were found not to result in significant improvements.

Site 2 is the one site at which the $(1, 1, 0, 0)$ model gives a substantially poorer fit than the separate-months model, and at which higher-order models are found to improve the fit. In fact, the $(2, 2, 0, 0)$ model is the best-fitting continuous-time model within this family, allowing for parsimony. However, even this fit is significantly poorer than the separate-months model. A summary of the

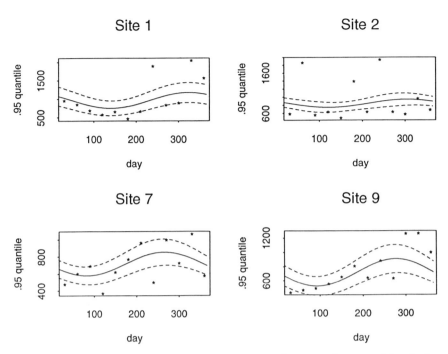

Figure 4.6 The 95% quantiles of estimated GEV distribution as function of day for four data sites. Asterisks correspond to separate monthly fits; the continuous curve relates to the (1, 1, 0, 0) harmonic model, with pointwise 95% confidence intervals shown as dashed curves.

likelihoods of the models fitted to this, and the other three sites, is given in Table 4.1.

The principal aim of this study is to obtain a general picture of the temporal behaviour of extreme rainfall, rather than using models for the specific purpose of extrapolating to obtain probabilities of rare events. Consequently, although the (1, 1, 0, 0) model is not particularly satisfactory at site 2, and the quality of the fit

Table 4.1 Negative log likelihoods of seasonal models (see text for notation)

	Site number			
Model	1	2	7	9
Monthly	3026.35	2990.79	2985.69	2937.36
(0, 0, 0, 0)	3088.98	3045.23	3029.95	2973.43
(1, 1, 0, 0)	3046.80	3028.13	3012.28	2954.23
(1, 1, 1, 1)	3045.35	3026.10	3011.32	2953.76
(2, 2, 0, 0)	3043.74	3020.75	3006.18	2950.90
(3, 3, 0, 0)	3040.79	3019.03	3000.47	2948.43

Table 4.2 Constant shape parameter ξ, dependence parameter α and extremal index θ in $(1, 1, 0, 0)$ models, standard errors in parentheses

	Site number			
Parameter	1	2	7	9
ξ	0.09988 (0.063)	-0.0437 (0.049)	-0.0402 (0.057)	0.0789 (0.062)
α	0.9437 (0.014)	0.9332 (0.015)	0.9760 (0.012)	0.9489 (0.014)
θ	0.903	0.891	0.927	0.909

at the other three sites has not been fully investigated other than by comparison of likelihoods and quantiles, we shall proceed with this model as a reasonable summary of the seasonal variability in extreme rainfall at each site. A summary of the constant shape and dependence parameters fitted at each site is given in Table 4.2.

For each site, the tail of the marginal distribution is seen to be near to exponential (corresponding to $\xi = 0$), and the dependence between successive observations is seen to be fairly weak (recall that $\alpha = 1$ corresponds to independence). This latter point confirms the observation made about Figure 4.4. A more direct way of interpreting the strength of dependence of the extremes is in terms of the extremal index θ of the series. A loose definition of θ is that it is the reciprocal of the mean number of observations in a spell that is extreme. Leadbetter *et al.* (1983) give precise definitions. Using calculations in Smith (1992), it is possible to calculate the value of θ for any given value of α within the logistic model described in Section 4.3. The corresponding values of θ from the fitted $(1, 1, 0, 0)$ models are also included in Table 4.2. Again, this reinforces the view that temporal dependence at extreme levels is very weak at all sites.

Despite the exploratory nature of this analysis, it was still felt to be necessary to question the appropriateness of the model assumptions. In particular, thresholds have been chosen at which it is assumed the asymptotic limits—marginal and joint—are reasonable approximations. Rather than examining diagnostics in detail, which would in any case be complicated by the non-stationarity, we have tried fitting the various models at a range of different thresholds and looked for stability in the model outputs. For example, Figure 4.7 gives quantile curves for the $(1, 1, 0, 0)$ model at site 9 based on 25 and 50 exceedances per month respectively. The results are similar in each case; certainly within the bounds of sampling variability. Various output diagnostics of this type confirmed that the models are not particularly sensitive to threshold choice; therefore all subsequent analysis is based on 25 exceedances per month.

It is almost obligatory for an extreme value analysis to include return level curves—plots of quantiles of the annual maximum distribution against probability level on a convenient scale. Obtaining the precise value of extreme quantiles is not the explicit aim of this analysis, but nevertheless return level curves are another useful diagnostic. Because of the non-stationarity and

Extremal Quantiles By Number of Monthly Exceedances

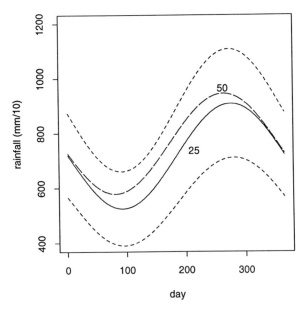

Figure 4.7   The 95% extremal quantiles as function of day based on the $(1,1,0,0)$ harmonic model fitted to data at site 9 using 25 and 50 exceedances within each set of corresponding months respectively. Outer curves are 95% confidence limits based on the 25 exceedances curves.

temporal dependence, it can be shown that the distribution of the annual maximum is given by

$$G(x) = \prod_{j=1}^{N} \exp\left\{-\frac{\theta(j)}{N}\left[1 + \xi(j)\frac{x - \mu(j)}{\sigma(j)}\right]^{-1/\xi(j)}\right\}, \qquad (4.10)$$

with the parametrisation defined by (4.5) and where $\theta(j)$ is the extremal index corresponding to day j. Solving $G(x_p) = 1 - p$ gives the quantile x_p with the exceedance probability p. A plot of x_p against $-\log[-\log(1-p)]$ based on the fitted $(1,1,0,0)$ model at site 9, with approximate 95% confidence intervals, is given in Figure 4.8. Also shown on this plot are the 10 largest annual maxima from site 9, plotted against their empirical exceedance probabilities on the same scale. It is reassuring that the fitted model, which aims to capture the seasonal variation in simple form, produces estimates for the annual maximum distribution that are consistent with just the annual maxima data.

The absence of any trend, as was apparent from the exploratory analysis, can be confirmed by appending linear time-dependent terms to the Fourier seasonal models. These extra parameters resulted in a non-significant increase

Quantiles of Annual Maximum Distribution

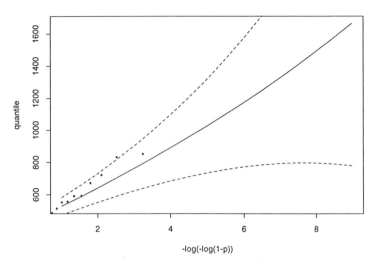

Figure 4.8   Return level curve of annual maximum distribution at site 9. Dashed lines correspond to 95% pointwise confidence limits. Asterisks are the 10 largest observed annual maxima plotted at appropriate empirical probability positions.

in likelihood. At site 9 for example, the inclusion of a linear trend in μ reduced the likelihood by just 0.03, while the corresponding reduction after including linear components for both μ and σ is 0.05.

As mentioned in Section 4.1, a longer data series for each site is available than has so far been analysed. Applying the above modelling procedures to each of these series results in much the same conclusion. Though there is some evidence of a poor fit at some sites during some months, overall the $(1, 1, 0, 0)$ models are found to be adequate summaries of the extremal behaviour.

4.4.2   Extending the analysis

Having identified the $(1, 1, 0, 0)$ model as being a reasonable fit at four of the sites, the analysis was extended to the remaining seven sites in the region. In each of these cases it was found that the continuous-time model fitted reasonably well by comparison with the corresponding 'separate months' model. The 95% extremal quantiles are plotted as functions of time in Figure 4.9 for each of the 11 sites. It is clear from this figure that both the severity and annual variability in extreme behaviour differs substantially across the region, while the modes of the seasonal behaviour appear reasonably consistent across sites. This all seems consistent with known meteorology, though conclusions should be drawn with care, since no allowance has been made for spatial dependence in the data.

Fitted .95 Extremal Quantiles By Site

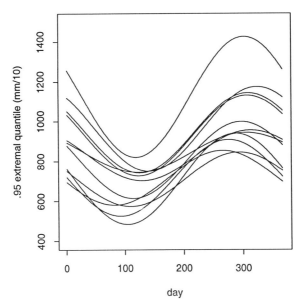

Figure 4.9   The 95% extremal quantiles based on the $(1, 1, 0, 0)$ harmonic model for all 11 data sites, plotted as function of day.

Despite the difficulties raised by having separate models based on spatially dependent data, it is desirable to assess, informally at least, possible explanations for the spatial variability in temporal behaviour which Figure 4.9 illustrates. Previous spatial analyses of extreme rainfall (Smith, 1994; Coles and Tawn, 1992) have identified that much of the spatial variation in extremal rainfall behaviour can be attributed to variation in SAAR, the Standardized Annual Average Rainfall.

In the present analysis we wish to allow for the effects on extremal behaviour of SAAR, and then assess whether there is any remaining structure to the spatial variation in temporal behaviour. This can be crudely examined in the following way. An overall measure of the severity of extreme rainfall at each site is obtained by averaging the extremal quantiles over an annual cycle. Figure 4.10 gives a plot of the averages of the 0.95 extremal quantiles based on the $(1, 1, 0, 0)$ model, against the corresponding value of SAAR for each site. The relationship appears reasonably linear, and the result of a simple linear regression is also shown on Figure 4.10. The predicted mean value of the 0.95 quantile from this regression is then subtracted from the corresponding quantile curve, and the resulting graph of standardized quantiles shown in Figure 4.11. In a sense then, Figure 4.11 displays the spatial variation in temporal behaviour after allowance

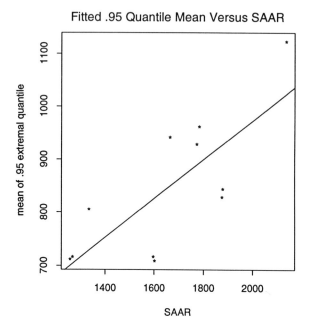

Figure 4.10   Mean of 95% extremal quantile against SAAR for each data site. The line corresponds to simple regression fit.

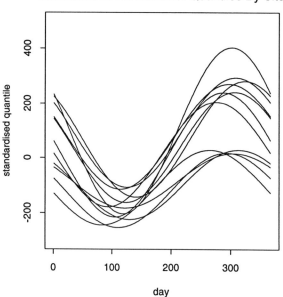

Figure 4.11.   Fitted 95% extremal quantiles, standardised by regression fit of SAAR, plotted as a function of day.

for the effects due to SAAR. There appears not to be much structure to this plot, except that four of the curves are substantially lower than the others around the mode corresponding to the most extreme conditions—the late autumn period. These curves correspond to the sites numbered 2, 6, 7 and 10, all of which are in the south-eastern corner of the region (cf. Figure 4.1). Thus there is some evidence that locations in this part of the region experience somewhat less extreme conditions in the autumn period than their annual rainfall levels would suggest are likely.

4.5   DISCUSSION

The derivation of simple continuous-time models within a Markov framework to explain the seasonal variation in extreme precipitation has highlighted a number of interesting features.

(i) In most respects, the seasonal variation in extremal behaviour at all sites can be accurately modelled using low-order Fourier models for the GEV and dependence parameters.
(ii) At each data site both the shape of the tail behaviour and the extent of temporal dependence appear homogeneous over the annual cycle. Furthermore, the extremal index at each site is close to 1, indicating that the extent of clustering at extreme levels is weak.
(iii) The severity of extremal behaviour and the extent of seasonal variation in extreme precipitation each appear to vary with location, but the modes of the annual cycle are seen to be approximately spatially homogeneous.
(iv) One quadrant of the region appears to suffer less extreme rainfall during the period of heaviest rain than would be expected given its average annual rainfall.

As stressed throughout, all spatial comparisons of models are extremely tentative because the models are fitted to spatially dependent data, and the spatial dependence has not been accounted for. A slightly more sophisticated approach would be to allow for spatial dependence when making spatial comparisons, perhaps along the lines of Smith (1994b). More desirable would be a fully integrated space–time description of the process. One possibility for this would be to extend the Markov-type model described in this paper, using high-dimensional multivariate extreme value models to describe both the spatial dependence of extremes and the short-term temporal dependence within a Markov process for temporal development. Currently, however, there are no multivariate extreme value distributions with the correct structural properties for such a model, so this remains an area for future research.

ACKNOWLEDGEMENTS

I am indebted to Richard Smith and Jonathan Tawn for the theoretical developments associated with the Markov model described above. The work was carried out using equipment provided by The Nuffield Foundation, whose support is gratefully acknowledged.

5

Wind Direction in Rain and Wind Storms

S. Nadarajah
University of Sheffield, UK

5.1 INTRODUCTION

This chapter is based on a study jointly undertaken by the University of Sheffield, Lancaster University and the Institute of Hydrology in Wallingford. The aim of the study is to assess the safety of reservoirs with respect to meteorological events that are jointly extreme in various factors, such as rainfall and wind speed. The safety requirements are assessed by examining the distribution of peak water level at the dam wall for a set of hypothetical reservoirs representative of a variety of real reservoir locations and conditions in the UK. A more complete introduction to the study and to the specific aspect of it considered here is contained in Anderson and Nadarajah (1993) and Tawn (1993). For simplicity of discussion, we only consider meteorological events over an hour and take the peak water level associated with an event as a function of the value of rainfall X_1, wind speed X_2, snowmelt X_3, wind direction X_4, catchment wetness X_5 and reservoir fullness X_6 at the hour of that event. The function depends in detail on the specific characteristics of the reservoir under study. The variables X_1, X_2 and X_3 are primary variables in the sense that it is their extremes that are almost certainly important in creating large water levels in the dam. The remaining variables are called secondary.

To simulate values of $(X_1, \ldots, X_6)$ for a given hypothetical reservoir, the study uses the following assumption: the estimate of the joint distribution of $(X_1, \ldots, X_6)$ from high-quality data from the Meteorological Office's station at Eskdalemuir— once the primary variable margins are standardised—is the same over all sites. The study identifies some qualitative grounds for expecting the assumption to be true, and aims to check its validity by using limited data from other sites. By

---

Statistics for the Environments: Water Related Issues. Edited by Vic Barnett and K. Feridun Turkman
© 1994 John Wiley & Sons Ltd.

combining the standardised joint distribution with the univariate marginal distributions—local to the reservoir—we can simulate values of $(X_1, \ldots, X_6)$ for a given reservoir. The crucial part of the study is therefore the estimation of the joint distribution of $(X_1, \ldots, X_6)$ from the Eskdalemuir data. The modelling of the primary variables, discussed in Anderson and Nadarajah (1993), is standard and follows approaches of Coles and Tawn (1991, 1994) and Joe et al. (1992). The novel component of the joint distribution modelling in the present problem is the subsequent modelling of the conditional distribution $(X_4, X_5, X_6)|(X_1, X_2, X_3)$. We simplify this by first testing for conditional independence between X_4, X_5 and X_6. The tests—based on the contingency-table approach for testing association between any number of factors—show that there is overwhelming evidence of conditional independence of X_4 and (X_5, X_6). However, the similarity of the X_5 and X_6 variables, in hydrological terms, leads to irrefutable dependence between them even when the values of the primary variables are known beforehand. Thus the joint conditional distribution $(X_4, X_5, X_6)|(X_1, X_2, X_3)$ reduces to a product of the distributions $X_4|(X_1, X_2, X_3)$ and $(X_5, X_6)|(X_1, X_2, X_3)$. In this chater we consider only the modelling of the former. In doing so, we transform to $U_i = F_i(X_i)$, where F_i is the d.f. of X_i, so that U_i is a $U(0, 1)$ random variable, $i = 1, 2, 3$, and estimate the conditional density

$$f_{X_4|U_1, U_2, U_3}(x_4|u_1, u_2, u_3) \tag{5.1}$$

in a standardised form. We let $\{(X_{i,1}, \ldots, X_{i,6}), i = 1, \ldots, n\}$ be a series of i.i.d. values of $(X_1, \ldots, X_6)$—representing independent events over the period of the Eskdalemuir data. These events are obtained by passing the data through a multivariate declustering scheme (see Nadarajah, 1994a) and so we shall refer to the series as the declustered series. Given a $X_{i,j}$ we estimate the corresponding value of U_i by $U_{i,j} = (R_{i,j} - 0.5)/n$, where $R_{i,j}$ is the rank of $X_{i,j}$ in the declustered data, counting from the lowest.

In Section 5.2 we first examine the effect of each primary variable on the variation of wind direction. This helps us spot important features in the marginal variations of wind direction and decide on an appropriate modelling approach. We then summarise some conclusions of formal tests on the model that simplify the estimation of (5.1) considerably. Throughout this paper the procedures leading to the estimation of (5.1) are performed separately for the summer (May to October) and winter (November to April) values of the data as a means to capture the temporal aspects of the variation of wind direction.

5.2   PRELIMINARY ANALYSIS

Figures 5.1 and 5.2 show kernel density estimates of X_4 directions when X_1 or X_2 exceed their various specified empirical quantiles. The wrapped normal density is used as the kernel.

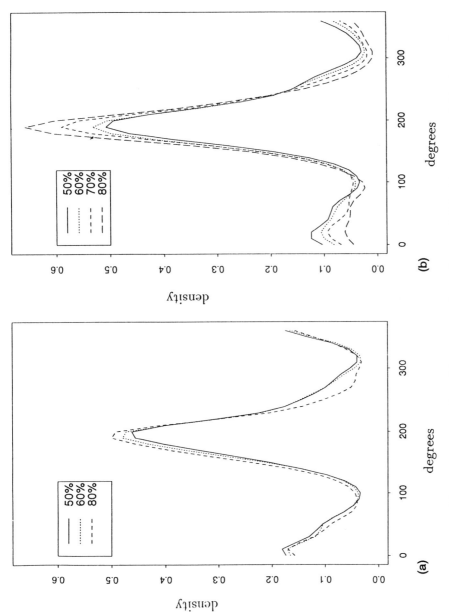

Figure 5.1 Kernel density estimates of wind direction conditional on rainfall for (a) summer and (b) winter values.

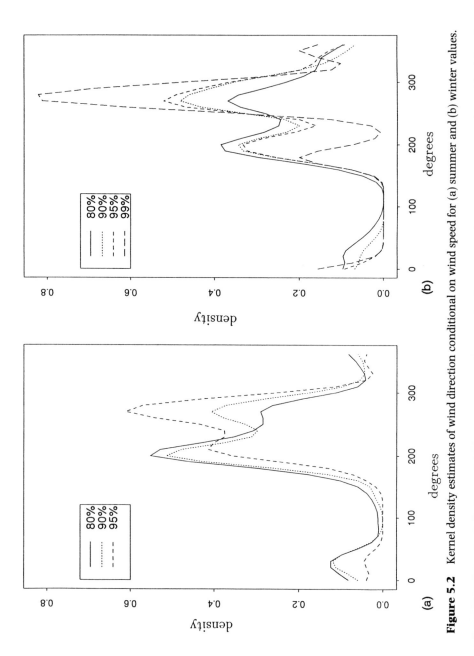

Figure 5.2 Kernel density estimates of wind direction conditional on wind speed for (a) summer and (b) winter values.

The densities in Figure 5.1 are bimodal, with the dominant mode around $190°$ and the other around $0°$. As larger values of X_1 are considered, the mode around $190°$ becomes increasingly dominant and the mode around $0°$ dies out, simplifying the density to an almost-unimodal shape. The heavier concentration around $190°$ is evidence of the fact that heavy rain storms approach from a south-south- west direction. We also note an anticlockwise movement in mean wind direction (possibly clearer in the summer season) as larger values of X_1 are considered. Further concentration of X_4 directions around the dominant mode appears larger in the winter than in the summer. This feature may be a consequence of the fact that wind direction is steadier with the frequent rainfall in winter and erratic with the occasional (but heavier) rainfall in summer.

It is evident from Figure 5.2 that strong winds mostly blow from westerly directions between $180°$ and $300°$. The densities show some trimodality, with two dominant modes around $180°$ and $270°$ respectively and the other around $0°$. The latter mode dies out as larger values of X_2 are considered. More importantly though, we see a tendency for winds corresponding to larger X_2 values to blow from points further around the compass than those corresponding to smaller X_2 values. This feature has also been observed in similar data from Sheffield, but not in data collected at sea. A possible explanation is that, at larger values of X_2, wind blows in a direction closer to the high-altitude geostrophic wind, which in UK latitudes is at an angle up to $+40°$ to the wind over land, but at a much smaller angle to wind over sea, because of smaller frictional effects there. The possibly related feature of the veering of wind direction at increasing altitudes (at which wind speeds are higher too) is well known to meteorologists as the Ekman spiral (see e.g. Haltiner and Martin, 1957, Chapter 14).

Similar densities for the X_3 variable are not produced here. However, they show that a large part of the behaviour of X_4 directions for varying values of X_3 could be explained by the corresponding variation in X_1 and X_2 values—an indication that there is little association between snowmelt and the variation of wind directions.

The veering of wind directions with increasing values of X_2 and the steadiness of wind directions with large values of X_1 (two of the main features observed in the foregoing discussion) provide a clear motive for parametric modelling of the conditional distribution of X_4. The increasing simplicity of the kernel density estimates with larger values of each primary variable suggests that the conditional distribution in (5.1) for events in the upper tail of rainfall, wind speed and snowmelt could be described by a unimodal parametric form. One such parametric form is the von Mises density given by

$$g(x_4; \alpha, \kappa) = \{2\pi I_0(\kappa)\}^{-1} e^{\kappa \cos(x_4 - \alpha)}$$

where $0 < \alpha = \alpha(u_1, u_2, u_3) \leqslant 2\pi$, $\kappa = \kappa(u_1, u_2, u_3) > 0$ and $I_0(\kappa)$ is the modified Bessel function of the first kind of order zero. The von Mises density is unimodal and the mean direction of X_4 is α, which is also the mode of the density. The parameter κ is referred to as the concentration parameter, since larger values of κ lead to greater clustering of X_4 around the mean direction. Chi-square tests of goodness of fit and discrimination with other unimodal and multimodal densities—when each primary variable is at its extreme—establish the von Mises density as a reasonable model. To economise on computations and to simplify interpretation, we assume throughout the rest of this chapter that the validity of the von Mises model holds over the entire range of $\{(U_{i,1}, U_{i,2})\}$. Formal GLR (generalized likelihood ratio) tests—based on the model—show that snowmelt has no significant effect on the variation of wind direction, that the variation of α with respect to the values of rainfall and wind speed is additive (on a circle) and that the variation of κ with respect to the values of rainfall and wind speed contains a significant interaction term. (For details of the tests and their diagnostics see Nadarajah (1994b).) The model assumption is now reduced to

$$f_{X_4|U_1,U_2}(x_4|u_1, u_2) = g(x_4; \alpha(u_1, u_2), \kappa(u_1, u_2)), \tag{5.2}$$

where $\alpha(u_1, u_2) = \beta + \gamma_1(u_1) + \gamma_2(u_2) \pmod{2\pi}$, and γ_1 and γ_2 are functions on $[0, 1]$ corresponding respectively to rainfall and wind speed effects.

Equation (5.2) provides a means for describing the variations of α and κ jointly in relation to values of rainfall and wind speed. An analogous description of the marginal variations of α and κ can be given by

$$f_{X_4|U_i}(x_4|u_i) = g(x_4; \alpha_i^{(m)}(u_i), \kappa_i^{(m)}(u_i)), \quad i = 1, 2. \tag{5.3}$$

Although (5.2) and (5.3) do not provide consistent models, we use the latter just as a diagnostic and an interpretative tool for the fit of the joint models (see Sections (5.3 and 5.4). We begin Section 5.3 by first identifying covariate models for α and κ and then giving their fitting procedure.

5.3   COVARIATE MODELS

Simple descriptive models are sought for the dependence of α and κ on u_1 and u_2. The models should be physically realistic, should broadly represent the main features of the present data and should be suitable for use with data from other sites. Natural choices are to take the $\gamma_i(u_i)$ functions to be low-order polynomials in u_1 and u_2 respectively, and log $\kappa(u_1, u_2)$ to be a low-order bivariate polynomial. Specifically, we might take

$$\left.\begin{aligned}
\gamma_i(u_i) &= \sum_{j=2}^{k_i} \gamma_{i,j-2} u_i(u_i^{j-1} - j), \quad i = 1, 2, \\
\kappa(u_1, u_2) &= \exp\left(\kappa_0 + \sum_{i=1}^{l} \kappa_i u_1^i + \sum_{i=1}^{l} \kappa_{l+i} u_2^i + \sum_{i=2}^{l} \sum_{j=1}^{i-1} \kappa_{2l+(i-1)(i-2)/2+j} u_1^j u_2^{i-j} \right),
\end{aligned}\right\} \tag{5.4}$$

where $k_1, k_2, l \geqslant 1$, $u_1, u_2 \in [0, 1]$ and $\gamma_i(u_i) \in (0, 2\pi]$. The models for $\gamma_i(u_i)$ are simply k_ith-degree polynomials satisfying the condition that $\gamma_i'(1) = 0$ $(i = 1, 2)$, which reflects a feature seen in diagnostics associated with the test for additivity (not reproduced here) that α becomes almost constant at high levels of u_1 or u_2. This special feature is not to be taken as an essential part of the present approach. The positiveness of $\kappa(u_1, u_2)$ is maintained by the exponential term and the parameters $\gamma_{i,j}$ are constrained so that $\gamma_i(u_i) \in (0, 2\pi]$ $(i = 1, 2)$. The maximum number of parameters required for specifying the conditional distribution in (5.1) is $k_1 + k_2 + \frac{1}{2}l(l-1) + 2l$.

The estimates of $\{\gamma_{i,j}\}$ and $\{\kappa_i\}$ in the joint model are obtained by maximising the likelihood for joint realisations of (U_1, U_2, X_4):

$$\prod_{i=1}^{n} g(X_{i,4}; \alpha(U_{i,1}, U_{i,2}), \kappa(U_{i,1}, U_{i,2})).$$

A standard approach to selection of the degree of polynomial models in normal theory linear regression is outlined in Rawlings (1989, Chapter 14), and we follow this here. Note, however, that this is only one of several procedures, and, as discussed in Section 5.4, may not in retrospect be judged ideal for the present application.

An evaluation of the marginal variations $\alpha_1^{(m)}(u_1)$ and $\alpha_2^{(m)}(u_2)$, from the joint models above for $\alpha(u_1, u_2)$ and $\kappa(u_1, u_2)$, is of special interest in that it may be used to quantify the descriptive features of Section 5.2 (in particular the veering of wind direction with increasing wind speed). Quantities of special interest are $\alpha_i^{(m)}(1)$ and $\alpha(1, 1)$: the limiting mean wind directions respectively at the upper end points of X_i $(i = 1, 2)$ and (X_1, X_2). It is easily shown by using the maximum-likelihood properties of a von Mises density and equations (5.2) and (5.3) that

$$\cos \alpha_i^{(m)}(u_i) = \int_{B_i} C_c(u, v)\, du\, dv / D_i, \tag{5.5}$$

$$\kappa_i^{(m)}(u_i) = A^{\leftarrow}\left(D_i \Big/ \int_{B_i} h(u, v)\, du\, dv \right), \tag{5.6}$$

where

$$D_i = \left\{ \left[\int_{B_i} C_c(u, v)\, du\, dv \right]^2 + \left[\int_{B_i} C_s(u, v)\, du\, dv \right]^2 \right\}^{1/2},$$

$$C_c(u, v) = h(u, v)A(\kappa(u, v)) \cos \alpha(u, v),$$

$$C_s(u, v) = h(u, v)A(\kappa(u, v)) \sin \alpha(u, v),$$

$A(u) = I_0'(u)/I_0(u)$, $B_1 = \{u_1\} \times [0, 1]$, $B_2 = [0, 1] \times \{u_2\}$ and h is the joint density of U_1 and U_2. The above equations express the marginal variations $\alpha_i^{(m)}(u_i)$ and $\kappa_i^{(m)}(u_i)$ in terms of the joint variations, $\alpha(u_1, u_2)$ and $\kappa(u_1, u_2)$, under the assumption that the joint models in (5.2) describe the variations in the data exactly. Once the models for $\alpha(u_1, u_2)$ and $\kappa(u_1, u_2)$ are fitted, the marginal variations can be evaluated by using the empirical density of $\{(U_{i,1}, U_{i,2})\}$ as an estimate of the joint density $h(u, v)$. We shall refer to these estimates of the marginal variations as the joint model estimates. The standard errors of these estimates are obtained by the delta method (Rao, 1973).

5.4   RESULTS, DISCUSSION AND CONCLUSIONS

The covariate models in equation (5.4) are fitted to the declustered series $\{(X_{i,1}, X_{i,2}, X_{i,4})\}$. Table 5.1 defines the selected models for α and κ by giving the parameter estimates. The standard errors of all estimates, throughout this section, are given in parentheses. Various assessments of the adequacy of the models show no evidence of a lack of fit. For details, again see Nadarajah (1994b).

Figures 5.3 and 5.4 are plots of the fitted surfaces $\alpha(u_1, u_2)$ and $\log(\kappa(u_1, u_2))$. The model formulae in Table 5.1 show that $\alpha(u_1, u_2)$ for summer is explained by the rainfall effect alone, while the $\alpha(u_1, u_2)$ for winter is explained equally by both main effects. The rainfall and wind speed components respectively exhibit a backing and a veering effect. In both seasons the insignificant variation of mean direction in the joint upper tail of (X_1, X_2) is explained smoothly and well by the $\gamma_i(u_i)$ components. The model estimates of $\alpha(1, 1)$ for summer and winter are respectively $190°(4°)$ and $191°(5°)$. A GLR test of equality of these limiting mean directions—performed by fitting a von Mises density to the wind directions in $[0.9, 1] \times [0.9, 1]$—shows no evidence against equality. In the sequel, whenever

Table 5.1   Parameter maximum-likelihood estimates with standard errors

	Summer		Winter	
	Parameter	**Estimate**	**Parameter**	**Estimate**
Concentration	κ_0	$-0.5\,(1.7)$	κ_0	$-8.6\,(2.0)$
	κ_1	$-4.5\,(2.8)$	κ_1	$17.5\,(5.0)$
	κ_2	$-0.7\,(0.9)$	κ_2	$-8.9\,(3.5)$
	κ_4	$1.6\,(7.3)$	κ_4	$11.1\,(2.9)$
	κ_5	$-13.5\,(11.4)$	κ_5	$-1.0\,(1.2)$
	κ_6	$13.5\,(6.1)$	κ_7	$-22.9\,(5.7)$
	κ_7	$26.7\,(8.3)$	κ_9	$14.6\,(4.2)$
	κ_8	$-22.4\,(6.2)$		
Mean	β	$-1.7\,(0.3)$	β	$-3.9\,(0.4)$
direction	$\gamma_{1,0}$	$-4.2\,(2.6)$	$\gamma_{1,0}$	$-11.2\,(2.2)$
(rad)	$\gamma_{1,1}$	$2.7\,(1.2)$	$\gamma_{1,1}$	$6.2\,(1.1)$
			$\gamma_{2,0}$	$-11.9\,(3.0)$
			$\gamma_{2,1}$	$4.8\,(1.3)$

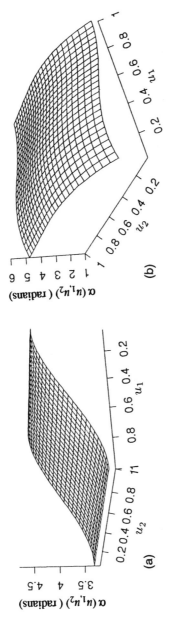

Figure 5.3 Fitted mean direction surface for $\alpha(u_1, u_2)$ for (a) summer and (b) winter values.

88

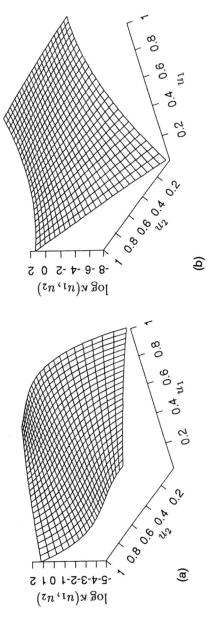

Figure 5.4 Fitted concentration surface for log $\kappa(u_1, u_2)$ for (a) summer and (b) winter values.

a test of equality of this kind is used, we shall only mention the interval in which the von Mises densities are fitted.

The concentration values $\kappa(u_1, u_2)$ generally increase as u_1 and/or u_2 approaches 1. Tests of equality of concentration values in the upper portions of the surfaces (including those where either rainfall or wind speed is small) show no evidence against equality.

Now consider the marginal variations $\alpha_i^{(m)}(u_i)$ and $\kappa_i^{(m)}(u_i)$. One set of estimates of these functions is obtained by partitioning the U_i range equally and then fitting a von Mises model to the X_4 directions in each cell of the partition. We shall refer to these estimates as the marginal estimates. Another set of estimates are the joint model estimates obtained from equations (5.5) and (5.6). The comparisons of marginal and joint model estimates of $\alpha_i^{(m)}$ and $\log \kappa_i^{(m)}$, leading to a diagnostic on the fit of the joint models, are illustrated in Figures 5.5–5.8 for different margins and seasons. In each figure the continuous lines join the joint model estimates and upper and lower bounds of a 95% confidence interval. Figure 5.5 and 5.7 correspond to the rainfall margin, while Figures 5.6 and 5.8 correspond to the wind speed margin. The plotted points in each figure correspond to the marginal estimates. Clearly the joint model estimates capture the general features evident in the variation of marginal estimates (and hence the features observed in Section 5.1)—an indication of the adequacy of the joint models.

The $\alpha_1^{(m)}(u_1)$ directions stabilise to a constant after a movement in the anti-clockwise direction (see Figures 5.5 and 5.7). The marginal estimates of $\alpha_1^{(m)}(u_1)$ in [0.95, 1] for summer and winter are respectively 192°(15°) and 197°(5°). Testing equality of these stabilised directions shows no evidence against equality. Testing inequality of $\kappa_1^{(m)}(u_1)$ values, for all values of u_1 close to 1, shows that the concentrations of directions for extreme rainfall are larger in winter.

The $\alpha_2^{(m)}(u_2)$ directions veer in both seasons but the veering in the winter is steeper and begins earlier (see Figures 5.6 and 5.8). The marginal estimates of $\alpha_2^{(m)}(u_2)$ in [0.95, 1] for summer and winter are respectively 247°(7°) and 259°(7°). Although the directions through which $\alpha_2^{(m)}(u_2)$ veers appear different from summer to winter for fixed values of u_2 away from 1, testing equality of the given 'veered to' directions shows no evidence against equality. Tests of equality of $\kappa_2^{(m)}(u_2)$ values, for all values of u_2 close to 1, show that the concentrations of directions for extreme wind speeds, in both seasons, are more or less the same.

A disturbing feature of the models in Table 5.1 is the number of parameters apparently needed to represent the data. It is possible that the selection procedure in Rawlings (1989) led to overfitting. The correct balance may not have been quite achieved between fidelity to the present data and simplicity and wider applicability of the model. In general, however, the overall strategy of simple covariate modelling based on (5.2) does appear capable of capturing the marginal and joint features of the variation of wind direction for the 1-hour events, including increased concentration with higher extremes and veering of wind direction with increased wind speed. Hence we can expect it to be capable of both estimating the distribution in (5.1) for d-hour events of primary variable values

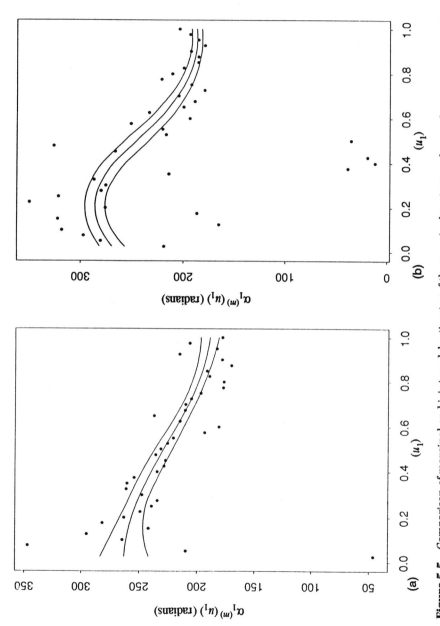

Figure 5.5 Comparison of marginal and joint model estimates of the marginal variation of mean direction corresponding to rainfall margin $\alpha_1^{(m)}(u_1)$, for a summer and (b) winter values.

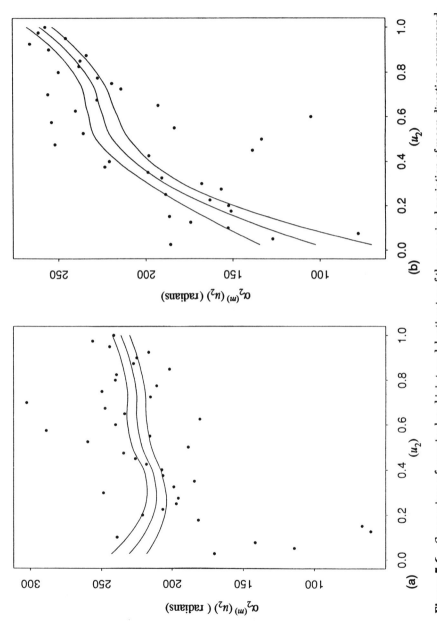

Figure 5.6 Comparison of marginal and joint model estimates of the marginal variation of mean direction corresponding to wind speed margin $\alpha_2^{(m)}(u_2)$, for a summer and (b) winter values.

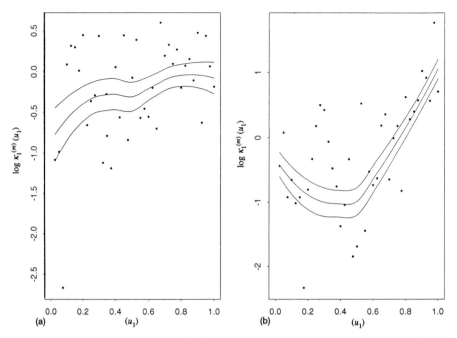

Figure 5.7 Comparison of marginal and joint model estimates of marginal variation of concentration corresponding to rainfall margin, $\log \kappa_1^{(m)}(u_1)$ for (a) summer and (b) winter values.

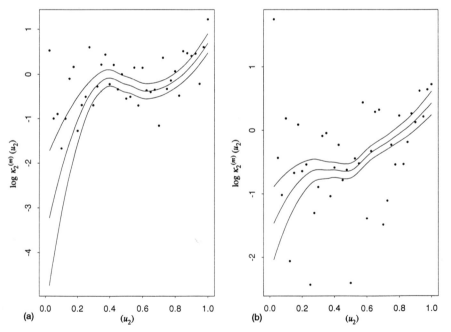

Figure 5.8 Comparison of marginal and joint model estimates of the marginal variation of concentration corresponding to wind speed margin, $\log \kappa_2^{(m)}(u_2)$ for (a) summer and (b) winter values.

from the Eskdalemuir data when $d > 1$ and representing the distribution for other sites.

ACKNOWLEDGEMENTS

This study of reservoir flood safety is supported by the UK Department of the Environment. The author gratefully acknowledges the guidance of Clive Anderson and Jonathan Tawn. Data for Eskdalemuir were supplied by the Meteorological Office Edinburgh Climate Office. Some of the work used equipment provided by SERC under the Complex Stochastic Systems Initiative.

6

Some Statistical Analyses of the Cluster Point Processes of Hurricanes on the Coasts of Mexico

J. I. Castro and V. Pérez-Abreu
Centro de Investigación en Matemáticas México

6.1 INTRODUCTION

Meteorologists have considerable interest in modelling, among other variables, the duration in days of hurricanes as well as their times of occurrence through year. Such models are useful in the study of general patterns of this meteorological phenomenon, and are relevant, for example, in the prevention of floods in some countries where the period of maximum intensity of hurricane occurrences coincides with the rainy season. This is the case in Mexico, where hurricanes arise from both the Pacific and Atlantic Oceans. This chapter presents a statistical analysis for the cluster point processes of hurricanes, based on those hitting the coasts of Mexico during the years 1966–1992.

A cluster point process N is a useful model for describing hurricanes or other meteorological or environmental phenomena over a period of time or an area Γ. The model comprises the cluster size X, which represents the duration in days of hurricanes, the times or sites $\{\tau_i\}$ of occurrences and the number of hurricanes $N(A)$ occurring in any period of time $A \subset \Gamma$. A traditional assumption made in these models is the Poisson nature of the point process N as well as the homogeneity of its intensity function $\Lambda(t)$. In this chapter we propose a new graphical method to check the Poisson nature of a point process of occurrences as well as its homogeneity. Our technique is based on the empirical probability-generating

Statistics for the Environment 2: Water Related Issues. Edited by Vic Barnett and K. Feridun Turkman
© 1994 John Wiley & Sons Ltd.

functional of a sample of independent point processes, which is also introduced in this work. On the other hand, to identify the distribution for counts of the cluster size X, we use an exploratory technique recently presented in Nakamura and Pérez-Abreu (1993a). Overall, we conclude that, for both coasts, the associated hurricanes follow a cluster point process model with a negative binomial distribution for the cluster size and a non-homogeneous Poisson process for the occurrences. We then choose suitable parametric intensity functions for the processes and compute maximum-likelihood estimates for all the parameters in the model. The intensity functions for the oceans are distinct.

The organization of this chapter is as follows. Section 6.2 presents a preliminary analysis of the cluster size distribution and identifies the corresponding distribution hurricane occurrences for each coast. Section 6.3 introduces a new method for identifying a non-homogeneous Poisson process of hurricane occurrences, based on the probability generating functional of a point process. Section 6.4 suggests parametric intensity functions for the identified non-homogeneous Poisson processes, and gives maximum-likelihood estimates of all parameters in the cluster point process model. Finally, Section 6.5 presents the conclusions of the statistical analysis.

6.2   DISTRIBUTION OF HURRICANE DURATIONS

In this section we identify a model of hurricane cluster sizes, namely, the distribution of the duration in days of these meteorological phenomena.

Since we are dealing with a distribution for counts, it is convenient to use a technique for the exploratory data analysis based on the empirical probability-generating function (Nakamura and Pérez-Abreu, 1993a). If X has a discrete distribution $\{p(k); k = 0, 1, 2, \ldots\}$ and probability generating function

$$\phi(t) = \mathrm{E}(t^X) = \mathrm{E}(\exp X \log t), \quad 0 \leqslant t \leqslant 1, \tag{6.1}$$

then, given a random sample $X_1, \ldots, X_n$, the empirical probability generating function (epgf) is defined as

$$\hat{\phi}_n(t) = \frac{1}{n} \sum_{j=1}^{n} t^{X_i}, \quad 0 \leqslant t \leqslant 1. \tag{6.2}$$

The empirical transform $\hat{\phi}_n$ has proved to be useful in the statistical analysis of distributions for counts as shown by Kocherlakota and Kocherlakota (1986), Kemp and Kemp (1988), Nakamura and Pérez-Abreu (1993a–c), and references therein. In particular, Nakamura and Pérez-Abreu (1993a) presents a technique for the exploratory data analysis for discrete distributions using the epgf. More precisely, it is suggested there to plot $Y_n(t) = \log \phi_n(t)$ against t for the preliminary statistical analyses of discrete distributions. Figure 6.1 shows this plot for the

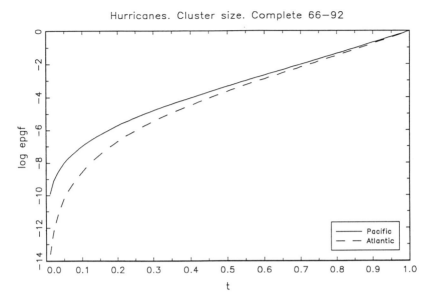

Figure 6.1

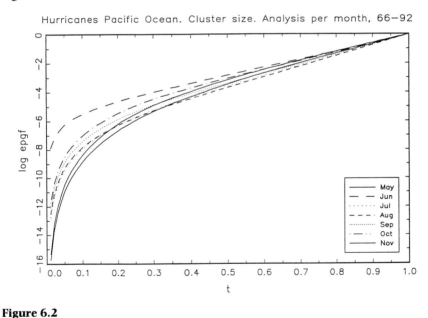

Figure 6.2

cluster sizes of hurricanes that occurred in each of the periods $\Gamma = [1$ May, 30 November] for 1966–1992, on the Pacific and Atlantic coasts of Mexico. It is concluded that both distributions, although distinct, belong to the same zero-truncated family. Figures 6.2 and 6.3 show plots of $Y_n(t)$ for each of the seven

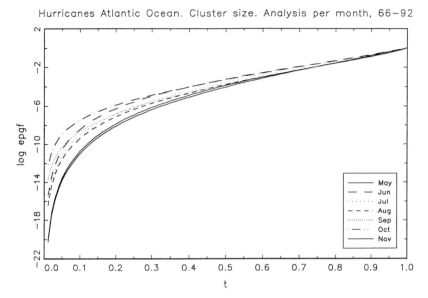

Figure 6.3

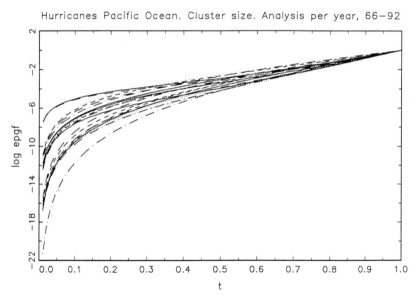

Figure 6.4

months corresponding to all the years in the Pacific and Atlantic Oceans respectively. From these, it can be concluded that the distributions belong to the same family with different parameters. Moreover, we observe that the distributions in both oceans follow the same monthly pattern.

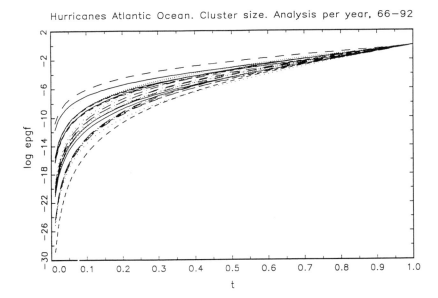

Hurricanes Atlantic Ocean. Cluster size. Analysis per year, 66–92

Figure 6.5

Figures 6.4 and 6.5 show plots of $Y_n(t)$ for each of the years considered in the Pacific and Atlantic Oceans respectively. From these, it can be concluded that there is no clear seasonal behaviour in the time series of distributions through the analysed years. From now on, we shall assume that the distribution of cluster size is the same for all months and years.

A possible model of hurricane durations in days is the truncated negative binomial model (Johnson *et al.* 1993). In order to verify this, we first estimate, as we indicate below, the additional number of observations equal to zero that are needed to have a sample from a non-truncated negative binomial model. Let us consider the following form of the negative binomial distribution with parameters N and P:

$$p(k) = \binom{N+k-1}{N-1}\left(\frac{P}{Q}\right)^k\left(1-\frac{P}{Q}\right)^N, \quad k = 0, 1, 2, \ldots, \tag{6.3}$$

where $Q - P = 1$. The corresponding zero-truncated negative binomial model is

$$p^*(k) = (1 - Q^{-N})^{-1}\binom{N+k-1}{N-1}\left(\frac{P}{Q}\right)^k\left(1-\frac{P}{Q}\right)^N, \quad k = 1, 2, \ldots. \tag{6.4}$$

The moment estimators of N and P under (6.3) satisfy the equations (Johnson *et al.*, 1993)

$$\tilde{N}\tilde{P} = \bar{x}, \quad \tilde{N}\tilde{P}(1 + \tilde{P}) = s^2,$$

that is, $\tilde{P} = s^2/\bar{x} - 1$ and $\tilde{N} = \bar{x}^2/(s^2 - \bar{x})$, where $\bar{x}$ and s^2 are the sample mean and variance respectively. If n_0 is the number of zeros and $n + n_0$ is the complete sample size in the non-truncated negative binomial sample, it can be shown that, for large n,

$$\tilde{P} \approx \frac{\sum_i x_i^2 - \left(\sum_i x_i\right)^2 \Big/ (n + n_0)}{\sum_i x_i} - 1.$$

From the zero-truncated data, we estimate P by the method of moments as (Johnson *et al.*, 1993) $\tilde{P}^* = s_*^2/[\bar{x}_*(1 - f_1)] - 1$, where s_*^2 and $\bar{x}_*$ are computed with the original data from the truncated distribution and f_1 is the relative frequency of the number of observations equal to one in the original sample. Hence an estimator of the number of zeros n_0 is given by

$$\hat{n}_0 = \frac{\left(\sum_i x_i\right)^2}{\sum_i x_i^2 - (1 + \tilde{P}^*)\sum_i x_i} - n. \tag{6.5}$$

Figure 6.6 shows plots of $Y_n(t)$ for both oceans, including the estimated zeros. Since these curves are convex, from Nakamura *et al.* (1993a) we conclude that both data sets come from a negative binomial model, and therefore the

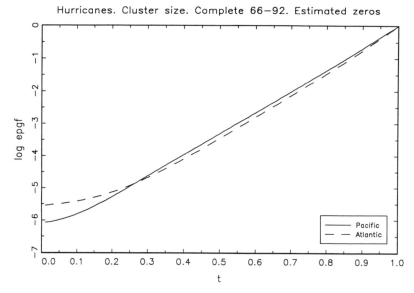

Hurricanes. Cluster size. Complete 66–92. Estimated zeros

Figure 6.6

distribution of cluster size follows a truncated negative binomial distribution with different parameters. Maximum-likelihood estimators for the parameters of the latter distribution are presented in Section 6.4.

6.3   EXPLORATORY ANALYSIS OF THE POISSON NATURE OF A POINT PROCESS

In this section we explore the nature of the point process N modelling the time of hurricane occurrences $\{\tau_{ij}\}$ in the coasts of Mexico during the period $\Gamma = [1$ may, 30 November] of the year. As one of the main contributions of this paper, we first introduce a new method for a graphical check of the Poisson nature of a general point process as well as of the homogeneity of a Poisson process. The proposed technique is based on the empirical counterpart of the probability generating functional (pgfl)—an important concept in the theory of point processes, defined by

$$G[h] = E\left[\prod_{i=1}^{N(\Gamma)} h(\tau_i)\right] = E\left[\exp\left\{\int_\Gamma \log[h(\tau)]N(d\tau)\right\}\right],$$ (6.6)

where h belongs to a suitable class of test functions V, namely, $0 \leqslant h(s) \leqslant 1$ and $h = 1$ outside a set of compact support (Cox and Isham, 1980; Daley and Vare-Jones, 1988). The pgfl $G[h]$ provides a complete description of the point-process N, and it is a generalisation of the probability-generating function of a discrete random variable.

An important cluster point process that occurs in this context is the non-homogeneous Poisson process with cumulative intensity function $\Lambda(t)$, $t \in T$, and cluster size with probability generating function ϕ. Its corresponding pgfl is given by (Cox and Isham, 1980)

$$G[h] = \exp\left\{-\int_\Gamma [1 - \phi(h(\tau))]\Lambda(d\tau)\right\}.$$ (6.7)

For independent copies $N_1, \ldots, N_n$ of an arbitrary point process N with corresponding location points $\{\tau_i^j; j = 1, \ldots, n\}$, the empirical probability generating functional (epgfl) $\hat{G}_n$ is defined as

$$\hat{G}_n[h] = \frac{1}{n}\sum_{j=1}^{n}\prod_i h(\tau_i^j), \quad h \in V.$$ (6.8)

It is easily seen that for each $h \in V$, $E(\hat{G}_n[h]) = G[h]$, and, by the law of large numbers, $\hat{G}_n[h] \to_{n \to \infty} G[h]$ almost surely. That is, $\hat{G}_n[h]$ is an unbiased consistent estimator of $G[h]$.

The epgfl (6.8) might be used to draw inferences about a point process N, analogously as way in which the empirical Laplace functional is applied (Karr, 1986). Here we only introduce some ideas for the exploratory analysis of point processes, generalising the concepts for discrete distributions presented in (Nakamura et al., 1993a). Let h be a function on Γ such that $(1 - h) \in V$, and let $0 < c < 1$. Then $1 - ch \in V$, $0 < c < 1$, and the following function is well defined:

$$Y_c[h] = \log G[1 - ch], \quad 0 < c < 1.$$

For a non-homogeneous Poisson process in $\Gamma \subset R^k$ with (cumulative) intensity function $\Lambda(\tau)$, the pgfl is (Cox and Isham, 1980)

$$G[h] = \exp\left\{-\int_\Gamma [1 - h(\tau)]\Lambda(d\tau)\right\}.$$

Hence, for each $1 - h \in V$,

$$Y_c[h] = -c\int_\Gamma h(\tau)\Lambda(d\tau)$$

is a straight line on $0 < c < 1$. This suggests that, if the point process is non-homogeneous Poisson, the plot of $\hat{Y}_c[h] = \log \hat{G}[1 - ch]$ should be a straight line for each function h such that $1 - h \in V$. Moreover, if the Poisson process is homogeneous in $\Gamma \subset R^k$, that is, $\Lambda(A) = |A|\lambda$ for some $\lambda > 0$ where $|A|$ is the Lebesgue measure of $A \in B(R^k)$, then

$$Y_c[h] = -c\lambda\int_\Gamma h(\tau)\,d\tau.$$

Thus functions h with identical Lebesgue integral $\int_\Gamma h(\tau)\,d\tau$ produce the same straight line $Y_c[h]$. Because of this, we suggest that $Y_c[h]$ be plotted for several functions h several with identical Lebesgue integral to test the homogeneity of a Poisson process.

The above technique has two main disadvantages. First, we have the problem of selecting possible functions for the analysis as well as their number. In this direction, after several simulation studies, we recommend the use of following four functions, where $\Gamma = [0, T]$:

$$h_1(\tau) = \frac{\tau^2}{|T|^2}, \tag{6.9}$$

$$h_2(\tau) = \frac{T^{\frac{1}{2}} - \tau^{\frac{1}{2}}}{T^{\frac{1}{2}}}, \tag{6.10}$$

$$h_3(\tau) = \begin{cases} 1 - (2/T)\tau & (\tau < \tfrac{1}{2}T), \\ (2/T)\tau - 1 & (\tau \geqslant \tfrac{1}{2}T), \end{cases} \tag{6.11}$$

$$h_4(\tau) = \begin{cases} (2/T)\tau & (\tau < \tfrac{1}{2}T), \\ 2 - (2/T)\tau & (\tau \geqslant \tfrac{1}{2}T), \end{cases} \tag{6.12}$$

which correspond to increasing, decreasing and two symmetric functions respectively. The functions $h_3(\tau)$ and $h_4(\tau)$ have the same Lebesgue integral on $[0, T]$. Secondly, the behaviour of $\hat{Y}_c[h]$ as c approaches unity has large variability. Therefore we suggest that $\hat{Y}_c[h]$ be plotted only in the interval $0 < c < 0.8$.

We now return to the sample of $n = 27$ point processes N_i in $\Gamma = [1\text{ May}, 30\text{ November}]$, $i = 1, \ldots, 27$, of occurrences of the strating data of the hurricanes over the Pacific and Atlantic coasts of Mexico, 1966–1992. We assume that the cluster point processes for these years follow the same model and that they are independent. Figures 6.7 and 6.8 show plots of $\hat{Y}_c[h_i]$ ($i = 1, \ldots, 4$) for these two sets of data. From these pictures, we conclude that both processes are of Poisson nature, even if the plot corresponding to h_4 is convex. Simulation studies have shown that the latter occurs if the sample size is not large enough and that for most of the non-Poisson alternative models the corresponding plot is concave. Since the plots of $\hat{Y}_c[h_3]$ and $\hat{Y}_c[h_4]$ are different, we conclude that these processes are non-homogeneous during the year.

The proposed technique is also useful in seeking possible outliers in a sample $N_1, \ldots, N_n$ of an arbitrary point process. Specifically, we suggest the evaluation of

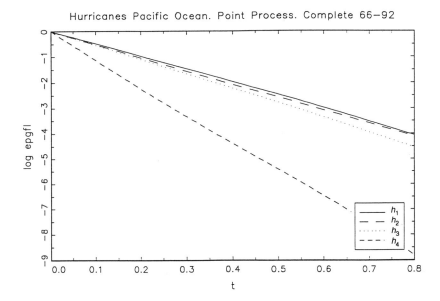

Figure 6.7

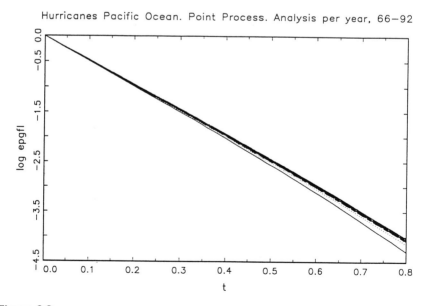

Figure 6.8

Figure 6.9

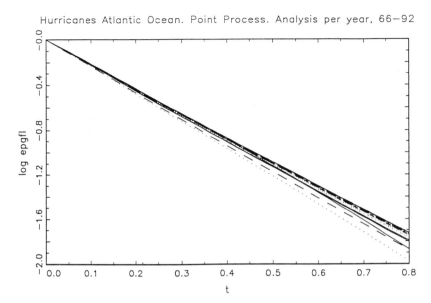

Figure 6.10

the effect of each point process on the function $\widehat{Y}_c[h]$ by a leave-one-out procedure, namely, plotting $\widehat{Y}_c[h]$ n times, leaving one point process out each time. If the N_ith process comes from a different family, the corresponding plot must be distinct from all the others. For example, Figures 6.9 and 6.10 show the functions $\widehat{Y}_c[h]$ for the Pacific and Atlantic cases respectively. From these plots, we conclude that all the years produce point processes of hurricanes of the same class, which is consistent with our assumption. Although the latter example produces non-homogeneous Poisson processes, the leave-one-out procedure can be applied to arbitrary point processes. It is important to remark, however, that this procedure would not have identified a smooth trend over the years.

6.4 MAXIMUM-LIKELIHOOD ESTIMATES OF THE PARAMETERS

To identify possible intensity functions for the non-homogeneous Poisson processes of hurricane occurrences, we plot the empirical cumulative intensity function

$$\widehat{\Lambda}(\tau) = \frac{1}{n} \sum_{j=1}^{n} \sum_{i=1}^{N_j(\Gamma)} 1_{\{\tau_i^j < \tau\}},$$

in Figures 6.9 and 6.10 for the Pacific and the Atlantic Oceans respectively. These plots suggest the following parametric models. For the Pacific ocean we propose

Table 6.1   Maximum-likelihood estimates

Pacific Ocean		Atlantic Ocean	
Parameter	**Estimation**	**Parameter**	**Estimation**
η	17.0003	η_1	20.9878
β	119.9015	η_2	145.1972
α	2.7301	η_3	521.5846
P	0.5325	β_1	313.5868
N	13.0000	β_2	294.2341
		β_3	679.6574
		α_1	2.0001
		α_2	4.8091
		α_3	4.8297
		P	1.5003
		N	5.0000

$$\Lambda_p(\tau) = \eta \left\{ 1 - \exp\left[-\left(\frac{\tau}{\beta}\right)^\alpha \right] \right\} \quad (1 \text{ May} \leqslant \tau \leqslant 30 \text{ Nov.}, \alpha, \beta, \eta > 0), \quad (6.13)$$

while for the Atlantic ocean

$$\Lambda_a(\tau) = \begin{cases} \eta_1 \left\{ 1 - \exp\left[-\left(\frac{\tau}{\beta_1}\right)^{\alpha_1} \right] \right\} & (\tau \in [1 \text{ May}, 31 \text{ July}]), \\[2mm] \Lambda_a(31 \text{ July}) + \eta_2 \left\{ 1 - \exp\left[-\left(\frac{\tau}{\beta_2}\right)^{\alpha_2} \right] \right\} & (\tau \in [1 \text{ Aug.}, 30 \text{ Sept.}]), \\[2mm] \Lambda_a(30 \text{ Sept.}) + \eta_3 \left\{ 1 - \exp\left[-\left(\frac{\tau}{\beta_3}\right)^{\alpha_3} \right] \right\} & (\tau \in [1 \text{ Oct.}, 30 \text{ Nov.}]). \end{cases}$$
$$(6.14)$$

Having identified the cluster point process of hurricane occurrences as a non-homogeneous Poisson process with intensity Λ and with a truncated negative binomial distribution $p^*(k)$ for the cluster size, the likelihood function for the model is

$$L = \exp\left[-n\Lambda(T) \right] \prod_{j=1}^{n} \prod_{i=1}^{N_j(T)} \lambda(\tau_i^j) p^*(M_i^j),$$

where M_i^j is the cluster size corresponding to τ_i^j and $\lambda(s) = \Lambda'(s)$. Maximum-likelihood estimators for this model based on the above data are shown in Table 6.1. Corresponding plots for the empirical and maximum-likelihood estimates of the intensity functions are given in Figures 6.11 and 6.12.

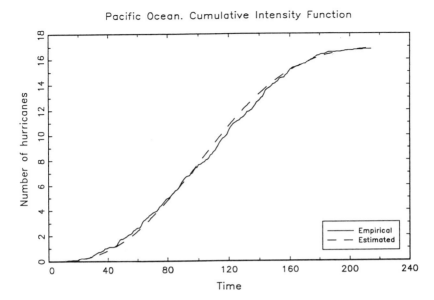

Figure 6.11

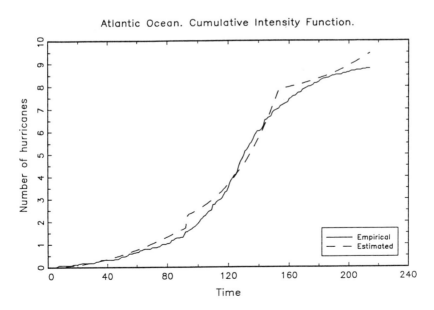

Figure 6.12

6.5 CONCLUSIONS

Two interesting conclusions have emerged from the present study. First, from Figures 6.11 and 6.12, we conclude that a cluster point process model with a truncated negative binomial distribution for the cluster size and a non-homogeneous Poisson process with intensity function (6.13) (respectively (6.14)) is consistent with the data of hurricanes hitting the Pacific (respectively Atlantic) coast of Mexico during the years 1966–1992. Secondly, graphical methods based on the empirical probability-generating functional of a sample of point processes are useful techniques for checking the commonly used Poisson nature of the process. The use of these tools in exploratory analyses of other meteorological or environmental problems where point processes arise looks promising.

ACKNOWLEDGEMENTS

The authors would like to thank Meteorologist Alberto Hernández Unzon for his help in preparing the data and Dr Miguel Nakamura for very useful discussions. This research was partially supported by CONACYT Grant 1858-E9219.

7

Bootstrap Confidence Intervals for the Estimation of Seeding Effect in an Operational Period

R. Nirel
The Hebrew University of Jerusalem, Israel

7.1 INTRODUCTION

This work was motivated by the problem of estimation of the effect of operational cloud seeding, on rain amounts in Israel. Rain enhancement activity in Israel started in 1960. Two controlled experiments were designed to test the null hypothesis of no effect of cloud seeding on rain amounts in the target area. These experiments indicated a statistically significant increase in rain amounts of about 13% (Gabriel, 1967, 1970; Gagin and Neumann, 1981). These encouraging results led to the initiation in 1976 of an operational seeding project in the north, which is still in progress. Whereas in the experimental periods seeding days were allocated randomly, in the operational period all days with suitable cloud conditions were seeded. However, an unseeded control area was kept at all times.

The seeding effect in the operational period has been estimated by a double-ratio-type statistic, based on target and control precipitation in historical and operational periods (Nirel, 1993). Construction of confidence interval for the double-ratio is not straightforward. The distribution of the double-ratio is known

Statistics for the Environment 2: Water Related Issues. Edited by Vic Barnett and K. Feridun Turkman
© 1994 John Wiley & Sons Ltd.

to be non-symmetric and is quite complicated under parametric assumptions (Flueck and Holland, 1976). Gabriel and Feder (1969) showed, for the rain data of the first Israeli experiment, that asymptotic normality of the double-ratio statistic was attained for samples greater than a few hundred observations. In rain enhancement experiments, permutation tests, based on the random allocation of seeding days, were highly recommended for inference on the effect estimate (Turkey *et al.*, 1978; Gabriel, 1979).

This chapter suggests that the non-parametric bootstrap method be used to construct a confidence interval for the seeding effect in the operational period. Various methods for construction of bootstrap confidence intervals for complicated statistics have been introduced in recent years, as discussed comprehensively by DiCiccio and Romano (1988) and Hall (1988). Hall showed that Efron's (1982) percentile interval is erroneous in the sense that the percentiles used in the upper and lower critical points are confused. Hall recommends the studentised interval, based on a pivotal statistic. Although the analytical properties of different intervals have been extensively analysed, the experience gained by application of these methods has been limited.

The objective of this chapter is to draw an empirical comparison between the optimal studentised interval and two other bootstrap confidence intervals for a double-ratio-type estimator, and recommend a preferable method, when possible. We show, in contrast to Hall's recommendation, that the percentile interval for the double-ratio is closer to the optimal interval than other intervals. An insight into this phenomena is provided by a comparison between the ratio and the double-ratio estimators.

The organisation of this chapter is as follows. Section 7.2 reviews methods for the construction of bootstrap confidence intervals and presents criteria for comparison between methods. Section 7.3 describes the rain data and the method of effect estimation in the operational period. In Section 7.4, three strategies of bootstrapping the effect estimate are suggested and discussed. Properties of ratio and double-ratio statistics, relevant to confidence interval estimation, are examined. The minimum number of bootstrap simulations is evaluated. A comparison between four confidence intervals, presented in Section 7.5, shows that the double-ratio is more sensitive to the choice of method than the ratio estimator. This result is found to characterise distributions with a constant coefficient of variation.

7.2   BOOTSTRAP CONFIDENCE INTERVALS

7.2.1   Definitions

The bootstrap is a resampling method for construction of the distribution of a statistic, by repeated sampling with replacement from the set of n observations (Efron 1979). Let $x = (x_1, \ldots, x_n)$ be n independent and identically distributed observations from an unknown distribution F_r. Let $\theta = t(F_r)$ be the functional of interest, with variance $\sigma = \text{var}\,(\theta)$, and let $\hat{\theta}$ be an estimate of θ based on $\boldsymbol{x}$. Non-parametric bootstrap inference on θ is based on B samples drawn from $\boldsymbol{x}$ with replacement. For each sample in the set $\{\boldsymbol{x}_b^*\}$ $(b = 1, \ldots, B)$ the estimate $\hat{\theta}_b^*$ is computed. The bootstrap distribution of the $\hat{\theta}_b^*$ values is constructed by

$$\hat{F}^*(u) = P^*(\hat{\theta}^* \leqslant u\,|\,\boldsymbol{x}) = \#(\hat{\theta}_b^* \leqslant u)/B.$$

Efron (1982) introduced the 'intuitive' percentile confidence interval (PER) for θ:

$$[\hat{\theta}_{\text{PER}}(\tfrac{1}{2}\alpha),\ \hat{\theta}_{\text{PER}}(1 - \tfrac{1}{2}\alpha)] = [\hat{F}^{*-1}(\tfrac{1}{2}\alpha),\ \hat{F}^{*-1}(1 - \tfrac{1}{2}\alpha)]. \tag{7.1}$$

Limitations of the percentile methods have been pointed out. Efron (1982) showed that the asymmetry and coverage probability of the percentile interval for the correlation coefficient were not accurate. Shenker (1985) discussed the error in coverage probability of the PER interval for the variance of a normal random variable.

Two approaches for improvement of the percentile method have been suggested. The first corrects for the bias in the interval's critical points by a suitable shift of these points (Efron, 1987; DiCiccio and Tibshirani, 1987). The second, introduced by Efron (1982) and Beran (1987), is based on a pivotal statistic (a statistic whose distribution is known, although the parameter θ is not known).

Several bootstrap confidence intervals will now be introduced. The notation and terminology follow Hall (1988). First, let us define a reference confidence interval for future comparisons. Suppose $\hat{\sigma}$ is an estimate of the unknown σ, and K is a known distribution of $(\theta - \hat{\theta})/\hat{\sigma}$. Then the exact (EXA) confidence interval for θ satisfies:

$$[\hat{\theta}_{\text{EXA}}(\tfrac{1}{2}\alpha),\ \hat{\theta}_{\text{EXA}}(1 - \tfrac{1}{2}\alpha)] = [\hat{\theta} - \hat{\sigma}\,K^{-1}(1 - \tfrac{1}{2}\alpha),\ \hat{\theta} - \hat{\sigma}\,K^{-1}(\tfrac{1}{2}\alpha)]. \tag{7.2}$$

Note that the upper percentile appears on the lower critical point and vice versa. Define also the bootstrap distributions:

$$\hat{K}^*(x) = P^*((\hat{\theta}^* - \hat{\theta})/\hat{\sigma}_b^* \leqslant x\,|\,\boldsymbol{x}), \qquad \hat{H}^*(x) = P^*((\hat{\theta}^* - \hat{\theta})/\hat{\sigma}^* \leqslant x\,|\,\boldsymbol{x}), \tag{7.3}$$

where $\hat{\sigma}^* = \sum_{b=1}^{B}(\hat{\theta}^* - \bar{\theta}^*)^2/B - 1$ is the bootstrap variance estimate based on the set $\{\hat{\theta}_b^*\}$, $i = 1, \ldots, B$, and $\hat{\sigma}_b^*$ is a variance estimate based on a single sample $\boldsymbol{x}_b$. $\hat{\sigma}_b^*$ is estimated either by a trustworthy asymptotic estimate of σ or by re-bootstraping each bootstrap sample $\boldsymbol{x}_b^*$. Note that $(\hat{\theta}^* - \hat{\theta})/\hat{\sigma}_b^*$ is approximately pivotal.

Bootstrap confidence intervals are based on $\hat{H}^*$ and $\hat{K}^*$ and on bootstrap variance estimates. The following intervals are compared.

(i) The 'hybrid' interval (HYB) is defined by substituting the percentiles of K in (7.2) by the percentiles of $\hat{H}^*$, and $\hat{\sigma}$ by the bootstrap estimate $\hat{\sigma}^*$, i.e.

$$[\hat{\theta}_{\mathrm{HYB}}(\tfrac{1}{2}\alpha), \hat{\theta}_{\mathrm{HYB}}(1 - \tfrac{1}{2}\alpha)] = [\hat{\theta} - \hat{\sigma}^* \hat{H}^{*-1}(1 - \tfrac{1}{2}\alpha), \hat{\theta} - \hat{\sigma}^* \hat{H}^{*-1}(\tfrac{1}{2}\alpha)].$$

(ii) The percentile interval (defined in (7.1)), named by Hall the 'backwards' (BAC) interval, is defined by

$$[\hat{\theta}_{\mathrm{BAC}}(\tfrac{1}{2}\alpha), \hat{\theta}_{\mathrm{BAC}}(1 - \tfrac{1}{2}\alpha)] = [\hat{\theta} + \hat{\sigma}^* \hat{H}^{*-1}(\tfrac{1}{2}\alpha), \hat{\theta} + \hat{\sigma}^* \hat{H}^{*-1}(1 - \tfrac{1}{2}\alpha)].$$

It is named the BAC interval because of the confusion of upper and lower percentiles at the critical points.

The equivalence of the BAC and PER intervals is demonstrated for the upper critical point:

$$1 - \tfrac{1}{2}\alpha = \hat{F}^*(u)$$
$$= P^*(\hat{\theta}^* \leqslant u)$$
$$= P^*((\hat{\theta}^* - \hat{\theta})/\hat{\sigma}^* \leqslant (u - \hat{\theta})/\hat{\sigma}^*)$$
$$= \hat{H}^*(u - \hat{\theta})/\hat{\sigma}^*)$$
$$\Rightarrow H^{*-1}(1 - \alpha/2) = (u - \hat{\theta})/\hat{\sigma}^*$$

and
$$u = \hat{\theta} + \hat{\sigma}^* \hat{H}^{*-1}(1 - \tfrac{1}{2}\alpha)$$

(iii) The 'studentized' interval (STD) is defined by replacing the percentiles of K in (7.2) by the percentiles of $\hat{K}^*$, and $\hat{\sigma}$ by $\hat{\sigma}^*$:

$$[\hat{\theta}_{\mathrm{STD}}(\tfrac{1}{2}\alpha), \hat{\theta}_{\mathrm{STD}}(1 - \tfrac{1}{2}\alpha)] = [\hat{\theta} - \hat{\sigma}^* \hat{K}^{*-1}(1 - \tfrac{1}{2}\alpha), \hat{\theta} - \hat{\sigma}^* \hat{K}^{*-1}(\tfrac{1}{2}\alpha)].$$

(iv) For θ with an asymptotic normal distribution and asymptotic variance $\hat{\sigma}_{\mathrm{as}}^2$, the normal (NOR) interval is defined by

$$[\hat{\theta}_{\mathrm{NOR}}(\tfrac{1}{2}\alpha), \hat{\theta}_{\mathrm{NOR}}(1 - \tfrac{1}{2}\alpha)] = [\hat{\theta} - \hat{\sigma}^* z(1 - \tfrac{1}{2}\alpha), \hat{\theta} - \hat{\sigma}^* z(\tfrac{1}{2}\alpha)].$$

where Φ is the normal c.d.f. and $z(\alpha) = \Phi^{-1}(\alpha)$.

7.2.2   Criteria for comparison between intervals

The differences between intervals are assessed by the following criteria.

(i) *Accuracy of the critical points.* A critical point is said to be 'first-order correct' if it differs from the corresponding critical point of the EXA interval (7.2) by $O_p(n^{-\frac{1}{2}})$, and is 'second-order correct' if the difference is $O_p(n^{-1}) = O_p((n^{-\frac{1}{2}})^2)$.

Hall showed that for the ('smooth function model', which assumes that θ and σ^2 are smooth functions of $E(\boldsymbol{x})$, the STD method is second-order correct while the BAC and HYB methods are first-order correct.

(ii) *Symmetry of interval.* The symmetry of the interval around $\hat{\theta}$ is defined by the index r/l:

$$\frac{r}{l} = \frac{\hat{\theta}_M(1 - \tfrac{1}{2}\alpha) - \hat{\theta}}{\hat{\theta} - \hat{\theta}_M(\tfrac{1}{2}\alpha)}.$$

where $\hat{\theta}_M$ is an estimate of θ based on a method M.

(iii) *Ease of use.* Although not a mathematical consideration, computational intensity as well as stability of an interval are major considerations in practice. The most convenient method is the PER interval, while the most computationally intensive is the STD interval based on a 'double bootstrap' variance estimate (see (7.3)).

7.3   DATA AND METHOD OF EFFECT ESTIMATION

7.3.1   Data

The analysis presented in this chapter is based on daily rainfall amounts for the seasons 1950–1990 (1950 stands for the 'rain season November 1949 to April 1950') in the northern target area and coastal control area. The data includes 1204 unseeded days in the period 1950–1975 and 1382 seeded days in the period 1970–1990 (of which 202 days were allocated to be seeded in the second Israeli experiment). Only days with at least 0.1 mm rain in the control area are included in the analysis. This selection criterion, introduced in the analysis of the first Israeli experiment, enables objective elimination of days with no clouds or no suitable clouds over the target area.

Extra care was taken in the uniformity of definitions over time. Thus, the unit of time was defined as a 24 h period beginning at 8:00 a.m. (the seasons 1963–1964 were eliminated because another definition was used). Also, even though the boundaries of the State of Israel changed in the period 1950–1990, the mean rain amounts in the target area was computed from a fixed network of about 70 rain stations that were within the boundaries of Israel throughout this period. Since changes in the seeding line occurred in the period 1960–1990, the mean rain amount in the control area were based on rain stations not seeded at any time.

7.3.2   Method of effect estimation

In rain enhancement projects with target and control areas, the effect of cloud seeding is frequently estimated by a double-ratio (DR) statistic (Davis, 1969; Flueck, 1986). The double-ratio is the ratio of two ratios: the ratio of the mean rainfall in the target area on seeded and unseeded days is compared to the same ratio in the control area. Gabriel and Rosenfeld (1990) suggested an improved double-ratio statistic adjusted for regression (DRR). In the DRR the control rainfall of the DR statistic is substituted for a linear regression prediction of target rainfall, adjusted for control rainfall. The DRR statistic was adapted to the estimation of the seeding effect in the operational period (Nirel 1993). The statistic, named the DRG, is defined by

$$\text{D}\hat{\text{R}}\text{G} = \frac{\hat{R}_s}{\hat{R}_u} = \frac{\bar{Y}_s/\bar{\hat{Y}}_s}{\hat{\bar{Y}}_u/\bar{\hat{Y}}_u}, \tag{7.4}$$

where $\bar{Y}_s$ is the mean rainfall in the target area on seeded days (in the operational period) and $\bar{Y}_u$ is the mean rainfall in the target area on unseeded days (in an historical period preceding the operational period). $\bar{\hat{Y}}_s$ and $\bar{\hat{Y}}_u$ are defined similarly for the predicted rainfall in the target area.

The predicted target rainfall used in the DRG is based on a logarithmic model with a shift parameter,

$$\ln(y_i + \mu) = \beta_0 + \beta_1 \ln(x_i + \mu) + \epsilon_i, \quad E(\epsilon_i) = 0, \quad \text{Var}(\epsilon_i) = \eta^2, \tag{7.5}$$

where y_i is the target rainfall, x_i the control rainfall, μ the shift parameter, β_0 the intercept, β_1 the slope and ϵ_i the uncorrelated identically distributed error terms.

The re-transformation to original scale is an asymptotic version of the estimate suggested by Bradu and Mundlak (1970):

$$\hat{y}_i = \exp(\hat{\beta}_0 + 0.5\hat{\eta}^2)(x_i + \hat{\mu})^{\hat{\beta}_1} - \hat{\mu}. \tag{7.6}$$

Estimating the shift parameter in (7.5), for all unseeded days, produces a local maximum-likelihood estimate at $\hat{\mu} = 0.3$. The rest of the parameters are then estimated by the least-squares method, with estimates $\hat{\beta}_0 = 0.047$, $\hat{\beta}_1 = 0.899$ and $\hat{\eta}^2 = 0.377$.

Note that for a linear regression model of Y on X, if the estimation of the regression coefficients and of $\bar{\hat{Y}}$ is based on the same observations then $R_u = 1$. But since $\text{D}\hat{\text{R}}\text{G}$ is based on the logarithmic model (7.5) and the retransformation (7.6), usually R_u will not be equal to one, mainly because of deviations of the residual distribution from a normal distribution. In this case R_u represents a 'calibration factor' of the model.

7.4   BOOTSTRAPPING RATIOS

The bootstrapping of complicated statistics is not always straightforward. As Efron and Gong (1983, page 43) observed,

> The main point here is that 'bootstrap' is not a well defined verb, and that there may be more than one way to proceed in complicated situations.

When bootstrapping of the DRG is considered, a few strategies seem plausible.

(i) In a 'thorough' bootstrap
   (a) sample n observations from the vector of unseeded days (x_{ui}, y_{ui}), $i = 1, \ldots$, n, with replacement;
   (b) estimate the parameters in model (7.6) for the sample in (a);
   (c) estimate $\hat{y}_{ubi}$ and $\hat{R}_{ub}$;
   (d) sample m observations from the vector of seeded days (x_{is}, y_{is}), $i = 1, \ldots, m$, with replacement;
   (e) estimate $\hat{R}_{sb}$ and $\hat{DRG}_b$;
   repeat B times.
(ii) In a 'full' bootstrap, proceed as in the 'thorough' bootstrap, but without stage (b), i.e. use the parameter estimates of the model (7.5) for estimating $\hat{y}_{ubi}$ in (c) and proceed as before. This strategy assumes the parameters of the model are 'known'.
(iii) In a 'partial' bootstrap estimate R_u from the original sample, and proceed with stages (d) and (e). This strategy assumes that the calibration factor and the model parameters are 'known'. This strategy actually bootstraps a *ratio* statistic (and not double-ratio).

The 'thorough' bootstrap is much more complicated than the other strategies. One has to consider if the accuracy gained by it is worth the computational cost. Let us examine the change in a single full bootstrap estimate $\hat{DRG}_{1b}$, by addition of stage (b) to the procedure, resulting in the 'thorough' bootstrap estimate $\hat{DRG}_{2b}$. Dropping the subscript b, examine the ratio $\hat{DRG}_1/\hat{DRG}_2$. Denote the parameters of the model (7.5) appearing in DRG_i by subscript i. Since the same samples are used for both estimates, $\bar{Y}_s$ and $\bar{Y}_u$ cancel. Therefore, for a fixed estimate of the shift $\hat{\mu} = 0.3$,

$$\frac{\hat{DRG}_1}{\hat{DRG}_2} = \frac{\bar{\bar{Y}}_{1s}/\bar{\bar{Y}}_{2s}}{\bar{\bar{Y}}_{1u}/\bar{\bar{Y}}_{2u}}$$

$$\approx \frac{\exp(\hat{\beta}_{01} + 0.5\hat{\eta}_1^2) \sum_{i=1}^m [(x_i + \hat{\mu})^{\hat{\beta}_{11}}] \exp(\hat{\beta}_{02} + 0.5\hat{\eta}_2^2) \sum_{j=1}^n [(x_j + \hat{\mu})^{\hat{\beta}_{12}}]}{\exp(\hat{\beta}_{02} + 0.5\hat{\eta}_2^2) \sum_{i=1}^m [(x_i + \hat{\mu})^{\hat{\beta}_{12}}] \exp(\hat{\beta}_{01} + 0.5\hat{\eta}_1^2) \sum_{j=1}^n [(x_j + \hat{\mu})^{\hat{\beta}_{11}}]}$$

$$= \frac{\sum_{i=1}^m [(x_i + \hat{\mu})^{\hat{\beta}_{11}}] \sum_{j=1}^n [(x_j + \hat{\mu})^{\hat{\beta}_{12}}]}{\sum_{i=1}^m [(x_i + \hat{\mu})^{\hat{\beta}_{12}}] \sum_{j=1}^n [(x_j + \hat{\mu})^{\hat{\beta}_{11}}]}.$$

The ratio $\hat{DRG}_1/\hat{DRG}_2$ is mainly influenced by the change in the slope estimate. The symmetrical structure of the ratio suggests that it is robust to small changes in the slope estimate.

For the rain data, the ratio $\hat{DRG}_1/\hat{DRG}_2$ was computed for the slope estimates $\hat{\beta}_{11} = 0.90$ (s.e. $(\hat{\beta}_{11}) = 0.014$ for 1204 unseeded days), and $\hat{\beta}_{12} = \pm 0.1$ and ± 0.2. The above ratio was examined for two data sets: the complete data set (2586 days), and the period of the second Israeli experiment (440 days). The results show that for the larger set, addition of 0.02 to the slope estimate changes the effect estimate by a factor of 1.0002. For the smaller set, the slope changes by a factor of 1.001. As the values of the full bootstrap estimates $\hat{DRG}_b$ vary from 0.98 to 1.16, the accuracy gained by the 'thorough' bootstrap does not seem to justify its use.

Consider next the usefullness of the full bootstrap. The ratio of the full bootstrap estimate and the partial bootstrap estimate exceeds the factor 1.03 for about 10% of the bootstrap samples. Hence the full bootstrap strategy cannot be given up for the partial strategy. However, since there is a methodological interest in the ratio estimate, the rest of the paper will focus on both the full (DRG) and partial (R) bootstrap procedures.

7.4.1   Number of simulations

The non-parametric bootstrap involve two stages of approximation. First, the empirical distribution F_n is substituted for the 'true' distribution F. Secondly, a finite number B of bootstrap simulations is performed. Thus the number of bootstrap simulations determines the accuracy of the second stage.

It is the aim of this section to determine the number of simulations that yields an adequate accuracy of the seeding effect estimate. Our experience shows that a relatively simple Fortran program is quite efficient for several thousand simulations. The following analysis is based on 'simple' bootstrap (and not on balanced or importance sampling: see Davison *et al.*, 1986; Hall, 1989, Efron, 1990, Johns, 1988 and Do and Hall, 1991).

The number of simulations was determined as follows: 3000 simulations of the ratio and 5000 simulations of the double-ratio were performed on all rain data. The resulting intervals were called the 'reference' bootstap intervals (increasing B to 10 000 did not alter the fourth decimal place of the critical point). The estimates of these simulations were divided into groups of $B = 100, 300, 500$ and 1000 simulations, and were compared to the 'reference' interval.

The results show that for $B = 100$ and 300, the intervals may differ from the 'reference' bootstrap interval by 0.01 for R, and by 0.02 for the DRG. The asymmetry index is erratic, changing by factors of up to 1.5 from the respective reference indices. As the seeding effect is estimated by DRG = 1.07, this level of accuracy is not acceptable. For $B = 500$ and 1000, the differences between the 'reference' and simulated intervals are in the third decimal place and the asymmetry index fluctuates around 1 for the BAC and HYB intervals (see Figure 7.1).

The conclusion from the above discussion is that a few hundred (or even one thousand) simulations do not provide an adequate accuracy for the rain data. Therefore, to be on the safe side, construction of confidence intervals in the following analysis will be based on (at least) $B = 2000$ and $B = 3000$ simulations for R and DRG respectively.

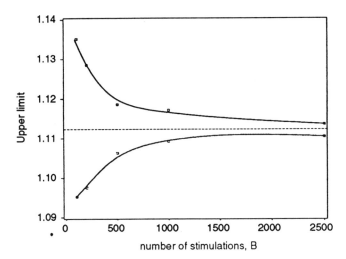

Figure 7.1 Minimum and maximum values of the upper critical points of the STD interval for the DRG, as a function of the number of simulations B, for n (seeded) = 1382 and n (unseeded) = 1204. The reference line corresponds to the critical points obtained by $B = 5000$ simulations. The full line below the reference line corresponds to the minimum critical point of all intervals estimated by B simulations; the values and the line above it correspond to the maximum critical points.

7.4.2   Bootstrap estimates of properties of ratios

In this section bootstrap estimates of properties of DRG and R, relevant to confidence interval construction, will be discussed. The analysis was based on the set of all observations, as well as on randomly selected subsamples of $n = 200$, 500 and 800 observations. Estimates based on all observations were taken for the 'true' value of the parameter.

The salient properties of the effect estimates are as follows.

(i) *Bias estimate.* The bias of $\hat{R}$ is negligible for all sample sizes examined. The mean bias of $\hat{DRG}$, for $n = 200$, is estimated by -0.002, with a maximum bias of -0.005. This means that for samples smaller 200 days, the bias is substantial, for the size of effect detected in the present analysis.

(ii) *Difference between bootstrap and asymptotic variance estimate (difs).* This parameter reflects the reliability of $\hat{\sigma}_{as}$ for construction of the STD interval. For $n = 200$, $\bar{difs} = 0.0005$, about 0.8% of $\bar{\delta}^{*}$, and attains a maximum of 3% of the variance estimate. *difs* assumes negative and positive values. Overall, for the sample sizes examined, the asymptotic variance estimate seems reasonable.

(iii) *Relationship between effect estimate and its standard error.* Regression of the standard error of $\hat{DRG}$ on $\hat{DRG}$ shows that as the sample size decreases, the slope of the regression, δ, and the percent of variance explained by the model, R^2, increase. For a sample size $n = 200$, the slope is estimated by $\hat{\delta} = 0.069$,

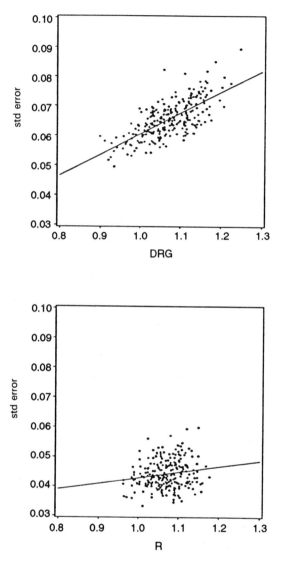

Figure 7.2 Relationship between estimate and its standard error, based on $g = 250$ subsamples of n(seeded) = 200 and n(unseeded) = 200 days. $B = 3000$ simulations were performed for each subsample.

and the percentage explained by the regression model by $R^2 = 0.51$. For the ratio estimate the corresponding estimates are $\hat{\delta} = 0.02$ and $R^2 = 0.3$. Thus for the DRG, the standard error of the estimate is proportional to the estimate, while for R, the s.e. is relatively stable, and is only slightly dependent on the estimate. The differences between R and DRG are evident in Figure 7.2. The implication of this property on the selection of method will be discussed in the next section.

7.5   COMPARISON OF CONFIDENCE INTERVALS FOR RATIOS

In this section the methods described in Section 7.2 are applied to the DRG and R statistics. The estimated 95% normal (NOR) confidence interval for the DRG, based on all observations in the period 1950–1990, is given by $[1.020, 1.119]$, and $\hat{DRG} = 1.07$. The differences between the critical points of this interval and those of the other intervals (BAC, HYB and STD) are smaller than 0.003. Thus, samples of about 2000 observations use of a bootstrap confidence interval is not required.

The differences between the intervals are apparent for smaller samples. The mean asymmetry of the STD interval for the DRG, based on 200 days, is 1.10. The differences between the critical points of the HYB, BAC and STD intervals are demonstrated in Figure 7.3. The distribution of the differences $\hat{\theta}_{BAC} - \hat{\theta}_{STD}$ and $\hat{\theta}_{HYB} - \hat{\theta}_{STD}$ is summarised in boxplots, for samples of 200 seeded and 200 unseeded days. The reference line represents the position of the STD interval. The most outstanding result in Figure 7.3 is the sensitivity of the DRG to the choice of interval. The BAC intervals for the DRG are shown to be fairly close to the STD interval, whereas all HYB intervals lie to its left. Some of the differences between the HYB and STD critical points exceed 0.02. For R, no systematic ordering of the intervals is evident. Both the BAC and HYB intervals fluctuate around the STD interval.

7.5.1   Explanation for the difference between DRG and R

The difference between DRG and R, observed in Figure 7.3, raises the question as to the source of this difference. It was seen in Figure 7.2 that the standard error of the DRG estimate is proportional to the effect estimate, while the standard error of the R estimate is more or less constant. Remember also that the STD interval is based on the percentiles of the studentised distribution $\hat{K}^*$, while the HYB and BAC intervals are based on the percentiles of $\hat{H}^*$.

The following lemma relates the differences between the DRG and R to the differences between $\hat{H}^*$ and $\hat{K}^*$.

Lemma

(a) For statistics with constant variance σ^2 over the entire range of parameter values θ, $\hat{K}^*(x) \approx \hat{H}^*(x)$ for all x.

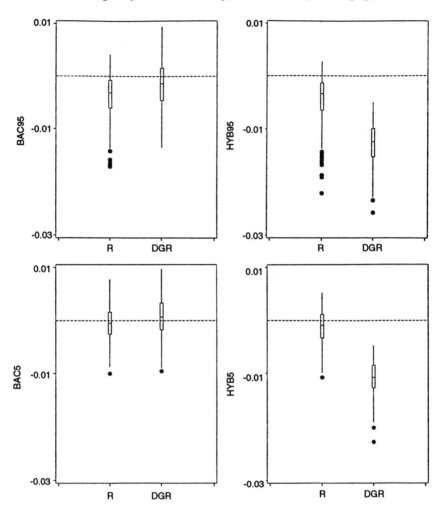

Figure 7.3 Differences between the critical points $\hat{\theta}_{BAC} - \hat{\theta}_{STD}$ and $\hat{\theta}_{HYB} - \hat{\theta}_{STD}$ for ratio (R) and double-ratio (DRG) statistics, based on $g = 250$ subsamples of $n(\text{seeded}) = n(\text{un-}$ seeded$) = 200$ days, $\alpha = 0.10$, and $B = 3000$. The bottom and top edges of the box are located at the sample 25th and 75th percentiles. The center horizontal line is drawn at the median. The whiskers extend to a distance of at most 1.5 interquantile ranges.

(b) For statistics with constant coefficient of variation $c = \sigma/\theta > 0$, $\hat{K}^*(x) > \hat{H}^*(x)$.

Proof

(a) Without loss of generality, assume $\hat{\sigma}^* = 1$; since $\hat{\sigma}_b^* \approx \hat{\sigma}^*$ for all b,

$$\hat{H}^*(x) \approx \hat{K}^*(x).$$

(b) If $\hat{\theta}^* \leqslant \hat{\theta}$ then $\hat{\sigma}_b^* \leqslant \hat{\sigma}^*$, and similarly with inequality signs reversed. Therefore, assuming $\hat{H}^*$ and $\hat{K}^*$ are monotonic and continuous,

$$P^*((\hat{\theta}^* - \hat{\theta})/\hat{\sigma}^* \leqslant x) = P^*((\hat{\theta}^* - \hat{\theta}) \leqslant \hat{\sigma}^* x)$$
$$\leqslant P^*((\hat{\theta}^* - \hat{\theta}) \leqslant \hat{\sigma}_b^* x) \quad \text{for } x \leqslant 0$$

and

$$\leqslant P^*((\hat{\theta}^* - \hat{\theta}) \leqslant \hat{\sigma}_b^* x) \quad \text{for } x \geqslant 0 \text{ too.}$$

Note that generalisation to any monotone relationship between θ and σ is immediate.

It follows from the lemma that

(i) for R, the distributions H and K approximately coincide, whereas for the DRG the right tail of H is longer than that of K, and the left tail of K is longer than that of H;
(ii) from (i), the STD, HYB and BAC intervals for R are close to each other, while the HYB interval for the DRG is located to the left of the STD interval;
(iii) the BAC interval for the DRG is closer to the STD interval, because H and K have the opposite asymmetry, and because of the 'confusion' of the upper and lower percentiles in the definition of the BAC interval—this confusion 'corrects' for the wrong asymmetry of H.

An example for the percentiles of $\hat{H}^*$ and $\hat{K}^*$ for samples of 200 days is presented in Table 7.1. An asymmetry index is defined by the sum of the $\frac{1}{2}\alpha$ and $1 - \frac{1}{2}\alpha$ percentiles. It is seen that for R, H has a slight negative skewness while K is positively skewed. For the DRG, the asymmetries of H and K are opposite and with similar absolute values. The closeness of the magnitude of the asymmetry, accounts for the closeness of the BAC and STD intervals in this case.

Table 7.1 Comparison of $\hat{K}^*$ and $\hat{H}^*$ for R and DRG, with $n(\text{unseeded}) = n(\text{seded}) = 200$, $g = 250$ and $B = 3000$

$1 - \frac{1}{2}\alpha$	$\hat{H}^{*-1}(1 - \frac{1}{2}\alpha)$	$\hat{K}^{*-1}(1 - \frac{1}{2}\alpha)$	$\text{asy}(\hat{H}^*)$	$\text{asy}(\hat{K}^*)$
DRG				
0.95	1.7747	1.5809	0.1745	−0.1864
0.975	2.1040	1.8708	0.2491	−0.2620
R				
0.95	1.6596	1.6356	0.0141	−0.1067
0.975	1.9893	1.8708	0.0352	−0.1376

$\text{asy}(H) = \hat{H}^{*-1}(1 - \frac{1}{2}\alpha) + \hat{H}^{*-1}(\frac{1}{2}\alpha)$ and $\text{asy}(K) = \hat{K}^{*-1}(1 - \frac{1}{2}\alpha) + \hat{K}^{*-1}(\frac{1}{2}\alpha)$.

Table 7.2 A case study of the differences between the percentiles of $\hat{H}^*$ and $\hat{K}^*$

Sample	$\hat{H}^{*-1}(0.05)$	$\hat{H}^{*-1}(0.95)$	$\hat{K}^{*-1}(0.05)$	$\hat{K}^{*-1}(0.95)$
#27	− 1.6301	1.7211	− 1.9454	1.51123
#41	− 1.6454	1.5881	− 1.7051	1.59564

7.5.2   Stability of the STD interval

Theoretical considerations ensure that if the STD interval is based on a true pivot then the method is more accurate than the BAC and HYB methods. The results in Section 7.4.2 reassure us that generally the asymptotic variance estimate is reliable; however, for some samples, its value is seen to be extreme. This problem may become more pronounced as the sample size decreases. In this section the caution required in the use of the STD interval is pointed out.

A case study of two samples identified as #27 and #41 shows that for the sample #27 the BAC and STD intervals for the DRG are quite different, while for sample #41 they are similar. A study of the percentiles of $\hat{H}^*$ and $\hat{K}^*$ in Table 7.2, shows that the left tail of $\hat{K}^*$ for sample #27 is extremely long. The effect estimate for this sample is also extreme, with $\hat{DRG} = 1.15$. Increasing the number of simulations from 3000 to 5000 does not change the results.

It is not clear how to operate for extreme cases. One may try to double bootstrap and get another variance estimate, or robustify K by some censoring. A simple diagnostic tool for the identification of extreme cases is the mean of the studentised values $(\hat{\theta}^* - \hat{\theta})/\hat{\sigma}_b^*$, whose expected value (if pivotal) is 0. When K is highly non-symmetric, this value will be relatively far from zero.

7.6   CONCLUSIONS

The investigation of bootstrap methods for construction of confidence intervals for ratio estimates, and the comparison between the double-ratio (DRG) and ratio (R) statistics, leads to the following conclusions.

(i) For samples of over 800 observations, a confidence interval based on the asymptotic normal distribution seems accurate enough. For smaller samples, the differences between the bootstrap BAC, HYB and STD intervals may be substantial.

(ii) The DRG statistic is more sensitive than R to the choice of method. In particular, the BAC interval for the DRG is found to be closer to the STD interval than the HYB interval, although the choice of the percentiles in the critical points of the BAC interval seems erroneous (Hall 1988, page 928). This phenomena is shown to characterise distributions with constant coefficient of variation. For these distributions, the BAC interval has the correct asymmetry.

(iii) The asymmetry index of the confidence interval for the DRG, for samples of about 200 observations, is about 1.13.

(iv) The STD interval is not robust, in the sense that for some samples the studentised distribution can have an unduly long tail. It is therefore recommended that the STD interval be used with care.

8

Inter-site and Inter-duration Dependence on Rainfall Extremes

D. W. Reed and E. J. Stewart

Institute of Hydrology, UK

8.1 INTRODUCTION

Hydrologists concern themselves with extreme rainfall in several contexts. One is the consideration of the spatial and temporal characteristics of severe storms, and perhaps their physical structure and the wider pattern of rainfall and weather within which they occur. Radar data can play a leading role in such studies, particularly those Doppler and dual-polarisation radars that provide information on the vertical structure of storm clouds (Doviak and Zrnic, 1984). The context is essentially the understanding and represention of severe storm cells and their rainfall fields.

The human impact of extreme rainfall is governed by where communities lie. Because the greatest impact is generally in the form of flooding, there is an important intermediary: the river catchment. Critical locations within a drainage system are known from past inundation or through topographic survey. Thus the relevant input is extreme rainfall affecting specific catchment areas.

For the hydrologist engaged in flood forecasting, the required steps are to monitor the rainfall field as it evolves in time, to anticipate its further development as the storm cell or rain band traverses the district, and to extract—from the observed and forecast fields—the rainfall information relevant to flood forecasting on the catchments draining to the critical sites. Through use of an appropriate rainfall-runoff model (see e.g. Moore, 1993) a real-time forecast of flow can be obtained and flood warnings issued where appropriate.

Statistics for the Environment 2: Water Related Issues. Edited by Vic Barnett and K. Feridun Turkman
© 1994 John Wiley & Sons Ltd.

8.1.1   The need for statistical analysis of rainfall extremes

For the hydrologist engaged in sizing flood defence structures, there is a complication. Flood alleviation works generally have to be designed to satisfy cost–benefit and performance criteria. The target level of service is usually expressed as an annual risk of inundation, with a value of 0.02 or 0.01 being the typical design standard for flood defence from main rivers in urban areas.

The difficulty is that the rainfall giving rise to flooding on a particular catchment is not unique to that catchment. Flooding may arise from a modest storm cell that is (unusually) stationary and (unusually) centred on the subject catchment, from a severe storm that passes across much of the catchment in a 'normal' fashion, or from an exceptionally severe storm that affects only part of the catchment.

Because it is the rainfall experienced by the catchment that is of interest—not the rainfall associated with the life of a particular storm cell or rainband—the statistical analysis required is that of rainfall depths recorded in arbitrary time periods but over fixed areas.

It is highly desirable that the method of synthesising rainfall frequency be fully generalised, to provide depth–duration–frequency estimates for any site, any duration and any return period. At present, it is impractical to carry out this generalisation other than in terms of point rainfall. Thus, in many design applications, it remains necessary to apply an 'areal reduction factor' to transform the T-year rainfall estimate (for a typical point within the catchment) to the required T-year catchment rainfall (see e.g. Stewart, 1989).

8.1.2   UK rainfall data

Most long-term records are of rainfall at a point, most often obtained as daily readings of a standard rain-gauge or hourly observations derived from chart logging of tilting syphon rainfall recorders. Radar observations of rainfall are of a different character, being spatial averages of instantaneous rainfall over an elemental area. Rainfall depths inferred from radar are often rather inaccurate, and radar data are therefore best used in conjunction with conventional rain-gauge data. Their spatial form is, of course, particularly suited to modelling applications requiring catchment or gridded rainfall data.

Although the UK has a history of organised rainfall data collection from the mid nineteenth century, the proportion of records available in computer form is relatively modest. Comprehensive records of daily data for several thousand sites are available from 1961 onwards, with longer-term daily data computerised for about 400 sites. Substantial computerised records of hourly data are available for about 118 sites, with start dates of 1970 and 1980 being typical (Stewart and Reynard, 1993).

A study of extremes can only be as good as the record lengths allow. The approach to be taken in new research for the National Rivers Authority is to augment the above computerised records with manuscript lists of extreme events, abstracted in support of Jenkinson's original generalization of rainfall frequency for the Flood Studies Report (NERC, 1975) and of a re-evaluation of 5-year 1-hour rainfall depths (May and Hitch, 1989).

Where extremes have been previously abstracted from tabular records, the data are usually in the form of annual maximum accumulations in fixed observing periods (e.g. one day ending at 0900 GMT, or one hour ending on the hour). However, where extremes have been abstracted directly from autographic records, these are more often in the form of true 60-minute or 120-minute maxima (i.e. 'sliding' or variable maxima).

8.1.3   Why adopt an annual maximum approach?

The peaks-over-threshold approach to the analysis of extremes has attracted new attention in recent years (Davison and Smith, 1990), but historically it has been applied much more often in flood frequency analyses than in rainfall frequency analyses. In the UK this distinction may have its origin in the numerous gauging stations for which rather short records of flood peaks were available at the time of the Flood Studies Report. However, the persistence of the idea that POT methods are less relevant to the analysis of rainfall frequency may reflect the greater acceptance of pooled analyses in such applications.

Another factor is that abstraction of rainfall extremes is more readily thwarted by missing data or uncertainty in the quality of the record. In the abstraction of peaks-over-threshold flood data from water level charts, it is usually possible to infer whether a flood may have been missed during a break in recording. Checks can be made of flows recorded before and after the break, of flows further down the river system and of rainfall measurements in the catchment area. Quality control for rainfall extremes is, in contrast, very much less certain. The rainfall process is erratic and intermittent, and it is hazardous to assume from the absence of an extreme at a neighbouring site that there was none at the subject site. Some relief has been given by the introduction of operational weather radars in the 1980s (May, 1988), but quality control of rainfall data remains problematic.

Whereas missing periods disrupt the POT rainfall series directly, it is possible to be rather more pragmatic in forming the annual maximum series, e.g. accepting maxima as valid where they derive from nine or more months' data. That maxima for several thousand site-duration-years have been abstracted in previous studies does, of course, further encourage retention of an annual maximum approach.

8.1.4   Pooling

The motivation for pooling data is to secure more consistent estimates of extremes. Because the required return periods are often high in comparison with the available record lengths at a particular site, an approach based on extrapolation of a single-site frequency curve is likely to be highly sensitive to the sample, the choice of distribution and the method of fitting. The limitations of a small sample are also likely to be seen in frequency curves for rainfall extremes of different duration that intersect anomalously. In such circumstances, when an accurate estimate of the required quantile is unattainable, good practice looks to supportable methods that provide consistent estimates at neighbouring sites, neighbouring durations and neighbouring return periods.

Some form of standardisation is usually appropriate prior to pooling data. The approach generally taken is to divide by an index variable, often the mean of annual maxima. As a consequence, the pooled frequency curve subsequently derived is dimensionless. This is usually termed a growth curve, and defines the size of the T-year extreme relative to the index value.

One approach to pooling is to seek an average growth curve across a region. This is relatively simple to arrange, but has the drawback that clear inter-site differences in rainfall growth at modest return periods (e.g. a quarter of the record length) may be suppressed in the drive to attain consistency at high return periods (e.g. four times the record length).

A further drawback is that pooling data by region leads to hard boundaries, at which the T-year growth factor jumps from one value to another. For flood peak data, pooling by hydrometric area may be a pragmatism made acceptable by a fractional (i.e. weighted) allocation scheme (e.g. Wiltshire, 1986). However, it is difficult to find any justification—other than administrative convenience—for fixed regional pooling of rainfall extremes.

Reed and Stewart (1989) and Burns (1990) introduce more flexible ways of pooling data, based on centring the region of pooling on the subject site. Reed and Stewart adopt the further refinement of adjusting the extent of pooling according to the segment of the rainfall growth curve being sought, so that the curve accords with locally recorded extremes at low return periods but is driven by regional extremes at high return period.

8.1.5   Dependence on rainfall extremes

While hydrologists have generally recognised the need to pool data if erratic extrapolation of frequency curves is to be avoided, the possible effect of dependence has been a recurrent worry. This is most obvious in the station-year method, where particular importance is attached to the largest standardised events in a region. Hosking and Wallis (1988) conclude that correlation among annual maxima leads to increased uncertainty in growth curve estimation. In essence, inter-site dependence is detrimental because it reduces the worth of a given number of station-years.

A consequence of inter-site dependence is that annual maxima from non-overlapping periods of record will in general be more valuable to a pooled analysis than those from concurrent records. This factor is generally ignored in traditional methods of pooling.

8.2 A CASE STUDY OF RAINFALL EXTREMES: MELBOURNE

8.2.1 Data

The study data comprise long-term hourly rainfall readings for six sites in Melbourne (Figure 8.1). Annual maximum values were abstracted for durations of 1, 2, 4, 6 and 12 h, together with their dates and times. A year was discounted only if more than 25% of the hourly data were missing. Given the general high quality of the data series, annual maxima could be ascribed in almost all site-years.

Record lengths are 'feathered'; for example, only one gauge has records earlier than 1930, while only two years (1971 and 1972) provide records for all six sites. To ensure that each year provided annual maxima for at least two sites, the study period was limited to 1930–1989 inclusive, providing 212 site-years of annual maximum values from 60 calendar years. Typically, there were 3.306 sites operating in any one year (geometric mean). Annual maxima for all five durations were available for each site-year, yielding a total of 1060 site-duration-years.

8.2.2 Choice of extreme value methods

The technique of L-moments (Hosking, 1990; see also Cunnane, 1989) is used to explore the statistical distribution of the extreme values. The first L-moment corresponds to the arithmetic mean, and provides the index by which to standardise annual maxima from different sites and/or durations, prior to pooling. Higher L-moments define measures of the variability, skewness and peakedness of extreme values; these are the L-moment ratios: L-CV, L-skewness and L-kurtosis (Hosking, 1990). The L-moments are linear combinations of probability-weighted moments; thus fitting by L-moments is synonymous with fitting by PWMs.

The generalised extreme value (GEV) distribution is used to represent the distributions of annual maxima. For $k \neq 0$:

$$x = u + a(1 - e^{-ky})/k, \tag{8.1}$$

where x is the variate and y the Gumbel reduced variate, i.e. $y = -\ln[-\ln F(x)]$. L-moment ratio diagrams (see e.g. Figures 8.3 and 8.4) provide a medium for checking the appropriateness of the distributional assumption.

The analyses reported use the biased PWM estimators recommended by Hosking *et al.* (1985). All the pooled applications standardise the annual maxima prior to pooling by dividing by the arithmetic mean. In arriving at 'regional' PWMs, the individual site PWMs are weighted according to record length (Wallis,

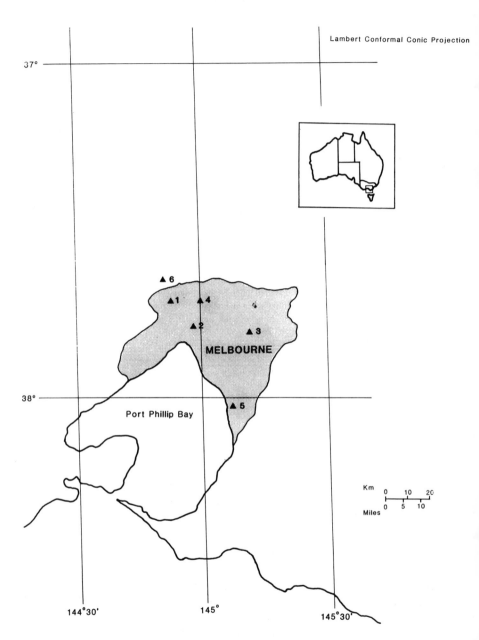

Figure 8.1   Location plan.

1980). The terminology used here is to refer to the network of sites or durations or, collectively, the network of site-durations.

8.2.3 A measure of dependence

The analysis assesses dependence in extreme values, i.e. the tendency for annual maxima at neighbouring sites or 'adjacent' durations to be correlated, and does so by reference to the network maximum SAM (Dales and Reed, 1989); SAM denotes standardised annual maximum. The network maximum SAM is simply the largest standardised annual maximum value from the network of sites or durations or site-durations, and is evaluated separately for each year of record.

The network maximum SAM is, by definition, greater than or equal to the single-site or single-duration SAM. Thus the distribution of network maximum SAMs lies above the typical (i.e. pooled) distribution of single-site or single-duration SAMs. The maximum of N independent and identically distributed (i.i.d.) random variables—each a GEV—is itself a GEV. Using the subscripts n and p to denote the parameters of the network maximum and pooled GEVs respectively, Dales and Reed (1989, Appendix 4) derive the relations

$$k_n = k_p = k \quad \text{(say)}, \tag{8.2}$$

$$a_n = a_p N^{-k}, \tag{8.3}$$

$$u_n + a_n/k = u_p + a_p/k. \tag{8.4}$$

Equation (8.2) states that the network maximum of N i.i.d. GEV distributions has the same shape parameter. Equation (8.4) shows that the network maximum distribution has the same bound as the parent distribution; for $k < 0$ the distributions share a common lower bound. For any parent distribution $F(x)$, the maximum of N i.i.d. variates lies a fixed distance $\ln N$ to the left of the parent on a variate-reduced variate plot. This follows from

$$y_n = -\ln[-\ln F^N(x)] = -\ln N - \ln[-\ln F(x)] = -\ln N + y_p,$$

so that

$$\ln N = y_p - y_n. \tag{8.5}$$

Here y_n and y_p denote reduced variate values corresponding to a given standardised value x in the distributions of the network maximum and parent respectively.

In practice, the standardised annual maxima being pooled from different sites or durations are partially dependent, and the network maximum distribution is found to lie a shorter distance, $\ln N_e$, to the left of the distribution fitted to the pooled SAMs, which estimates the underlying parent distribution. The measure N_e defines an effective number of independent sites or durations. Dales and Reed (1989), using a measure of bivariate dependence due to Buishand (1984), found the degree of dependence in rainfall extremes to vary only weakly with quantile x. This motivates a particular method of estimating the degree of dependence, based on the assumption that N_e is invariant with quantile.

If there is no dependence between the N sets of annual maxima being pooled, $N_e = N$; if there is complete dependence, $N_e = 1$. Following Reed and Stewart (1991), the degree of dependence in the network maxima is defined as

$$d = 1 - \frac{\ln N_e}{\ln N},$$ (8.6)

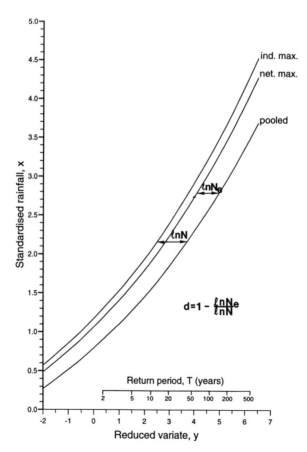

Figure 8.2 Derivation of the equivalent number of independent sites (or durations) N_e and the degree of dependence d.

so that $d = 0$ corresponds to no dependence and $d = 1$ to complete dependence. The measure d is the proportion of the distance that the (partially dependent) network maximum distribution is found to lie from the no dependence (i.e. $N_e = N$) case to the complete dependence (i.e. $N_e = 1$) case. Figure 8.2 provides an illustration.

8.2.4   Procedure used to estimate dependence

General extreme value distributions are fitted separately to the pooled and network maximum SAMs. The shape parameter is then constrained to an intermediate value $(k = \frac{1}{2}(k_p + k_n))$, and the GEVs re-fitted to derive new scale and location parameters of the pooled and network maximum GEVs. The implied bound of each distribution is noted, and the arithmetic mean taken. The remaining parameters $(a_p$ and u_p, a_n and $u_n)$ are then constrained to satisfy this common bound while also fitting the observed mean of each distribution. The motivation behind enforcing these constraints (i.e. (8.2) and (8.4)) is to harmonise the distributions fitted to the network maximum and pooled SAMs, prior to calculating the degree of dependence. The effective number of independent sites or durations is then estimated from (8.3), i.e.

$$N_e = (a_n/a_p)^{-1/k}, \tag{8.7}$$

and the degree of dependence calculated using (8.7) as shown in Figure 8.2.

   Whereas Dales and Reed (1989) consider spatial dependence in rainfall extremes, here both spatial and temporal dependence are explored. N_s denotes the number of sites and N_t the number of durations; the additional suffix e is used to denote the equivalent number of independent sites or durations. The indices d_s and d_t denote the degrees of spatial dependence and temporal dependence respectively.

   A premise of the approach is that annual maxima are identically distributed across different sites and durations, save only for a scaling factor. This assumption has to be checked for the Melbourne data set.

8.2.5   Rainfall growth characteristics for individual sites and durations

The annual maxima are considered in four ways:

  (i) individual sites, individual durations;
 (ii) pooled sites, individual durations;
(iii) individual sites, pooled durations;
(iv) pooled sites, pooled durations.

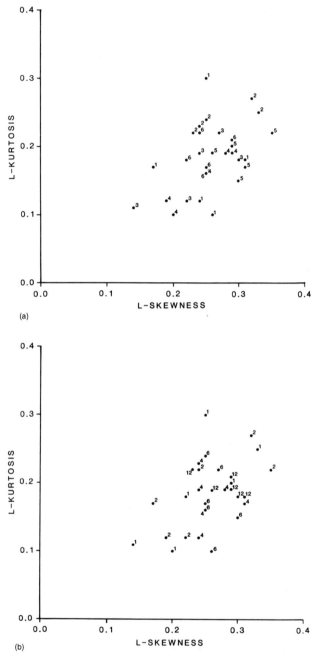

Figure 8.3 L-moment ratio diagrams for standardised annual maximum rainfall depths: (a) labelled by site; (b) labelled by duration.

Figures 8.3(a, b) show L-skewness and L-kurtosis for the complete set of six sites and five durations (i.e. 30 site-durations). The labels in Figure 8.3(a) indicate the site, whereas those in Figure 8.3(b) denote the duration. It is seen that the different sites and durations are relatively well mixed within the overall set. The general mix of points evident in Figure 8.3, and the absence of any notable outliers, suggests that the six sites and five durations can be treated as a homogeneous network of standardised annual maxima, to which a single growth curve can be fitted.

8.2.6 Pooled analyses of rainfall growth

Figure 8.4(a) shows L-moment ratio diagrams corresponding to the PWMs pooled across durations. Standardised annual maxima were available for all five durations for each site-year of record; thus $N_t = 5$. The labels in Figure 8.4(a) distinguish the site. The grouping is not especially tight, and there is some suggestion that site 5, and possibly site 2, exhibit greater skewness and kurtosis than the others. While this could reflect a coastal influence, some of the variation in Figure 8.4(a) may stem from the different periods of record available at each site.

The resultant (GEV) parameters for the rainfall growth curves for each site are summarised in Table 8.1. The R-statistic in the final column relates to the use of the G-point test (Wiltshire and Beran, 1987) of regional homogeneity to determine if the network of standardised annual maxima of different durations is homogeneous. The critical value of R for this sample size (i.e. five durations) is 9.49 at the 5% significance level; thus for none of the sites is there compelling evidence of inhomogeneity. However, on the basis of the R-statistic, homogeneity across durations is somewhat less certain for sites 1, 2 and 5 than for sites 3, 4 and 6.

Figure 8.4(b) shows L-moment ratio diagrams corresponding to the PWMs pooled across sites. Standardised annual maxima were available for an average

Table 8.1 Analysis of annual maxima: pooled by duration

| Site | GEV distribution parameters | | | R-statistic |
	u	a	k	
1	0.806	0.274	−0.119	6.59
2	0.793	0.271	−0.161	4.02
3	0.832	0.243	−0.103	0.46
4	0.800	0.285	−0.112	0.33
5	0.766	0.286	−0.198	2.61
6	0.763	0.330	−0.124	0.72

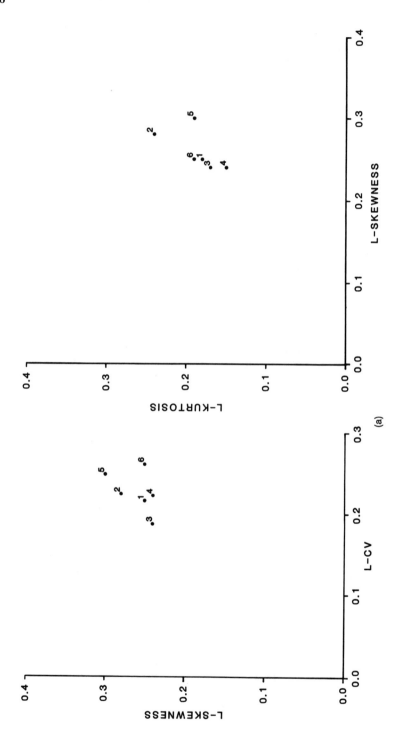

(a)

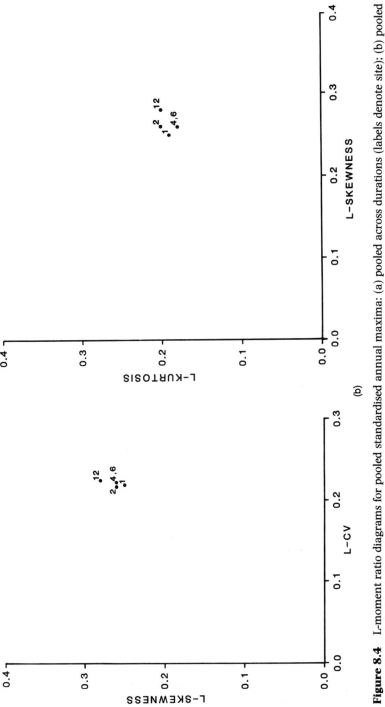

Figure 8.4 L-moment ratio diagrams for pooled standardised annual maxima: (a) pooled across durations (labels denote site); (b) pooled across sites (labels denote duration).

Table 8.2 Analysis of annual maxima: pooled by site

Duration (h)	GEV distribution parameters			R-statistic	Mean annual maximum value (mm)
	u	a	k		
1	0.801	0.279	− 0.121	10.12	15.29
2	0.801	0.275	− 0.129	4.20	21.41
4	0.799	0.277	− 0.132	3.20	29.07
6	0.797	0.277	− 0.137	5.83	34.23
12	0.793	0.271	− 0.160	5.36	44.67

(geometric mean) of 3.306 sites for each duration-year of record; thus $N_s = 3.306$. The numbers attached to data points in Figure 8.4(b) denote duration. It is seen that—with the possible exception of the 12-hour data—the points group closely, supporting the assumption that maximum rainfalls are identically distributed for the range of durations considered here.

The resultant GEV parameters for the rainfall growth curves for each duration are summarised in Table 8.2. Again, the G-point test provides no compelling evidence of inhomogeneity at any of the durations; the critical value of R for this sample size (i.e. six sites) is 11.07 at the 5% significance level.

The R-statistic confirms that homogeneity across sites is generally somewhat poorer than across durations, with homogeneity being supported least strongly for the network of 1-hour rainfall extremes (10.12 is close to the test value of 11.07 at which the G-point test would pronounce significant evidence of inhomogeneity). It is recognised that the power of the G-point test to detect inhomogeneity is very low for the sample sizes relevant here.

It is seen from Table 8.2 that the pooled growth curves for different durations are remarkably similar, further supporting the decision to pool across durations as well as across sites.

8.2.7 Depth–duration formula for the mean annual maximum rainfall

The mean values in the final column of Table 8.2 are the weighted cross-site means of the annual maximum values, i.e. weighted by record length. Logarithmic regression on duration yields

$$\text{RBAR}_{\text{CH}}(D) = 15.65 D^{0.431}, \tag{8.8}$$

with a factorial standard error of 1.025, where RBAR_{CH} is the mean annual maximum rainfall (mm) in D clock hours.

Table 8.3 Analysis of annual maxima: pooled across all sites and durations

GEV distribution parameters			
u	a	k	R-statistic
0.798	0.276	-0.136	29.07

8.2.8 Two-way pooled analysis of rainfall growth

Pooling the standardised annual maxima for all sites and durations yields the rainfall growth curve summarised in Table 8.3, also appearing in Figure 8.2. Applying the G-point test, the grouping of 30 site-durations is not judged inhomogeneous ($R = 29.07$ compared with a test value of 42.56 at the 5% significance level).

8.2.9 Dependence assessments

Spatial and temporal dependence are assessed using the approach illustrated in Figure 8.2. Table 8.4 presents the degree of spatial dependence found for each duration. That the degree of spatial dependence is found to increase with duration is consistent with the typically greater physical extent of storms giving rise to extreme rainfalls over longer durations.

The geometric mean of d_s from Table 8.4 is 0.263. Applying this to the typical number of sites operating in any year, the effective number of independent sites is judged to be

$$N_{se} = N_s^{1-d_s} = 3.306^{0.737} = 2.414.$$

Table 8.5 presents the degree of temporal dependence d_t found in rainfall extremes for each site. At all sites the degree of temporal dependence is marked. The geometric mean of d_t is 0.625; thus the effective number of independent durations is judged to be

$$N_{te} = N_t^{1-d_t} = 5^{0.375} = 1.829.$$

Table 8.4 Spatial dependence in rainfall extremes: Melbourne

Duration D (h)	1	2	4	6	12
Degree of spatial dependence d_s	0.163	0.238	0.268	0.307	0.393

Table 8.5   Temporal dependence in rainfall extremes: Melbourne

Site	1	2	3	4	5	6
Degree of temporal dependence d_t	0.599	0.567	0.608	0.630	0.675	0.678

The measures of spatial and temporal dependence can be combined to give an estimate of the degree of spatio-temporal dependence d_{st} in the overall data set of extreme values. The approximate relation

$$d_{st} = \frac{d_s \ln N_s + d_t \ln N_t}{\ln N_s + \ln N_t} \tag{8.9}$$

assumes that the effective number of independent site-durations N_{ste} is adequately estimated by multiplying the effective number of independent sites N_{se} by the effective number of independent durations N_{te}. For $N_s = 3.306$ and $N_t = 5$, the formula yields $d_{st} = 0.471$. This agrees well with the value $d_{st} = 0.468$ obtained by applying the method of Figure 8.2 directly to the entire pooled data set of standardised annual maxima.

8.3   DEPENDENCE MODELS AND THEIR APPLICATIONS

8.3.1   Generalised dependence models

Dales and Reed (1989) present a general model of spatial dependence in rainfall extremes, albeit calibrated for rainfall durations in the limited range of one to eight days. The model is of the form

$$d_s = d_s(N_s, \text{AREA}, D) \tag{8.10}$$

where D is the rainfall duration and AREA is a measure of the area spanned by the N_s sites. The model can be used to assess the degree of dependence in rainfall extremes of a given duration for any collection of sites in the UK.

The model is not without imperfection; in its present form, it is capable of yielding estimates of d_s outside the permitted range $(0, 1)$. Furthermore, it would be helpful to construct the spatial-dependence model separately at different reference durations, so that the manner in which D influences d_s is not prejudged.

Construction of a general model for temporal dependence in rainfall extremes has yet to be attempted. However, a candidate model is of the form

$$d_t = d_t(N_t, \text{RANGE}) \tag{8.11}$$

where RANGE is a measure of the temporal range spanned by the N_t durations.

8.3.2   Collective risk estimation

Dales and Reed (1989) use their spatial dependence model to develop estimates of the collective risk r of experiencing an event with return period T years at one or more of a network of critical sites:

$$r = 1 - \left(1 - \frac{1}{T}\right)^{N_{se}}. \tag{8.12}$$

They also demonstrate the corollary to spatial dependence that, when an exceedance occurs, it is quite likely to affect a number of sites.

Somewhat similar applications are envisaged for a generalised model of temporal dependence. If this is constructed thoughtfully, it should be possible to evaluate the collective risk of experiencing an extreme rainfall for a set of durations that is representative of *all* durations to which a particular installation is thought to be sensitive. Collective risk estimation using models of spatial and temporal dependence could also be of profound importance in generalizing rainfall rarity estimation for drought severity assessment.

8.3.3   Cost–benefit assessment of data extraction

A characteristic of studies of subdaily rainfall extremes is the difficulty in assembling relevant data in a cost-effective manner. Records of hourly rainfall data held in computer form are typically much shorter than those for daily data. Because of the very large number of data values involved (e.g. 175 320 for a 20-year record) in comparison with the number of extreme events (e.g. 20 for an annual maximum analysis of a 20-year record), computerisation of historical series of hourly data is unlikely to prove economic unless benefit can be assigned to other applications. A more cost-effective alternative is likely to be to seek out tabulated annual maxima for one or more subdaily durations.

In preparatory work for a major analysis of rainfall frequency in England and Wales (Stewart and Reynard, 1993), it has proved possible to recover many site-duration-years of data painstakingly abstracted from hourly tabulations and charts as part of the UK Flood Studies Report (NERC, 1975). Their format differs from gauge to gauge. Sometimes annual maxima have been abstracted for a wide range of durations, in other cases only for one, two or three subdaily durations.

Dependence models will be helpful in determining where further effort in data extraction can best be concentrated. Because of the many more numerous computerised records of daily data, the much easier task of extending annual maximum abstractions of daily data to earlier start dates (than for hourly data), and the easier task of abstracting 1-hour extremes (than say 12-hour extremes) from hourly data, it is likely that an efficient strategy will seek to concentrate new data abstractions at key durations, and will, of course, give greater weight to sites in localities for which few annual maxima are presently held. Another feature is that abstraction of annual maximum data for years of record that are poorly represented in the overall data set is likely to prove particularly valuable; there is usually no dependence, spatial or temporal, in annual maxima from different years!

One rider is that the pooling of annual maxima from different sites and durations requires confidence in the form of standardisation used. Techniques based on peaks-over-threshold analysis, or correlation with neighbouring gauges, could be used to improve estimates of the mean annual maximum value from short records. A pragmatic approach may be to discount short-record (e.g. less than 10 years) where they add little to the local network of site-durations.

8.4 GENERATING RAINFALL FIELDS THAT ARE SPATIALLY CONSISTENT: A PLEA FOR NEW RESEARCH

An attempt was made in Section 8.1 to set the context in which rainfall frequency estimates are required for flood defence design. This context is not fixed. While details of the Flood Studies Report rainfall-runoff method of flood frequency estimation have been progressively enhanced, it is now recognised that the 'design event' approach on which it is based has fundamental limitations. It continues in widespread use in the UK largely because of the requirement for design hydrographs when studying catchments with intricate configurations and/or storage reservoirs.

Elsewhere in Europe there is renewed interest in Gradex-type methods (see e.g. Duband *et al.*, 1988; Margoum and Oberlin, 1991), which seek to use rainfall frequency estimates directly in the extrapolation of flood frequency curves. Thus, in some respects, the tide is flowing against the use of catchment models.

In conjunction with the Ministry of Agriculture, Fisheries and Food, the Institute of Hydrology is promoting a programme of work to develop a radically different method of flood frequency estimation, based on continuous simulation modelling of catchments. This will avoid a key weakness of the existing rainfall-runoff method of flood estimation, which makes only nominal allowance for the influence of pre-storm catchment wetness on flood formation.

Also, hydrological modellers are keen to exploit the several high-quality spatial data sets now available, for example digital terrain data (spot heights at 50 m grid interval), improved land cover classification (derived from satellite imagery), and more detailed soil classification (Boorman *et al.*, 1991). Many applications of flood frequency estimation arise as the result of land use change, past or planned. The new approach will permit the use of specialised models to represent particular effects, something that simply is not possible within the confines of the 'design event' approach of the FSR rainfall-runoff method.

If suitable continuous simulation models can be developed, it will be possible to make better use of detailed mathematical models of river flow to transcribe flood flows into water levels for the detailed design of flood defence structures. Because of necessarily coarse assumptions about suitable input hydrographs—particularly where two or more tributaries are involved—the potential of such models is yet to be fully realised for flood design.

However, two key obstacles have to be overcome if the continuous simulation approach is to provide a general method of flood estimation. First, hydrologists must develop catchment distributed models (see e.g. Beven, 1987) that represent *all* wetness conditions—normal and extreme. Secondly, a methodology is required to generate long-term catchment rainfall series that are spatially consistent. This might be achieved by introducing treatment of dependence into multisite stochastic models of point rainfall. Alternatively, statistical or mathematical models may be developed to generate rainfall fields on a regular grid, continuously through time. The problem is a relatively general one and would, we believe, reward the attention of non-hydrologists over the next three to five years.

ACKNOWLEDGEMENTS

The cooperation of the Australian Bureau of Meteorology, in making extensive records of hourly rainfall available, is warmly appreciated. Rainfall frequency research at the Institute of Hydrology has been funded by the Ministry of Agriculture, Fisheries and Food, the Department of the Environment, the National Rivers Authority, South West Water, and the Natural Environment Research Council (all UK organisations).

PART II

Sea Levels and Wave Energy

9

Trends in Sea Levels

J. A. Tawn, M. J. Dixon and P. L. Woodworth
Lancaster University and Proudman Oceanographic Laboratory, UK

9.1 INTRODUCTION

Over the last few years there has been significant public and scientific debate concerning the impact of climate change on the environment. In the UK the impact on sea levels has probably received the most attention. As part of this debate there have been a number of scientific papers (Houghton *et al.*, 1990; Woodworth, 1990; Pugh, 1990; Wigley and Raper, 1992) and reports by pressure groups (Barkham *et al.*, 1992) that make various predictions about future increases in mean sea level and flood frequencies. As a consequence of such predictions, the authorities responsible for managing coastal flood protection schemes have been forced to reassess the viability of their schemes. This has led to discussions such as whether a controlled retreat in certain low-lying coastal plains, for instance in the Wash and the estuarine regions of the Thames and Humber, is more cost-effective than improving coastal flood defences.

On a global scale, the issue of sea-level trends is even more emotive, with communities in low-lying countries, such as Bangladesh or the Republic of Maldives, threatened with losing their lands to the sea, since the cost of flood defences is largely beyond their resources. Other countries, for example the Netherlands and USA, face similar problems to areas of the UK.

This concern primarily stems from predictions of increases in mean sea level and flood frequencies due to global warming arising from increased greenhouse gas concentrations. Even without such changes induced by mankind, the mean sea level has not been a stationary process, as can be seen in Figure 10.1 in the following chapter of this volume (Vrijling, 1994). In fact, geological records show that over millions of years, sea levels have varied by several hundred metres (Pugh, 1987). Thus estimated increases of 0.3–1.5 m for the coming century, resulting from a combination of long-term trend and climate change impacts, are

Statistics for the Environment 2: Water Related Issues. Edited by Vic Barnett and K. Feridun Turkman
© 1994 John Wiley & Sons Ltd.

relatively small. Since such estimates are being used for future planning, it is a matter of some concern that the current mean sea-level trend induced by factors other than climate change is not, at present, accurately estimated. Much of the mean sea-level rise debate concerns the acceleration of a process for which we as yet do not know the current trend, and it is clear that evidence of accelerations is likely to be difficult to extract from sea-level data.

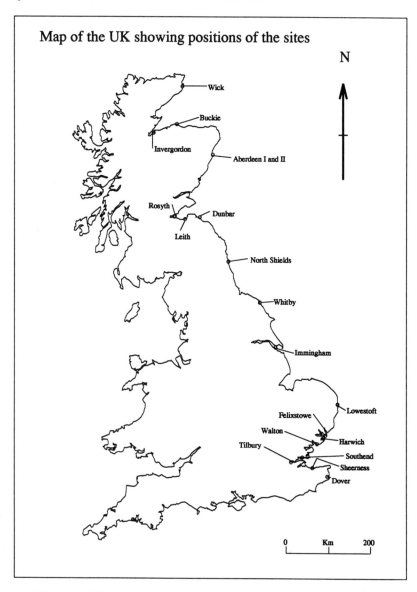

Figure 9.1   Map of the UK showing the positions of the 19 east coast data sites.

At a simplistic level, the problem is to estimate the trend and accelerations over recent years based on mean sea-level data. The data are annual estimates of mean sea level obtained from tide gauges, over a global network of sites, which have been in operation a maximum of 100–200 years, but more typically 10–30 years. Over this period, the long-term trend, to a first approximation, is linear, and so at first sight this seems to be a standard statistical problem, since simple regression methods can be used to estimate the linear trend and a quadratic term can be included to assess evidence for accelerations. However, such direct methods produce few helpful conclusions, because the trend itself is so small (≈ 0–3 mm yr^{-1} in the UK) relative to inter-annual and decadal variations of the process.

In this chapter the scientific issues associated with estimation of current sea-level trends are addressed by drawing on a combination of techniques from physical science and statistics. In Section 9.2 we give a background description of the physical processes involved and how they are observed, together with a discussion of the causes and impacts of current and future trends. In Section 9.3 the existing methods of analysis, from standard regression through to the use of hydrodynamical models, are described. Spatial statistical regression procedures are developed in Section 9.4 to fully exploit the limited information, and extensions of these to provide methods which are robust to errors in datum controls are outlined in Section 9.5. The impacts of mean sea-level changes are discussed in Section 9.6, where particular emphasis is given to issues faced by coastal engineers responsible for flood prevention schemes.

Although the principal change in mean sea level is likely to be on a global scale, there are physical reasons to expect regional variations as well (see Section 9.2), so a natural approach is to model first at a regional scale and then to combine regions to form a global perspective. Thus in this chapter we have taken the UK east coast as a region to illustrate the proposed methods. This region contains the areas of the UK most at risk from mean sea-level trends, and has a dense network of sites providing high-quality data. Figure 9.1 gives the positions of the 19 available data sites. The sites are irregularly positioned, being located for historical reasons at ports or more recently for storm surge monitoring, rather than for mean sea-level trend estimation. Mean sea-level data from these sites are shown in Figure 9.2. These are plotted on common scales to clarify the spatial similarity of the mean sea-level process and the variation in available data over sites. In particular, it is clear that there are a limited number of long-term series and a significant proportion of missing values.

9.2 BACKGROUND TO THE PHYSICAL PROCESSES

For any site, the observed sea-level process Y_t at time t is the composition of four physically distinct processes: mean sea-level M_t, astronomical tide T_t, surge S_t and surface waves W_t, such that

$$Y_t = M_t + T_t + S_t + W_t. \tag{9.1}$$

150

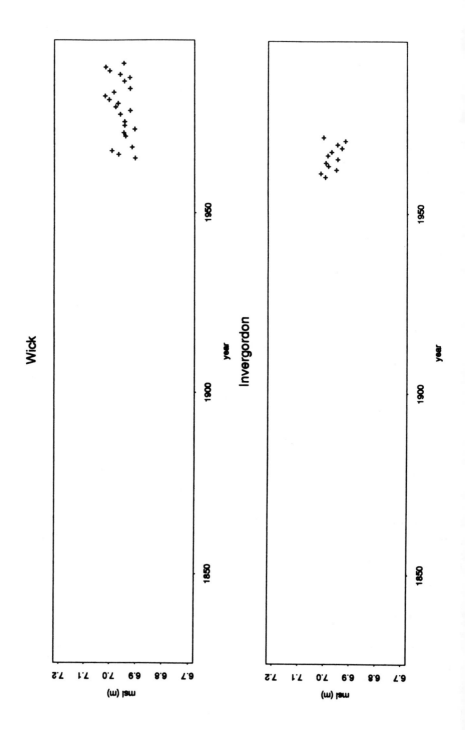

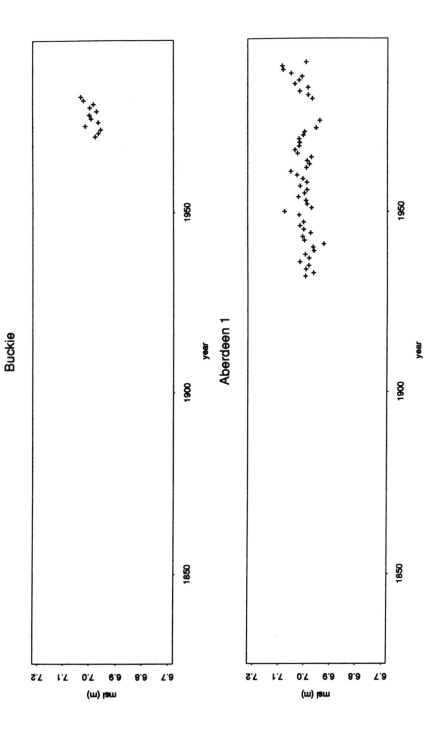

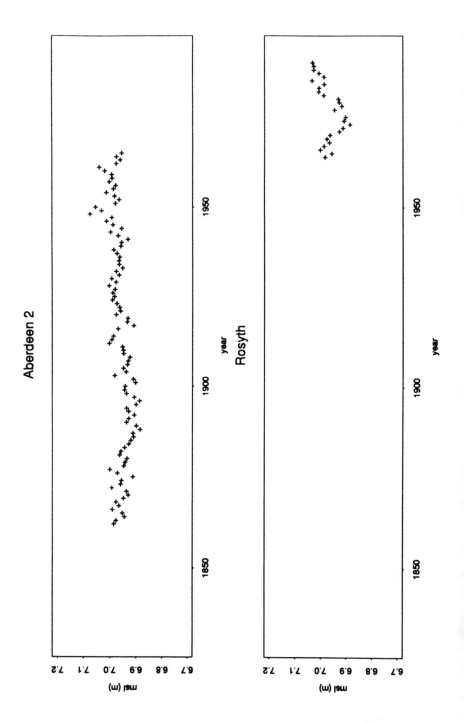

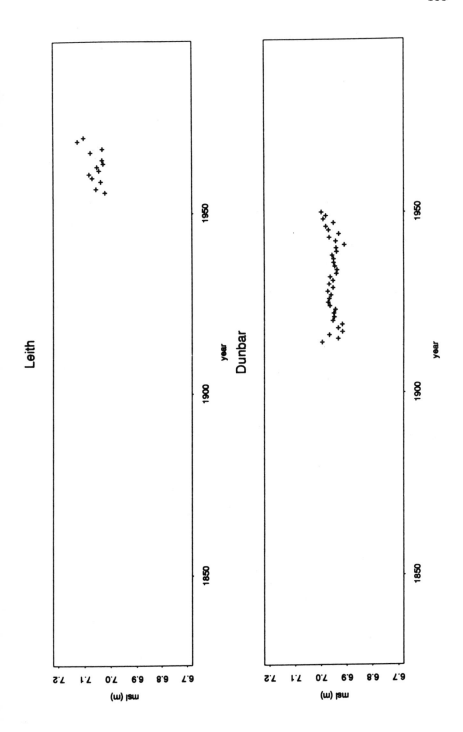

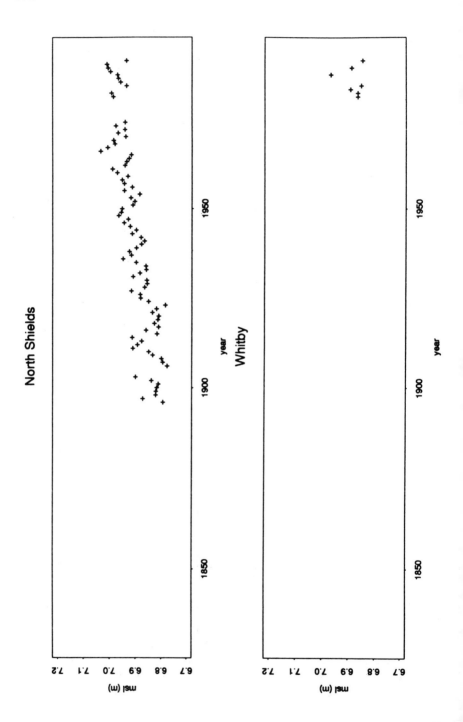

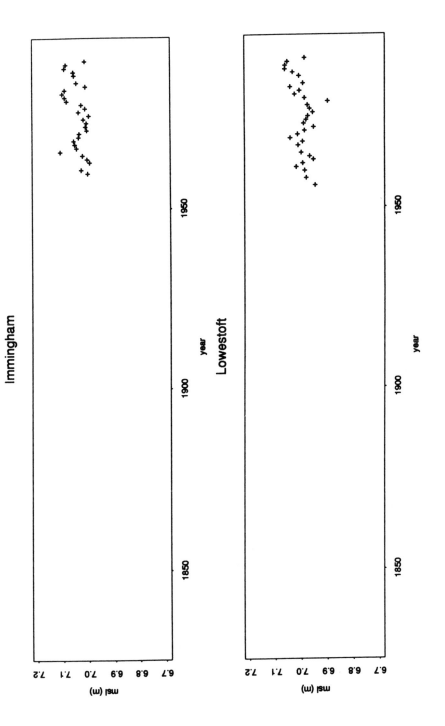

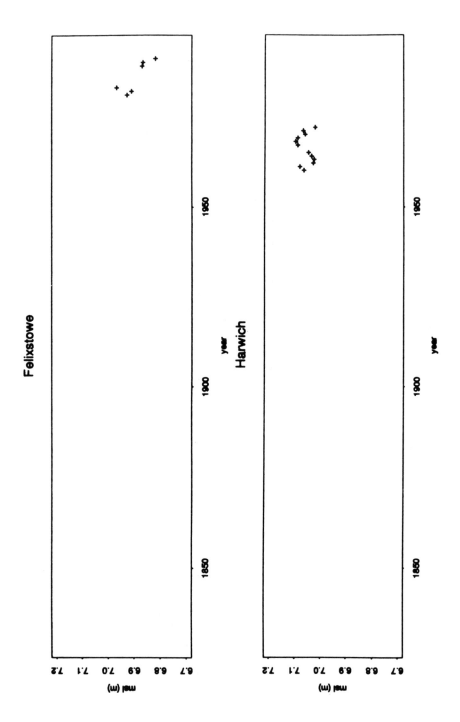

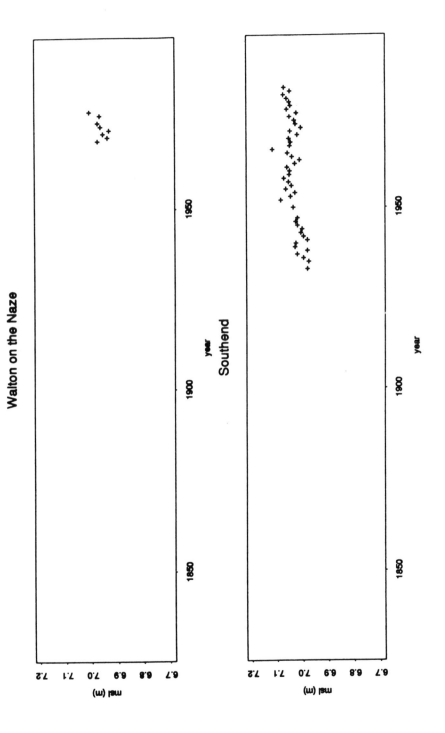

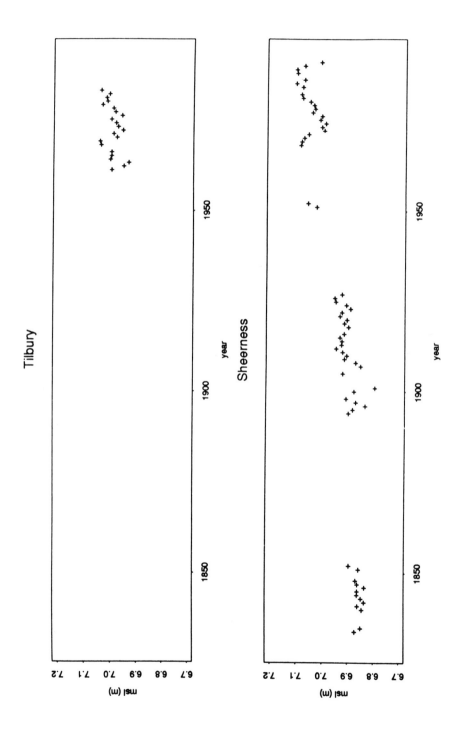

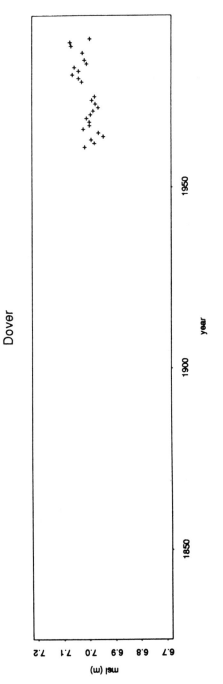

Figure 9.2 Mean sea-level data plotted using common axes for each of the 19 UK east coast data sites. The data are plotted relative to the Revised Local Reference level.

Here M_t is a low-frequency component, with variations on an annual scale; W_t is high-frequency, with variations of the order of a few seconds; S_t and T_t are intermediate in terms of frequency. As the meteorologically induced surge component can have a non-zero mean, the decomposition (9.1) is ill defined. A standard approach to remove this feature is to incorporate the mean surge into M_t, and take the surge to have zero mean. This approach is used here, so care must be taken in interpreting M_t, since it combines the mean sea level, in the absence of meteorology, with the long-term mean meteorological effects. The surge component, so defined, varies primarily over a scale of a few days, but also exhibits some inter-annual dependence. Similarly, the principal variation of the tide is over semi-diurnal and diurnal time scales, but has smaller-magnitude cycles, including the 18.61 years nodal cycle (Pugh, 1987). Historical interest in sea levels focused on the estimation and prediction of the deterministic astronomical tidal component, since this is typically the cause of most hourly variation. Consequently, sea-level measurements have generally been made to aid such analyses by filtering out the high-frequency wave component from the process at the observational stage. Tide gauges, such as that shown in Figure 9.3, can be used to achieve this by limiting the size of the orifice to dampen the high-frequency component of the process. Thus sea-level data are observations of the still-water process

$$M_t + T_t + S_t. \tag{9.2}$$

In theory, tide gauges could be modified to have much smaller orifices and thus filter out all characteristics of the sea-level process other than the low-frequency mean sea-level component of interest here. However, this is impractical for engineering reasons and undesirable because of the requirements of other applications for sea-level data. Thus M_t is obtained by analysis of the observed still-water levels to produce a series of annual estimates of M_t. This is achieved typically by either taking annual averages of observations made regularly in time, or by first using a low-pass numerical filter to largely remove the effects of semi-diurnal and diurnal tides before annual averaging (Pugh, 1987).

For some sites, the earliest annual mean sea-level estimates come from observations sampled at less frequent intervals (such as 3, 12 or 24 h) than the 1 h archiving typically used today. This is the case with data from Aberdeen—hence the distinction between the two Aberdeen series here (Woodworth, 1987)—although from Figure 9.2 it appears to make little difference in terms of inter-annual variation.

For each site, annual estimates of mean sea levels are obtained using some form of averaging over an annual series of still-water levels. Such data are widely available from the Permanent Service for Mean Sea-Level, based at the Proudman Oceanographic Laboratory, which holds a large database of annual mean sea levels from over 1300 sites worldwide (Pugh *et al.*, 1987; Woodworth *et al.*, 1990). Although these are mainly short records, over 500 series consist of at least

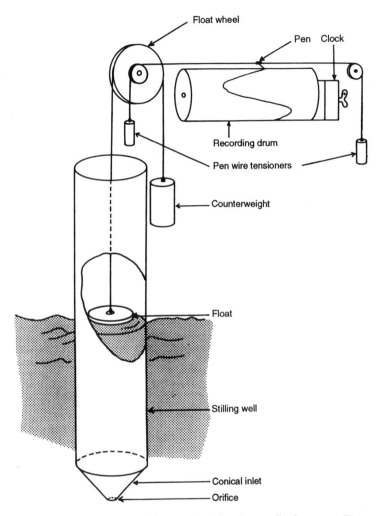

Figure 9.3 Schematic diagram of the traditional stilling well tide gauge. (Figure reproduced from IOC (1985).)

20 years of high-quality data. Each of these series has been transformed to have a common datum level, the Revised Local Reference. Our aim is to eventually extend the analysis of this chapter, which focuses on a single coastline of the UK, of 19 sites, to this high-dimensional global database.

The quality of the data in the database is generally high; however, there are likely to be some systematic errors in the mean sea-level data arising from the handling of the gauge, from timing errors due to consistent incorrect positioning of the chart on the drum of the tide gauge (see Figure 9.3) or from datum control errors. The first two sources of systematic error should be identifiable in preliminary data analysis before the mean sea-level data are entered into the database, whereas datum errors are likely to be small so may not have been identified. However, since small datum shifts may be highly influential in trend estimation, methods that protect against this potential source of error should be employed; this feature is addressed in Section 9.5.

Now consider the physical causes of variation for the mean sea-level process. From (9.2), the annual average still-water level is of the approximate form

$$\frac{1}{k} \sum_{t=1}^{k} (M_t + T_t + S_t) = \bar{M} + \bar{T} + \bar{S}, \tag{9.3}$$

where k is the annual number of observations. Thus, from (9.3), an annual mean sea-level observation consists of a signal, in terms of the long-term mean sea level $\bar{M}$, with the error being the annual average tide and surge, $\bar{T} + \bar{S}$. By definition of the processes involved, the long-term mean of $\bar{S} + \bar{T}$ is zero, so the error in $\bar{M}$ is zero-mean.

Over the observational period, $\bar{M}$ has risen owing to an increase in the volume of water in the oceans due to the melting of glaciers and polar ice and from the thermal expansion of water. These changes occur on a global scale; however, because of ocean circulations and water density variations, they are likely to appear as homogeneous trends over regional scales only (Thompson, 1981, 1986). As the tide has cycles of longer period than a year, for example the nodal tide with an 18.61 year period, a component of the error will be a deterministic cycle. The surge is induced by meteorological effects, principally air pressure through an inverse barometric effect, and the forcing of winds. Thus inter-annual variations in atmospheric conditions produce inter-annual variations in $\bar{S}$, which lead to the observed annual mean sea level having an inter-annual noise component (Figure 9.2). However, long-term trends in the atmospheric influence through changes in climatic patterns can lead to an additional meteorological component of the trend, which may be taken as homogeneous only over a regional scale. Thus the observed trend at a site may be due to a combination of global and regional trends. Even by viewing the problem spatially, separation of these effects is impossible using mean sea-level data only. Thus estimation of these confounded trends must first be undertaken on a regional scale, so here we restrict attention to the UK east coast.

The climatic variable that has been linked most with future mean sea-level rise is global temperature. Gornitz *et al.* (1982) claim that past mean sea-level data show a rise of 0.16 m, with an 18 year lag, per 1 °C rise in temperature. The IPCC (Houghton *et al.*, 1990) suggest that the rise since 1860 has been 0.5 °C, but that a 1.5–4.5 °C rise will result from a doubling of carbon dioxide concentrations, and hence could lead to 0.24–0.72 m mean sea-level rises. Although these estimates of mean sea-level rise are quite crude, they do provide some appreciation of the potential future impact of climatic change. Further background discussion is given by Peltier and Tushingham (1989), Pugh (1990) and Wigley and Raper (1992).

Finally, consider the reference level of mean sea-level data. As measurements of still-water levels are made using tide gauges, or some equivalent land-based measuring equipment, the reference level is the local land level. This, however, is not constant, since land levels also have trends owing from geological effects caused by shifting loads of ice and water on the Earth's crust, an isostatic rebound effect from the last ice age. Local land trends also occur owing to water extraction, e.g. at Venice (Pirazzoli, 1987), or earthquakes (Emery and Aubrey, 1991). Thus, even if the true mean sea-level signal, which is termed the eustatic trend, were globally homogeneous, the observed mean sea-level would have heterogeneous trends due to differential land-level trends (see Pugh, 1987, Figure 9.6). For the UK, the principal change is a gradual uplift in the North and subsistence in the South-East. This is consistent with knowledge about land movements inferred from geological data (Shennan, 1989). An alternative method of estimating land-level trends is to use geodynamic models of the Earth (Peltier and Tushingham, 1989; Trupin and Wahr, 1990). Both geodynamic models and geological data provide fairly crude estimates of current land-level trends and so more advanced methods, based on measurement using satellites, are starting to be used. Statistical methods that incorporate these measurements are discussed in Section 9.4.

9.3 EXISTING METHODS OF ANALYSIS

All the existing procedures for estimating mean sea-level trends are marginal analyses in that only data from the site of interest are used. In this section we outline these methods and illustrate the simplest version by application to the data shown in Figure 9.2. Based on the expression (9.3), four basic structures of model have been proposed for the observed annual mean sea level $Z_t(\boldsymbol{x})$ for site position $\boldsymbol{x}$ in year t. Let $M_t(\boldsymbol{x})$ denote the mean sea-level signal, and let $\epsilon_t(\boldsymbol{x})$ be a temporally independent and identically distributed normal error, with zero mean and variance $\sigma^2(\boldsymbol{x})$. Then, for all t and $\boldsymbol{x}$, the four models are as follows:

Model 1

$$Z_t(\boldsymbol{x}) = M_t(\boldsymbol{x}) + \epsilon_t(\boldsymbol{x});$$

Model 2

$$Z_t(\boldsymbol{x}) = M_t(\boldsymbol{x}) + N_t(\boldsymbol{x}) + \epsilon_t(\boldsymbol{x}),$$

where $N_t(\boldsymbol{x})$ denotes the deterministic nodal tide, with period 18.61 years;

Model 3

$$Z_t(\boldsymbol{x}) = M_t(\boldsymbol{x}) + N_t(\boldsymbol{x}) + \phi_0 P_t(\boldsymbol{x}) + \phi \nabla P_t(\boldsymbol{x}) + \epsilon_t(\boldsymbol{x}),$$

where $P_t(\boldsymbol{x})$ and $\nabla P_t(\boldsymbol{x})$ denote respectively the annual average air pressure and the annual average gradient in air pressure, i.e. annual average winds, and both ϕ_0 and ϕ are regression parameters;

Model 4

$$Z_t(\boldsymbol{x}) = M_t(\boldsymbol{x}) + H_t(\boldsymbol{x}) + \epsilon_t(\boldsymbol{x}),$$

where $H_t(\boldsymbol{x})$ denotes the annual mean of the predicted sea-level data using a hydrodynamic model.

These models have signal $M_t(\boldsymbol{x})$ corresponding to $\bar{M}$, and characterise the $\bar{S} + \bar{T}$ component of (9.3) by four levels of a hierarchical error model: in Model 1 the error model is taken to be white noise, while Model 2 accounts for inter-annual variations in the $\bar{T}$ term by the inclusion of long-period tides. Models 3 and 4 extend these forms to include inter-annual variations in $\bar{S}$. Specifically, Model 3 uses a linear model to represent the atmospheric influences of air pressure and winds on surges; whereas Model 4 exploits a sophisticated hydrodynamical model that relates surge levels to the atmospheric influences through nonlinear spatial numerical models. Since the physical phenomena that produce the inter-annual variability of $\bar{T} + \bar{S}$ are temporally dependent, models that fail to capture these error terms are liable to have time-dependent errors. This is a relevant issue, but since our main interest here is in spatial dependence, this is not considered further in this chapter.

Other aspects of these regression models can also be derived from the form of (9.3) and the physical background of the processes involved. The error term arises from an averaging operation, so arguments based on the central limit theorem are used to justify taking $\epsilon_t(\boldsymbol{x})$ to be normally distributed.

Within the class of models defined above, two principal forms for the mean sea-level component that have been explored are a linear model

$$M_t(\boldsymbol{x}) = \alpha(\boldsymbol{x}) + \beta(\boldsymbol{x})t, \tag{9.4}$$

and a model incorporating an acceleration term either via a piecewise-linear form, considered by Woodworth (1990) or a quadratic model

$$M_t(\boldsymbol{x}) = \alpha(\boldsymbol{x}) + \beta(\boldsymbol{x})t + \gamma(\boldsymbol{x})t^2. \tag{9.5}$$

These models for $M_t(\boldsymbol{x})$, which are based on the environmental and geological processes that govern trends, are simple and yet capture all the relevant features of the mean sea-level process.

At present only limited use of Models 3 or 4 can be made, since each requires high-quality meteorological data. In particular, for the UK, the application of Model 4 has only recently become possible owing to the availability of results of a hydrodynamical model run continuously for the past 30 years. Even though the $H_t(\pmb{x})$ covariate, produced using the hydrodynamical model, can explain much inter-annual mean sea-level variability, wide applicability of such models is unlikely, since hydrodynamical models, or the input atmospheric characteristics that are used to drive them, are not generally available for other coastlines. Furthermore, even for the UK, the 30 year period for which they are available is short relative to available records. However, it is clear that such models have a future role, since they provide covariates for the entire coastline that do not incorporate land-level trends, so are ideal covariates for inter-annual variations in trend modelling. At present, the only published account of the use of such long runs of the hydrodynamical models is Richards *et al.* (1993).

If interest lies in obtaining an estimate of the trend in mean sea level, as opposed to estimating future sea levels, then a potential drawback of Models 3 and 4 is that if there are trends in the atmospheric influences that induce mean sea-level trends then these are accounted for by the covariates and hence the mean sea-level trend will be under-estimated. Although Models 3 and 4 are superior, for simplicity here we restrict attention to Models 1 and 2. Thompson (1980) and Woodworth (1987) fitted these models with the linear trend (9.4), by maximum likelihood, and we illustrate their approach, with Model 1, using our UK east coast data.

Here and subsequently we define x to be the convex coastal distance, in kilometres from Wick, the most northerly site in our data set and t to be the time in years from a base year of 1800, and we take $\pmb{x}$ to be the scalar distance x. Then the data plots in Figure 9.2 are displayed in order of increasing distance x. Estimates of intercept $\alpha(x)$, trend $\beta(x)$ and standard deviation $\sigma(x)$ obtained from fitting Model 1 are shown in Figure 9.4, where estimated trends are typically between 0 and 3 mm yr^{-1}. There is some regularity in the estimated $\beta(x)$, with generally larger estimates in southern England and smaller values along the Scottish coast, i.e. for large and small x respectively. For many sites, estimates of the trend parameters, and the corresponding intercepts, are quite poor, but exhibit negative association, which is typical between these parameter estimates. The $\sigma(x)$ parameters vary little along the coast, which is somewhat surprising given the substantially different variabilities of tide and surge along the coast; however, this may reflect a common nodal tide that is not removed within Model 1. The two Aberdeen records ($x = 310$ km) have almost equal estimates of standard deviation despite the differences in sampling frequency discussed in Section 9.2.

As applied by Woodworth (1987), only a small subset of these sites is considered, owing to worries over quality and datum control, and the length of the record. In particular, the pairs of sites Aberdeen I and II, and Sheerness and Southend were each combined to provide an increased sample. Although this seems a potential source of bias, Figure 9.4 shows that there is sufficient spatial smoothness in estimates to pool local data. A potentially more efficient approach, though, is to preserve the spatial nature of the data by fitting explicit spatial models, a theme developed in Section 9.4.

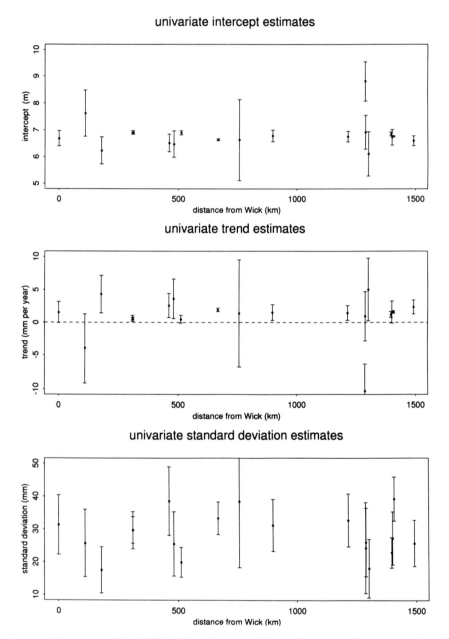

Figure 9.4   Marginal-model-based estimates, together with 95% confidence intervals, of $\alpha(x)$, $\beta(x)$ and $\sigma(x)$ plotted against coastal distance x from Wick. The parameters, defined by Model 1 with $M_t(x)$ given by (9.4), are estimated using an ordinary least-squares regression separately for each of the 19 sites.

As discussed in Section 9.2, the trend in mean sea level, as measured by tide gauges, is the sum of two components: eustatic sea-level trend and land-level trend. Within our region the eustatic sea-level trend should be approximately homogeneous over all sites; however, the land-level trend varies with site position, so we take the sea-level trend $\beta(x)$ to be

$$\beta(x) = \beta_l(x) + \beta_e, \tag{9.6}$$

where $\beta_l(x)$ is the land-level trend at position x and β_e is the eustatic sea-level trend. Shennan and Woodworth (1992) use (9.6) together with estimates of $\beta_l(x)$, obtained by Shennan (1989), to obtain a series of β_e estimates, which are effectively averaged to give a regional estimate of eustatic sea-level trend. Some authors estimate eustatic mean sea-level trends from spatial averages of estimates of $\beta(x)$; however, if when averaged over sites the land level trend is non-zero, the scope for bias here is substantial (see Barnett (1984) and Dixon and Tawn (1992) for further discussion). Whichever approach is used, owing to the spatial dependence of the data, estimates that exploit the spatial nature of the problem should provide an improved estimate, together with an unbiased estimate of the standard error (see Section 9.4).

Woodworth (1987) and Woodworth and Jarvis (1990) side-step the problem of estimating $\beta_l(x)$ or β_e directly by studying the process $Z_t(x) - Z_t(y)$ for sites at positions x and y, which are close together, a procedure they term buddy checking. From Model 4, with $M_t(x)$ given by (9.4),

$$Z_t(x) - Z_t(y) = [M_t(x) - M_t(y)] + [H_t(x) - H_t(y)] + [\epsilon_t(x) - \epsilon_t(y)]$$
$$\approx \alpha(x) - \alpha(y) + [\beta_l(x) - \beta_l(y)]t + \epsilon_t(x) - \epsilon_t(y), \tag{9.7}$$

since the hydrodynamical model is spatially smooth. Also, since the $\epsilon_t(\cdot)$ terms are positively correlated, the noise in (9.7) has small variance; thus accurate estimates of relative land-level trends are obtained, and these could be used to improve the estimated $\beta_l(x)$.

Finally, consider the quadratic model (9.5). Both Woodworth (1990) and Douglas (1992) fit this form for Model 1, in each case finding no evidence of accelerations; i.e. their estimates of $\gamma(x)$ do not differ significantly from zero and show no spatial pattern. Since land-level trends are linear over the observational period, and acceleration in eustatic sea levels, if present, will be homogeneous over the region, $\gamma(x)$ should be regionally homogeneous, i.e.

$$\gamma(x) = \gamma. \tag{9.8}$$

Thus this feature is not exploited by the existing analyses, but can be by spatial analyses. This is of potential importance, since evidence of accelerations from any one site is likely to be extremely weak—yet, viewed spatially, may be significant.

9.4 SPATIAL STATISTICAL METHODS

In Section 9.3 certain parameters of the models were identified as being spatially homogeneous, yet such features were not exploited since the data exhibited spatial dependence. In this section we examine the joint distribution of mean sea levels over data sites, and assess the benefits of treating the problem spatially through a simple analytical example together with a spatial application of Model 1 to the UK east coast data.

Marginally, Models 1–4 are applicable; thus the only change arises from the introduction of the joint distribution over the d sites of the errors $\{\epsilon_{t,i} = \epsilon_t(x_i):$ $i = 1, \ldots, d\}$, where x_i denotes the position of site i. Again the central limit theorem suggests that a multivariate normal distribution, with variance–covariance matrix $\Sigma = (\sigma_{ij})_{i,j=1,\ldots,d}$, is appropriate. Thus, marginally, for site i, Model 1 with the linear model (9.4) gives the result that the annual mean sea level $Z_{t,i} = Z_t(x_i)$ follows a normal distribution with mean

$$M_{t,i} = M_t(x_i) = \alpha(x_i) + \beta(x_i)t = \alpha_i + \beta_i t$$

and variance σ_{ii}.

The benefits of a spatial analysis arise from combining information over sites. It is most apparent when there are missing data, as in Figure 9.2, since there is a transfer of information between sites that partially compensates for the loss of data. Here we illustrate this using a simple bivariate case ($d = 2$) for Model 1 with linear form (9.4). Let the annual mean sea levels at the sites be $Z_{t_1,1}, \ldots, Z_{t_n,1}$ and $Z_{t_1,2}, \ldots, Z_{t_m,2}$, where $m < n$ and the t_i are year indices, so site 2 has missing data in relation to site 1. We take ρ to be the correlation between $Z_{t,1}$ and $Z_{t,2}$, i.e. $\rho = \sigma_{12}/(\sigma_{11}\sigma_{22})^{\frac{1}{2}}$, and $(\hat{\alpha}_k, \hat{\beta}_k)$ to be the maximum-likelihood estimates of (α_k, β_k) for $k = 1, 2$; then it can be shown that

$$\hat{\beta}_1 = \frac{S_{tn1}}{S_{tntn}}, \quad \text{Var}(\hat{\beta}_1) = \frac{\sigma_{11}}{S_{tntn}}$$

and that $\hat{\beta}_2$ satisfies

$$\hat{\beta}_2 = \frac{S_{tm2}}{S_{tmtm}} + r_m \frac{S_{t1}^*}{S_{tmtm}}, \quad \text{Var}(\hat{\beta}_2) = \frac{\sigma_{22}}{S_{tmtm}}\left[1 - \rho^2\left(1 - \frac{S_{tmtm}}{S_{tntn}}\right)\right]. \tag{9.9}$$

Here, for $k = 1, 2$,

$$S_{tjtj} = \sum_{i=1}^{j} (t_i - \bar{t}_j)^2, \quad S_{tjk} = \sum_{i=1}^{j} (t_i - \bar{t}_j)(z_{t_i,k} - \bar{z}_{j,k}),$$

$$S_{tk}^* = \sum_{i=m+1}^{n} (t_i - \bar{t}_m)(z_{t_i,k} - \hat{M}_{t_i,k}),$$

$$r_m = \left[\sum_{i=1}^{m} (z_{t_i,1} - \hat{M}_{t_i,1})^2\right]^{-1} \sum_{i=1}^{m} (z_{t_i,1} - \hat{M}_{t_i,1})(z_{t_i,2} - \hat{M}_{t_i,2}),$$

with

$$\bar{t}_j = j^{-1} \sum_{i=1}^{j} t_i, \qquad \bar{z}_{j,k} = j^{-1} \sum_{i=1}^{j} z_{t_i,k}, \qquad \hat{M}_{t,k} = \hat{\alpha}_k + \hat{\beta}_k t.$$

The trend estimate and standard error for the longer Z_1 series are unaffected by knowledge about the shorter Z_2 series, whereas the trend estimate for the Z_2 series is essentially the usual estimate adjusted, by the second term in (9.9), for transferred information from the Z_1 series. The elements of the second term are of interest: S_{t1}^* measures how unrepresentative the Z_1 series is over the additional observation period relative to during the overlapping observation period, r_m is a measure of dependence between the two series during the overlapping period, and S_{tmtm} is a measure of the information from the overlapping period. If either the additional observations are unrepresentative, there is no correlation in the overlapping series, the period of overlapping data is large, or the residual variance of the Z_1 series is large relative to the Z_2 series then there is limited information transfer in terms of the trend estimate value. However, if there is non-zero correlation then $\mathrm{Var}(\hat{\beta}_2)$ is reduced from the variance of the marginal trend estimator.

In the multivariate case the analytical expressions are more complex, and consequently more difficult to interpret, but information continues to be transferred (see Johnston, 1984; Zellner, 1962, 1963, 1972). We shall illustrate this by our application to the UK east coast, and this requires the likelihood for the data, accounting for missing values: we define δ_{ij} to be an indicator function of whether or not data are observed at site j, in year i, $A_i = \{j : \delta_{ij} = 1, j = 1, \ldots, d\}$, d_i to be the number of sites with data in year i, i.e. $d_i = |A_i|$, and n to be the number of years. Then the overall likelihood, for parameter $\boldsymbol{\theta}$, is

$$L(\boldsymbol{\theta}) = \prod_{i=1}^{n} L_i(\boldsymbol{\theta}), \qquad (9.10)$$

where

$$L_i(\boldsymbol{\theta}) = (2\pi)^{-\frac{1}{2}d_i} |\Sigma_{(i)}|^{-\frac{1}{2}} \exp\left[-\tfrac{1}{2}(\mathbf{z}_{(i)} - \boldsymbol{\mu}_{(i)})^{\mathrm{T}} \Sigma_{(i)}^{-1}(\mathbf{z}_{(i)} - \boldsymbol{\mu}_{(i)})\right],$$

i.e. L_i is the likelihood for an observation from a d_i-dimensional multivariate normal distribution. Here

$$\boldsymbol{\mu}_{(i)} = \{(M_{i,k_1}, \ldots, M_{i,d_i}) : k_j \in A_i, j = 1, \ldots, d_i\},$$

$$\mathbf{z}_{(i)} = \{(z_{i,k_1}, \ldots, z_{i,d_i}) : k_j \in A_i, j = 1, \ldots, d_i\},$$

$$\Sigma_{(i)} = (\sigma_{jk})_{j,k \in A_i}.$$

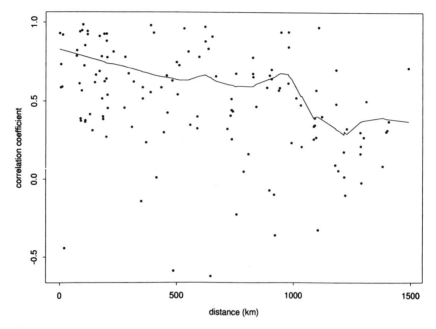

Figure 9.5 Intersite correlation plotted against intersite distance. The points correspond to the maximum-likelihood estimates from fitting the bivariate normal model (9.10) to each pair of sites for which there are observations that, overlap in time. The plotted curve is a simple smoother of these points that accounts for weighting given by the respective standard errors associated with each estimate. For clarity, estimates with large standard errors have been omitted from the plot, although they have been accounted for in the smoothing.

Owing to the spatial coherence of the mean sea-level process, the correlation between sites i and j, $\rho_{ij} = \sigma_{ij}/(\sigma_{ii}\sigma_{jj})^{\frac{1}{2}}$, should vary as a smooth function of the intersite distance $|x_i - x_j|$. Figure 9.5 shows estimates of ρ_{ij}, obtained by maximising (9.10) with respect to θ for $d = 2$, against the intersite distance for pairs of sites. The form of the smooth local average curve on the figure motivates our selection of

$$\rho_{ij} = \rho_0 \exp\left(-\rho_1 |x_i - x_j|\right), \qquad (9.11)$$

with $0 \leqslant \rho_0 \leqslant 1$ and $\rho_1 > 0$, i.e. an exponential correlation model with a nugget effect. With ρ_{ij} taken as (9.11) and all the other parameters unconstrained over sites, maximising the likelihood (9.10) gives the estimates shown in Figure 9.6 when $M_{t,i}$ is taken to be linear. Here ρ_0 and ρ_1 are $0.7(0.04)$ and $2.3 \times 10^{-4}(9.0 \times 10^{-5})$, but, despite this high correlation, the trend estimates are only slightly changed from those obtained by the univariate analyses (see Figure 9.7 for a comparison). In particular, the estimates for the longer series

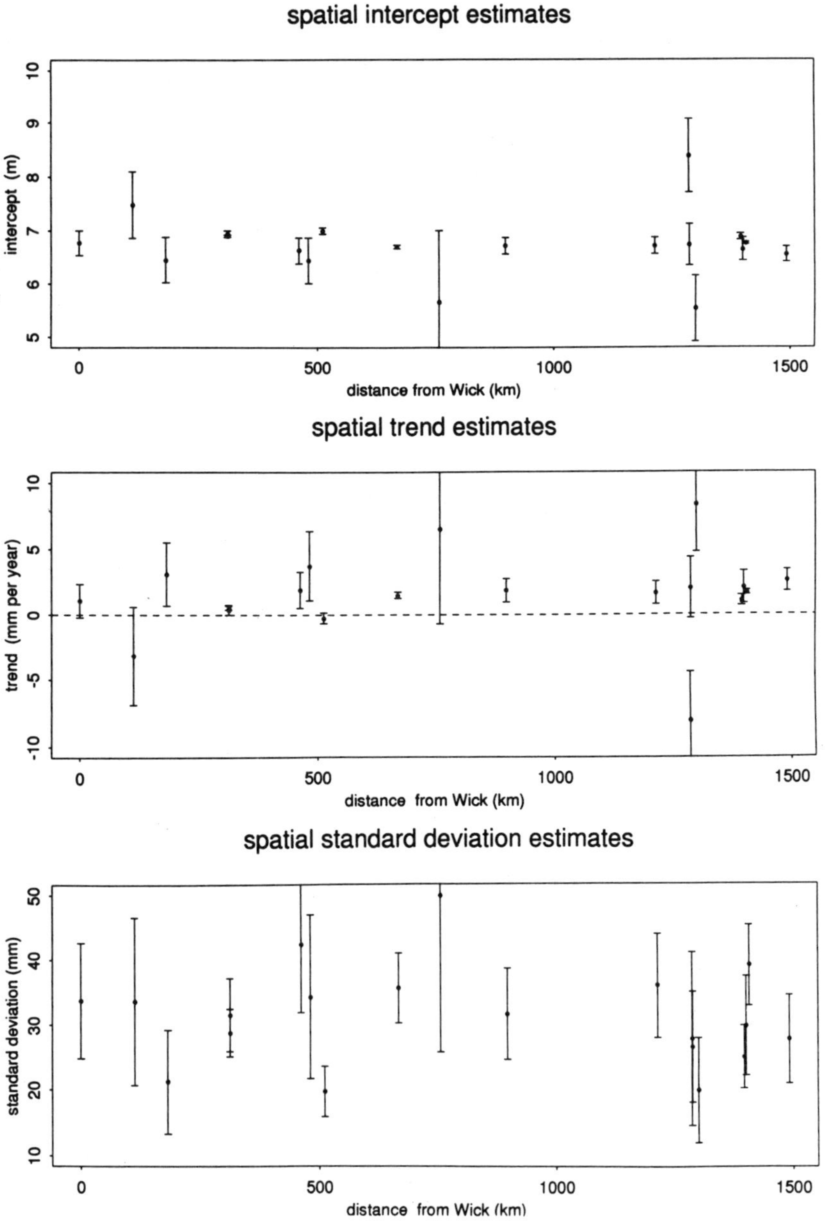

spatial intercept estimates

spatial trend estimates

spatial standard deviation estimates

Figure 9.6 Spatial-model-based estimates, together with 95% confidence intervals, of $\alpha(x)$, $\beta(x)$, $\sigma(x)$ plotted against coastal distance x from Wick. These estimates are maximum-likelihood estimates of the parameters of the multivariate normal model (9.10) obtained using the spatial correlation model (9.11).

remain largely unchanged, while the poor estimates from sites with short records become either more or less consistent with neighbouring estimates, depending on how typical their observed data are with the fitted correlation model (9.11) and data from neighbouring sites, as is seen from (9.10).

It is clear from Figure 9.6 that there is still limited spatial smoothness in the trend estimates, and that the intercept $\alpha(x)$ continues to be negatively correlated with $\beta(x)$. Since $\alpha(x)$ corresponds to the mean sea level at the base year, 1800, it must have a smooth form along the coast. Because of the complex form of the spatial variation of $\alpha(x)$, parametric models are not suitable, so some form of non-parametric model could be used, in which case the regression model would be of a generalised additive form (Hastie and Tibshirani, 1990) and estimation could proceed using techniques in Zeger and Diggle (1993). As we are not specifically interested in estimating $\alpha(x)$, instead we simply specify that it must vary smoothly along the coast by using a penalised likelihood

$$L_{\text{pen}}(\boldsymbol{\theta}) = L(\boldsymbol{\theta}) - \lambda \sum_{i=1}^{d} [\alpha'(x_i)]^2, \tag{9.12}$$

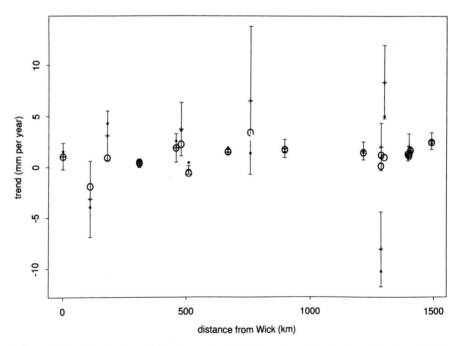

Figure 9.7 Marginal, spatial and penalised likelihood-based estimates of the trend $\beta(x)$, plotted against coastal distance x from Wick. Spatial estimates are denoted by $+$, and are given with associated 95% confidence intervals. Marginal estimates are shown by $\bullet$, and the penalised likelihood estimates by $\bigcirc$.

where $L(\theta)$ is given by (9.10), the second term penalises spatial roughness in the $\alpha(x)$ estimates, and λ is a tuning constant. Figure 9.7 shows estimates of $\beta(x)$ obtained from maximising the penalised likelihood (9.12), for correlation model (9.11), with $\lambda = 10^{-4}$. Clearly this has led to a significant smoothing of the trend estimates without imposing an explicit smoothness on them.

Now consider the estimation of the eustatic mean sea level trend. Following Shennan and Woodworth (1992), we use (9.6) with $\beta_l(x)$ estimated by Shennan (1989). Thus β_e is obtained by maximising the penalised likelihood (9.12), giving $\hat{\beta}_e = 1.17\,\text{mm yr}^{-1}$ with standard error $0.07\,\text{mm yr}^{-1}$. This estimate and the associated 95% confidence interval are shown on Figure 9.8, together with the penalised likelihood estimates, and confidence intervals, for $\beta(x)$ adjusted by estimates of $\beta_l(x)$. This figure suggests that the land level trends in the South East, i.e. large x, may be inaccurate, and consequently may have led to an underestimation of the eustatic mean sea-level trend. Our estimate of β_e is consistent with previous findings (Barnett, 1984; Woodworth, 1987; Peltier and Tushingham, 1989), but is more precise as a result of our use of spatial analyses. Accelerations

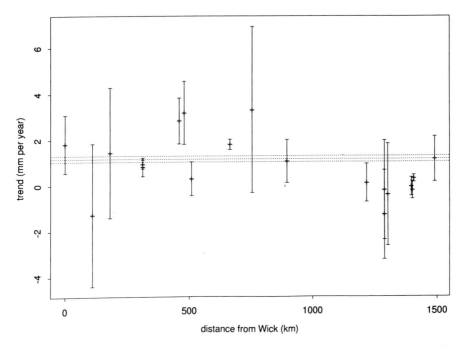

Figure 9.8 Estimated eustatic sea-level trend. The plot shows estimates, together with 95% confidence intervals, obtained using the penalised likelihood, where the estimates have been adjusted by the estimated land-level trend (Shennan, 1989). Also shown on the plot, by the dotted lines, are the estimate, and 95% confidence interval, of a common regional eustatic sea-level trend, obtained using the penalised likelihood.

can be studied similarly: following (9.8), the quadratic trend parameter, γ can be estimated using the penalised likelihood (9.12), giving $\hat{\gamma} = 9.9 \times 10^{-4}\,\mathrm{mm\,yr^{-2}}$, with standard error $14.5 \times 10^{-4}\,\mathrm{mm\,yr^{-2}}$. Thus, even though the spatial properties of the data are exploited in this analysis, no significant evidence of accelerations is obtained. Assuming that this current estimated linear trend continues throughout the next century, the model here gives an estimate of the increase in eustatic mean sea level by 2100 of 0.12 m which is much smaller than the original IPCC estimate (Houghton *et al.*, 1990) of 0.66 m and the revised estimate by Wigley and Raper (1992) of 0.50 m, which were obtained under climate-change scenarios. On the basis of Figure 9.8, our estimate appears to be a slight under-estimate, whereas estimates obtained using climate models appear high, but are much reduced from the original claims of 1–2 m, and this reduction appears to be continuing.

Maximum likelihood in all these models was computationally intensive owing to the high dimensionality of the parameter space and, as a consequence of missing data, the need to invert as many as n different variance–covariance matrices at each iteration in the maximisation routine. Evaluation of standard errors also proved troublesome numerically, with problems involving rounding errors that caused the Hessian matrix to be negative-definite.

A principal weakness of the proposed methodology presented in this section is our taking a crude estimate of land-level trend to be exact. These land-level trend estimates were obtained from geological data and are subject to large uncertainties. Current scientific work in this area focuses on the use of global positioning satellites to obtain accurate geodetic measurements of land-level movements. Thus improved estimates of $\beta_l(x)$ should be obtained. Since the measurements are subject of small random errors, observations are required over a number of years for each site. Letting $S_t(x)$ denote the measurement in year t at site x,

$$S_t(x) = s_0(x) + \beta_l(x)t + \varepsilon_t(x),$$

where $s_0(x)$ is the base-year level and $\varepsilon_t(x)$ is the noise term. So, as in Section 9.4, the likelihood $L_{\mathrm{sat}}(\beta(x_i); i = 1, \ldots, d)$, for the measurements over sites and years can be developed. Then the joint likelihood of mean sea-level observations and satellite-based geodetic measurements is

$$L_{\mathrm{pen}}(\boldsymbol{\theta})L_{\mathrm{sat}}(\beta(x_i); i = 1, \ldots, d) \tag{9.13}$$

since the two measurement processes are independent. In (9.13) the trend parameters of $\boldsymbol{\theta}$ are $\beta_l(x_i) + \beta_e$ for $i = 1, \ldots, d$, so information about $\beta_l(x)$ is obtained from both components of the likelihood function, and this leads to improved estimates of both $\beta_l(x)$ and β_e by comparison with the use of separate likelihood forms.

9.5 METHODS TO PROTECT AGAINST DATUM ERRORS

Throughout the analysis in Sections 9.3 and 9.4, the data for each site are assumed to be measured relative to a common datum level. However, datum errors, in the form of shifts, do occasionally occur during the upgrading of tidal gauges or through poor management of the gauge. If such shifts in level are not detected, they can lead to bias in trend estimates. Figure 9.9 illustrates this for the mean sea-level data at Portpatrick, a site on the Scottish west coast, where a datum shift appears to occur in 1976. If undetected, the trend would be $5 \, \text{mm yr}^{-1}$, as opposed to $1\text{–}2 \, \text{mm yr}^{-1}$ given by the data before and after 1976. Similarly, if for Sheerness (see Figure 9.2) datum errors occurred in the long periods with no data, this would be highly influential. Consequently, to avoid such errors, sites where there are worries about the datum control have been excluded from the analysis (Woodworth, 1987). As datum errors are quite rare in the UK, this approach is potentially inefficient, and this motivates the development of models that are robust to datum changes, yet remain able to detect small trends. Ideas for such models are outlined in this section. In essence, such methods must be changepoint resistant, since datum shifts lead to changepoints in the incremental process $Z_{t+1}(x) - Z_t(x)$.

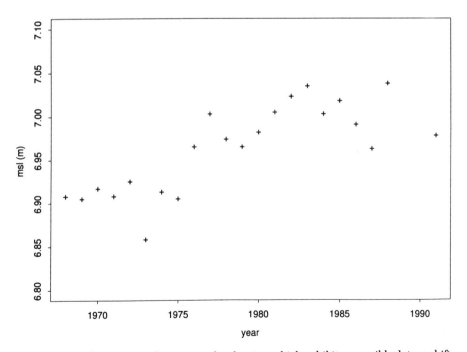

Figure 9.9 The Portpatrick mean sea-level series, which exhibits a possible datum shift in 1976.

A simple approach based on analysing the univariate series $\{Z_{t+1}(x) - Z_t(x)\}$, separately for each site, will be unlikely to detect change points owing to the large inter-annual variability of mean sea levels, $\varepsilon_{t+1}(x) - \varepsilon_t(x)$. Figure 9.9 illustrates why this is the case: the largest incremental rise occurs in 1989, and this does not appear to be a changepoint. This shows how difficult it is to detect small datum shifts using only the data from the site. Again this suggests that a spatial approach will be beneficial.

The buddy checking procedure of Woodworth (1987) could be used here, since if there is a shift c in year t^* at site x then (9.7) gives

$$Z_t(x) - Z_t(y) = \begin{cases} \alpha(x) - \alpha(y) + [\beta_l(x) - \beta_l(y)]t + \varepsilon_t(x) - \varepsilon_t(y) & \text{if } t < t^*, \\ c + \alpha(x) - \alpha(y) + [\beta_l(x) - \beta_l(y)]t + \varepsilon_t(x) - \varepsilon_t(y) & \text{if } t \geqslant t^*, \end{cases}$$

which amounts to analysing spatial increments, between pairs of sites, for changepoints. Buddy checking is likely to be an improvement over analysing temporal increments, since $\varepsilon_t(x) - \varepsilon_t(y)$ will have a smaller variance than $\varepsilon_{t+1}(x) - \varepsilon_t(x)$ owing to spatial dependence for x and y sufficiently close together.

These approaches to detect datum shifts have two drawbacks; first, they cannot exploit the complete spatial structure of the problem, since only multiple pairwise comparison can be assessed via buddy checking, and, secondly, knowledge of the maintenance history of the gauges cannot be incorporated. The first deficiency can be overcome using the following model: assume there exists a single changepoint of size c^* at time t^* and site x^*, and, referring to (9.12), define

$$Z_t(x^*) = \begin{cases} M_t(x^*) + \varepsilon_t(x^*) & \text{if } t < t^* \\ c^* + M_t(x^*) + \varepsilon_t(x^*) & \text{if } t \geqslant t^*, \end{cases} \tag{9.14}$$

and $Z_t(x) = M_t(x) + \varepsilon_t(x)$ for all other x. Here c^*, t^* and x^* are parameters of the model, so, for example, the model can be extended to have n_c changepoints at the expense of additional parameters $\{(c_i^*, t_i^*, x_i^*); i = 1, \ldots, n_c\}$. In essence, this extends the buddy checking procedure to all the sites, and incorporates a model for the shifts at the detects sites and times.

The issue of exploiting the additional knowledge of the times of potential datum errors is best handled through a Bayesian prior. Bayesian changepoint problems have been studied previously (Smith, 1975); however, here the prior for the time of the changepoint is informative, since it relates to the history of the gauge maintenance. The novel aspect in this case arises from the feature that when gauges are replaced, or have a fault, there will be intervals without a gauge, and hence no data. Therefore if a site has a period of missing data, we assign it a higher prior probability for a changepoint than during a period of continuous data. At first sight, this may appear contrary to the Bayesian paradigm, but of course it is not since missing data are non-informative in the likelihood formulation.

Suitable priors for changepoint times are yet to be formulated, but models should capture the above feature, together with independence over sites, and be

that changepoints are thought to be more likely further in the past. With the flexibility of Bayesian modelling using Gibbs sampling (Gelfand and Smith, 1990), many possible forms for the prior can be examined, and thus the positions of possible changepoints identified and fitted simultaneously within the Bayesian analysis, through the model (9.14). These trend estimates will be more robust than existing estimates, and no efficiency should be lost. Although application at a computational level is still required, on the basis of results thus far we believe significant improvements can be made.

9.6   IMPACTS AND EXTREMES

Large increases in mean sea level would drastically influence life over the coming centuries. Much land would be lost, and protection of coastal regions would have to be limited to small areas of most importance. Such severe changes are unlikely, but even modest changes in mean sea level will result in significant impacts on the coastline. The more obvious effects include penetration of salt water into the water table, damage to the salt marshes, and increased rates of coastal erosion (Bird, 1993).

The impact of deeper seas on the sea-level process will be through modification of tidal patterns, a reduced degree of interaction between the tidal and surge processes, damped variability of surges, and increased wave heights: these are likely to be relatively minor effects though. Potentially more important changes could arise from the changes in climatic circulation patterns associated with global warming, since these may lead to significant increases in surge variability.

A direct impact of positive trends in the sea-level process is an increased risk of flooding, which occurs owing to resulting changes in the distribution of extreme levels. Here interest focuses on mean sea-level trends, rather than eustatic sea-level trends, since these provide a direct measure of the impact on the coastline. A consequence of this is that the trend depends on the coastal position, with the largest trends along the east coast occurring in the South. In the UK public attention was drawn to the issue of extreme sea-level change by a report for the Friends of the Earth (Barkham *et al.*, 1992) which showed a significant increase in the risk of flooding. These results were based on the following model.

(i) The annual maximum sea level X_t in year t was taken to follow a generalised extreme value distribution

$$\Pr\{X_t \leqslant x\} = \exp\left\{ -\left[1 + \frac{\xi(x - a_t)}{b} \right]_+^{-1/\xi} \right\}, \qquad (9.15)$$

where $s_+ = \max(s, 0)$ and a_t, b $(b > 0)$ and ξ are location, scale and shape parameters respectively.

(ii) With t and $t + T$ the present and a future year respectively, the mean sea level is taken to have risen by an amount c, so that, with a linear increase, the trend would be c/T per year.

(iii) A rise in mean sea level by c increases extreme sea levels by the same amount; thus $a_{t+T} - a_t = c$ in (9.15).

On the basis of IPCC estimates (Haughton *et al.*, 1990), with $t = 2000$ and $t + T = 2100$, c was taken to be the combination of 0.66 m and the relevant land-level trend estimate from Shennan (1989). The standard method of displaying the relationship between extreme levels and the risk of them being exceeded is the return-level–return-period curve. If $x_{t,p}$ satisfies

$$\Pr\{X_t > x_{t,p}\} = p$$

then $x_{t,p}$ is termed the return level with return period $r = -1/\log(1-p)$; i.e. if the process were stationary, r would be the expected waiting time, in years, between exceedances of $x_{t,p}$. From (9.15) for $\xi \neq 0$,

$$x_{t,p} = a_t - b(1 - r^\xi)/\xi,$$

and when $\xi = 0$,

$$x_{t,p} = a_t + b \log r,$$

so when $x_{t,p}$ is plotted against r, on a logarithmic scale, as in the return-level–return-period curve, the relationship is linear. Figure 9.10 shows such curves for various ξ, and with $c = 0.3$ m: a compromise between the findings of Section 9.4 and Wigley and Raper (1992) for a site with no land-level trend. In the top part of the figure, where $\xi = 0$ and $b = 0.2$, which are typical parameters for the UK east coast (Graff, 1981; Coles and Tawn, 1990), we find that if current coastal flood defences were designed to the 100 year return level, a mean sea-level increase of 0.3 m would lead to a fourfold increase in the risk of flooding. However, if $\xi = 0$ and $b = 0.3$, i.e. the steeper line, then this increase would be threefold. Thus the impact of trends on risk depends on the existing statistics of extremes. These effects are significantly more pronounced in the lower part of the figure, when ξ is varied over a range consistent with UK data (Coles and Tawn, 1990). Here, in one case if flood defences were designed to the 1000 year return level the risk is increased over one-hundred-fold, whereas in the other case it is only twofold.

Berkham *et al.* (1992) conclude by identifying the sites that are at most risk from such increases in sea levels. There are some caveats of their analysis: first they used estimates of return levels obtained from Graff (1981) that are subject to large uncertainties, which cause the risk level at many sites to be misclassified; and secondly they did not discuss the fact that, although the flood defence would be breached more often, it would be by a limited amount. Improved estimates of return levels are given by Coles and Tawn (1990), which are still subject to estimation errors, so typically these are incorporated by over-design. Thus, even

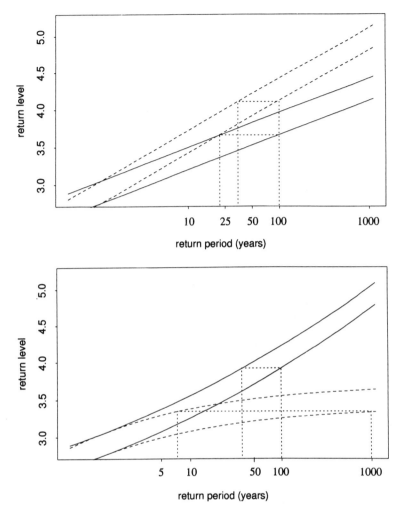

Figure 9.10   The impact of mean sea-level trend on extreme sea levels. The solid and dashed lines show the existing and the new return level curves, whereas the dotted lines act as a guide to assessing the change in frequency of flooding. Return levels are in metres, and $a_t = 2.75$ and $a_{t+T} = 3.05$ for each curve. In the top figure $\xi = 0$, while in the lower figure $\xi = 0.1$ for the steeper curve and -0.3 for the shallower curve. The parameter $b = 0.2$ in all cases except the steepest-gradient curve in the top figure, where $b = 0.3$.

after a rise of 0.3 m, in most cases the level of protection will be at least that intended.

   The model in Barkham *et al.* (1992) can be used to estimate extreme sea-level trends only if they are identical to mean sea-level trends. This is a restriction, since the two trends could differ, since the mean sea-level trend is just one component of the trend in extreme sea levels. Using observed annual maximum values, from a 62-site network, Dixon and Tawn (1992) estimated a_t directly from the distribution (9.15) with a linear model, $a_t = c_0 + [\beta_l(x) + c]t$. They used all the coastal data by exploiting the spatial extreme value model of Coles and Tawn (1990); see also Tawn (1993). Thus the model is the extreme value equivalent of (9.12) in Section 9.4, and, after adjusting for land-level changes, gives equivalent results to those of Section 9.4. The similarity of the mean and extreme sea-level trend estimates suggests that a joint analysis of means and maxima will improve estimation precision compared with that obtained from marginal analysis. In other contexts such ideas have been explored by Blackman and Graff (1978), Tawn and Mitchell (1993) and Anderson and Turkman (1991), the latter providing the basis of a suitable model framework. This is not pursued here.

Since the principal information required for coastal flood risk assessment concerns the extremes of the process, further analysis, such as Smith (1989), aimed specifically at estimating trends in extremes is needed. An additional motivation for concentrating on trends in extremes is that if trends occur only in the surge standard deviation then trends will occur in extreme sea levels but not in mean sea levels. For example, if the surge standard deviation increases linearly with time then both the location, a_t, and scale, b_t, parameters of (9.15) will also be linear, subject to a_t/b_t being constant. Such trends have not been tested for, but are important since they increase both the intercept and the gradient of the return level curve and so, for example, could lead to changes from the lower shallow curve to the higher steep curve in the top plot of Figure 9.10. Again the hydrodynamical model can help address questions concerning extremes (Flather, 1987) by enabling the impact of quite subtle changes in the meteorological conditions to be evaluated.

9.7 DISCUSSION

In the chapter a procedure has been proposed for obtaining improved trend estimates by using spatial statistical models that utilise knowledge of the physical processes involved. Although a precise estimate $(1.0, 1.3)$ mm yr^{-1} of the mean sea-level trend was obtained, it is subject to bias and false precision, since our findings were based on taking a crude estimate of geological trends to be exact. Better estimates of land-level trends through advanced geodetic measurements would help, but at present such information is not available. No significant evidence of an acceleration in the mean sea-level process is found on the basis of the data used here; however, it is clear that acceleration effects, such as those predicted by climatological models, may not be detected using only the mean sea-level data available at the present time.

Generally the results show that there is a need for careful monitoring and data analysis over the coming years. At the monitoring stage an important role for statisticians is to advise engineers and oceanographers on suitable choices for sampling schemes for the mean sea-level process. This area has received no statistical attention in this context, although some relevant techniques are available through experimental and spatial design (Cressie, 1991, Chapter 5). On the basis of the models and methods outlined in this chapter, questions such as 'Is the rate, or the precision, of sampling by tide gauges important?' and 'Where are the most informative locations for new gauges?' can at least be partially addressed. To make significant progress at the analysis stage, the statistician must exploit all information from the data. This, however, may be achieved only by fully utilising the oceanographers' knowledge, through the use of hydrodynamical models that can accurately predict the tidal and surge components, and hence the inter-annual variations of the mean sea-level process. Clearly there is considerable scope for future interdisciplinary work here in addressing these important environmental issues.

ACKNOWLEDGEMENTS

J.A.T. and M.J.D. were partly supported by the Ministry of Agriculture Fisheries and Food grant. The work used equipment provided by SERC under the Complex Stochastic Systems Initiative.

10

Sea-Level Rise:
A Potential Threat?

J. K. Vrijling
Delft University of Technology, The Netherlands

10.1 INTRODUCTION

Contrary to popular belief, the relative sea-level rise is not a new phenomenon, but a known fact from geological observations (Lamb, 1982) as shown in Figure 10.1. About 8000 BC, the northern edge of the presently shallow southern part of the North Sea called the Doggersbank (Figure 10.2) was dry and connected the British Isles with the continent. The rivers Thames, Meuse, Rhine and Elbe flowed over the Doggersbank and emerged at the northern edge. As the present depth of the Doggersbank is equal on average to MSL − 35 m, a sea-level rise of approximately 35 m must have occurred in the last 10 000 years. This rise opened the Strait of Dover and pushed the mouths of the rivers flowing into the North Sea back to their present position.

Besides this geological evidence, the records of the port of Amsterdam over the period from 1682 to 1930 show a relative sea-level rise of 0.17 m measured at the official benchmark. The phenomenon is called relative, since it is composed of a rise of the sea level and a sinking of the land. The velocity of the recorded rise appears to be an increasing function of time. Starting with a value of 0.04 m per century, the velocity had increased since 1850 to 0.17 m per century in 1930. In 1960 the Delta Committee, set up by the Government of the Netherlands in the aftermath of the 1953 disaster, chose a design value for this phenomenon of 0.2 m per century. In the last two decades accurate measurements of the relative sea-level rise have confirmed this estimate.

Statistics for the Environment 2: Water Related Issues. Edited by Vic Barnett and K. Feridun Turkman
© 1994 John Wiley & Sons Ltd.

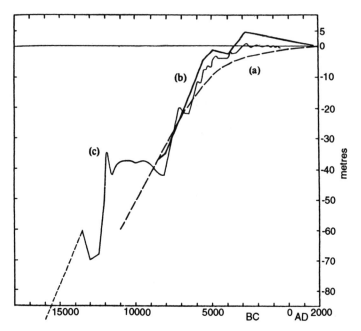

Figure 10.1   The rise in world sea level—three reconstructions: (a) a smooth mathe-matical fit; (b) a reconstruction of the main stages from dated shore lines in the Baltic; (c) a carefully calculated curve due to Mörner.

However, experts have recently increased their estimates of this figure for the future. In these studies several causes are mentioned that may increase the velocity of the relative sea-level rise. These include

 (i) melting of the polar ice pack;
 (ii) thermal expansion of the ocean;
(iii) tectonic movement of the Earth's crust;
(iv) settlement of alluvial soils due to dewatering.

Recent estimates of the future sea-level rise differ considerably. For the period 1990–2100 values ranging from 0.30 to 1.10 m have been mentioned. It should be noted, however, that these extreme values are not yet confirmed by actual observations.

If the sea-level rise accelerates it will pose a significant threat to low-lying countries and islands. The question is when and to what extent should these lands improve their sea defences.

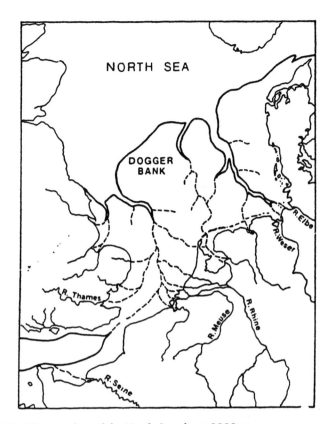

Figure 10.2 The coastline of the North Sea about 8000 BC.

10.2 THE HISTORY OF SEA DEFENCES IN THE NETHERLANDS

10.2.1 Terps and dikes

Long before the Christian era, people started to live in the low-lying unprotected northern parts of the Netherlands and Germany. These people, called Frisians, built mounds of clay (Vern, 1949) to protect their families and their cattle against storm-tides. On top of the mounds, named terps, farms and even villages were erected. So it is not surprising that some of these terps exceed the pyramid of Cheops in volume.

After visiting the low countries bordering the North Sea in AD 47, the Roman chronicler Pliny reported his impressions as follows: A poor country, that is flooded two times a day. So the inhabitants have to live on man-made hills, where they warm their frozen limbs on a fire of dried mud.

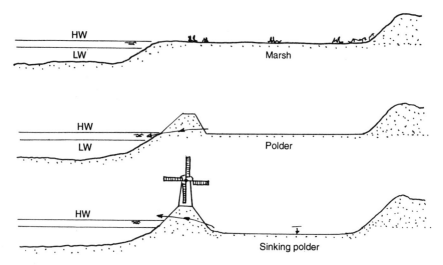

Figure 10.3   Gradual sinking of a drained polder.

In the ninth century the first dikes were built to protect the pastures against too-frequent flooding and to provide safe pathways. In the Frisian language roads are called dikes, even today.

This approach proved so successful that it developed a social framework in the thirteenth century. A certain degree of social organisation was necessary to share the costs of construction and maintenance equally and to reap the benefits of diking for the entire population. When the safety of the land was threatened by storm surges, special regulations came into force. Then no feuds were allowed—the 'dike peace' reigned. Breach of this peace was punished by death.

At first, the aim of diking was defensive; in a later stage the art of diking was used in an offensive way to reclaim land from the sea. In these cases a new marsh, which by accretion had reached the level of high water, was diked and added to the agricultural area (Figure 10.3). Because of settlement of the soil, drainage became more difficult with time. In the beginning the areas could be drained via discharge sluices by gravity, but in the course of time the application of wind energy became a neccessity.

10.2.2   Draining lakes

The newer polders have been reclaimed from lakes or estuaries, and consequently the ground level of the polder is considerably below the water level. In these cases part of a lake is diked or the entire lake is surrounded by a drainage canal. Next the water is pumped out and drained to the sea via a system of canals. This more advanced way of reclamation (Figure 10.4) started around 1500 when windmill technology was well developed.

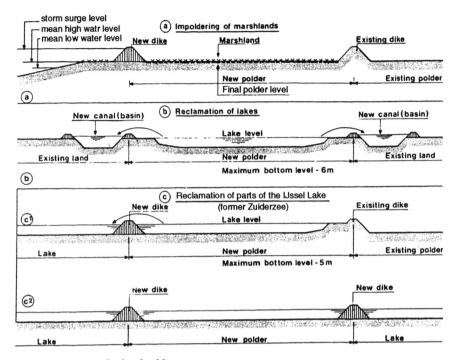

Figure 10.4   Methods of poldering.

Later, owing to the development of British mine pump technology, fossil energy became available to drain even larger and deeper polders. From that moment, the reclamation of large lakes like the Haarlemmermeer (location of Schiphol), Alexanderpolder and the polders in the former Zuiderzee (Figure 10.5) with a depth of up to 5 m below MSL became technically feasible. In these polders the process of 'sinking', caused by the drainage of the alluvial soil, plays a relatively less important role.

In some cases the aggressiveness of the offensive to reclaim led to calamities. For instance, the first large lake, the Beemster (Figure 10.5), which was drained by Leeghwater in 1610, was flooded a few months later. A storm surge from the Zuiderzee was the apparent cause. However, because of the insufficient knowledge of soil mechanics in those days, the dikes were constructed of peat, and this also contributed to the disaster. The perseverance of hydraulic engineers appears clearly from the fact that the polder finally fell dry in 1612.

10.2.3   Flood disasters and the reactions

All the time, the sea fought back, and many storm-flood disasters prove that it was a formidable opponent. One of the oldest floods reported struck in the night of

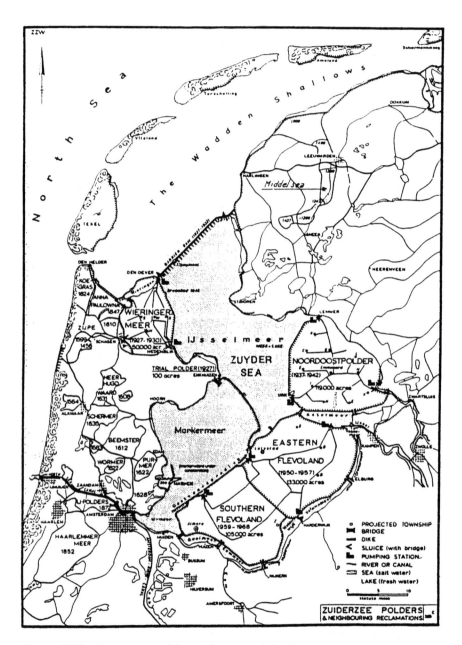

Figure 10.5   An overview of the polders around the former Zuiderzee.

14th December 1287. About 50 000 people seem to have lost their lives in this terrible winter storm. Many storm floods were to follow: 1421, 1570, 1682, 1686, 1717 and 1775. In 1825 an almost unbelievable disaster happened, when the sea took 3770 km² of land around the Zuiderzee (Figure 10.6) Seventy years later in 1894, these levels were exceeded again and large areas were flooded.

The flood of 1916, which hit the low-lying areas around the Zuiderzee again (Figure 10.7), was the stimulant needed to bring the plans made by C. Lely some years earlier into the decision phase. Lely proposed to close off the Zuiderzee estuary with a 32 km long dam and to reclaim land in the resulting lake. After a thorough study, in which the famous physicist Antoon Lorentz played an important part, the closure dike was built from 1927 to 1932. It reduced the length of the coastline by 300 km.

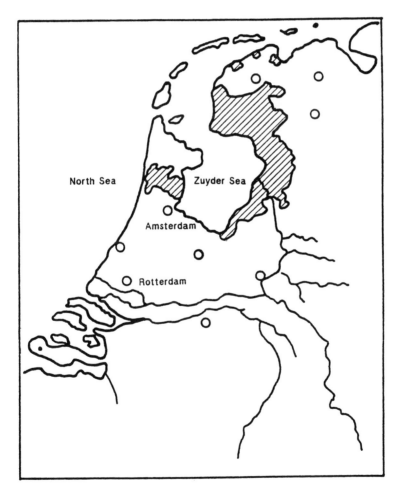

Figure 10.6   The storm surge of 1825.

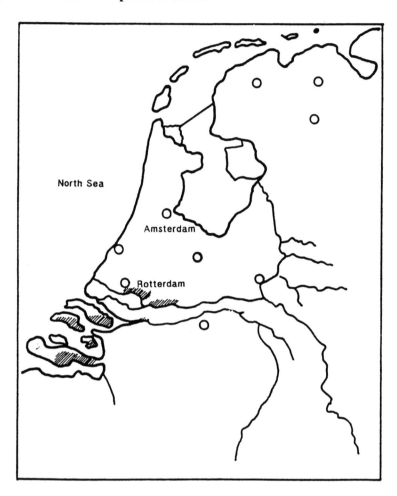

Figure 10.7   The storm surge of 1916.

Six years after the Second World War, on 31 January 1953, the Enclosure dike proved its value: it survived the storm surge without serious damage, although the waves, overtopped the crest on many places. However, this time another almost unbelievable disaster took place in the south-western part of Holland (Figure 10.8). Approximately 150 000 ha polderland were inundated through 90 breaches and 1800 people and thousands of cattle lost their lives.

After this flood, the Dutch government decided to take firm action. A proposal by Johan van Veen to shorten the coastline in the south-western part of Holland— a measure that had proved successful in the North—demanded attention. The Delta Committee was established with van Veen as secretary, and a few years later the Delta Plan (Figure 10.9) was born. According to this plan, all major estuaries in the Delta, except for the Western Scheldt, the seaway to Belgium,

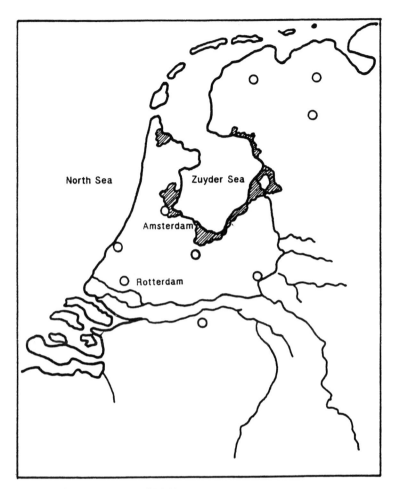

Figure 10.8   The storm surge of 1953.

were to be closed. These closures, to be executed in order of growing magnitude to gain experience, would reduce the length of the coastline considerably. The plan also included the construction of a large discharge sluice in the Haringvlietdam to let the water of the rivers Rhine and Meuse flow into the North Sea. All existing primary seadikes were to be checked against the newly developed Delta standard design rule and improved if necessary.

After a decade of hard and successful construction work, political resistance against the plan grew. And in the years after 1970 environmental pressure groups forced a decisive change to the Delta Plan. The closure of the Eastern Scheldt by a dam was amended to the construction of a barrier that would preserve the tidal movement essential for the estuarine ecosystem. During storms, the Eastern Scheldt barrier would be closed to protect the hinterland. The

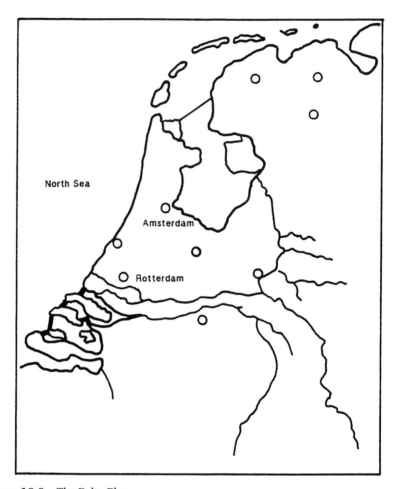

Figure 10.9 The Delta Plan.

reconstruction of river levees along the branches of the Rhine also ran into environmental resistance.

10.3   STRATEGY OF SEA DEFENCE SINCE THE DELTA COMMITTEE

10.3.1   Some aspects of sea defence

The above historical overview shows that the Netherlands have been plagued by inundations with a certain regularity—even more frequently than mentioned, since floods caused by the large rivers were omitted above. Here high discharges, often in combination with icejams, caused severe damage. Flood mitigation and water-transport requirements led in the nineteenth century to the taming (normalisation) of the Dutch rivers and the birth of the central government organisation Rijkswaterstaat.

The major cause of inundations by the sea is the combination of the astronomical tide and a storm-induced extreme wind set-up. Experience over the centuries indicated that always higher storm-surge levels and more extreme wave attack occurred than had been seen before. This can be understood in a statistical framework, where longer return periods show more extreme events.

A less readily analysed cause must be sub-optimal maintenance of the sea defences. When the memory of floods recedes, the importance of dike maintenance seems reduced—A tendency that may be enhanced by economic or social decline. In this respect the Second World-War certainly contributed to the 1953 disaster.

The perpetual (relative) rise of the sea level has complicated the defence of the polders during living memory. Although the phenomenon is clear, the contributions of the various causes are not easily established. The rise in the level of the ocean, the tectonic movement of the Earth's crust and the settlement of alluvial soils due to dewatering all contribute. The last cause plays a part in every polder. The oldest polders were reclaimed by diking, as soon as the marshlands had reached the high water level due to the sedimentation of clay. Initially the polder is drained by gravity through discharge- sluices in the dikes. In the course of time, the soil of the drained polder settles by the order of 1 m owing to lowering of the groundwater level. This process (Figure 10.3), in conjunction with the rise in sea level, finally leads to a situation where the polder has to be drained by means of wind, animal or fossil energy. And because the groundwater level is kept at some distance below ground level, the settlement process continues.

The increasing energy requirements for drainage has loder considerable technological changes, but this has not impaired the economic viability of the polders, as the relative wealth of these areas shows.

Today the mining of water, gas and oil contribute to the sinking process in no small measure. The problems in Venice, the northern part of the Netherlands and along the shores of Lake Maracaibo clearly show the consequences of the respective mining operations.

An extremely important aspect of the age-old, successful fight against the threat of the water is the social organisation. The diking of marshlands and the reclamation of polders required well-developed cooperation of the people living in and near the drained area. Since the early Middle Ages, the polders have been governed, with regard to water problems, by 'democratic' water boards. The distribution of votes relates to the area of the polder the inhabitants possess. The water board has to manage the maintenance of the dikes, the pumping stations, the sluices, the drainage ditches etc. To meet expenses, the water board has the right to tax the inhabitants of the polder.

10.3.2   Statistical and economic approaches

The present strategy of sea defence is based on the recommendations of the Delta Committee.

One of the major innovations of the Committee was a statistical approach to the design storm-surge level. Wemelsfelder (1949) had shown that the observed storm-surge levels followed an extreme value distribution. Statistical extrapolation thus seemed the right answer to the age-old experience that the highest known storm surge was always surpassed. The choice of the appropriate return period was a problem for the Committee. Although the question of the strengthening of the sea defences is a multifaceted one, it is readily schematised to an economic decision problem (von Pontzig, 1960).

The optimal return period or the optimal dike height is found by minimising the total cost TC, consisting of the investment in the seadefences $I(h)$ and the present value of the expected value of the loss in the case of a flood:

$$\text{TC}\,(h) = I(h) + \int^{T} P(h)We^{-rt}\,dt,$$

where h is the height of the sea defence, $P(h)$ is the probability of a flood exceeding h, W is the damage by a flood, r is the real rate of interest and T is the planning period. Using this model, the optimal values of h and $P(h)$ are readily found (Figure 10.10).

The result of the optimisation for the large polder of Central Holland was a level of MSL + 6 m and a return period of 125 000 years. As this in fact represented the storm-surge level that causes inundation, the Committee chose a *design* storm-surge level of MSL + 5 m and a return period of 10 000 years.

The basic design storm-surge level for other places along the Dutch coast was fixed on the basis of the return period of 10 000 years. If, however, the economic importance of the polder under consideration was less than that of Central Holland, an economic reduction according to an economic decision model was

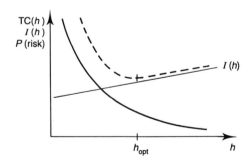

Figure 10.10   The economically optimal dike height.

applied. For the mainly agricultural south-western Delta, a return period of 4000 years was advised. The return period for the river levees was established at 3000 years.

It is assumed that the surges, the waves and the discharges result from a stationary process, and that therefore statistical extrapolation is meaningful. To account for the known non-stationarity, a sea-level rise of 0.20 m per century should be incorporated in the design. In this respect, a planning period of 50 years was advised for flexible structure and 100 years for structures that are difficult to adapt.

10.3.3   Application in practice

According to the recommendations of the Delta Committee the height of new dikes includes a provision of 0.10 m for sea-level rise during 50 years. The settlement of the subsoil should also be added to the dike height, and due account should be taken of other causes of sinking, such as dewatering and mining.

Measures to incorporate the effects of sea-level rise in the design of dunes are less straightforward since the positions and masses of these natural structures are the result of a long-term dynamic equilibrium between errosion and accretion. This equilibrium has along the Dutch coast been disturbed by (among others) the taming of the rivers (normalisation) and the partial closure of estuaries, all leading to coastal erosion. Here the effect of sea-level rise accounts for a small but steady erosion equal to the beach slope times the sea-level rise ($0.10\,\mathrm{m}/\frac{1}{50}-\frac{1}{100} =$ 5–10 m). Such effects may be neutralised by regular beach nourishments, needed to combat the overall beach recession.

For less readily adaptable structures like barriers, locks and sluices, the Delta Committee adopted a planning period of 100 years and a provision of 0.20 m for sea-level rise.

As stated before, the Delta plan (Delta commisie, 1975) proposed in 1960 the closure of the estuaries in the south-west and the improvement of all sea defences and river levees to the Delta standards. The closure of the estuaries was completed in 1986 with the opening of the Eastern Scheldt storm-surge barrier—an event signifying the successful compromise between safety and ecology.

10.3.4   Present state of affairs

Today after 40 years, the Delta plan is still not fully completed. Because of budgetary problems, the refurbishment of the Enclosure dike (constructed in 1932) and the locks in IJmuiden that give entrance to the harbour of Amsterdam were postponed.

The reconstruction of the river levees ran into political problems in the 1970s, as environmental pressure groups opposed these projects. Resistance was met where the reconstructed levees would cross and damage areas of great ecological value or scenic beauty. It appeared that the era of safe, slightly oversized, technical solutions was over and that environmental values had to be incorporated in the design. The political process reduced in 1977 the return period from 3000 to 1250 years to lower the design water level. It also coined the phrase 'sophisticated' dike design to provide more space for environmental values compared with the classical dike design. Fierce opposition was met where the levees passed the centuries-old villages bordering the river. In some villages old and picturesque houses (Figure 10.11) stand along the levee with such a density that the dike

Figure 10.11   Some scenic views of the dike village Sliedrecht.

cannot be discerned from an old street. Sometimes the century-old main shopping street is located on the dike.

The classical approach to dike reconstruction, which requires the total destruction of all buildings on the dike, is economically and politically impossible today.

Innovative approaches, that merge the water-retaining function into the scenery without compromising safety have to be developed. This was recently confirmed again by a government committee on river levees.

10.4 FUTURE STRATEGY ON SEA DEFENCE

10.4.1 No formal change of strategy

The Dutch Ministry of Public Works has not yet officially decided to allow for an increased rate of sea-level rise in the design and the management of sea defences. As the increased rate has not been observed, the predictions based on complicated simulations seem insufficiently reliable to form the basis for real sea defence policies. So, for most operational decisions, the above-mentioned strategy still has validity.

To guide research and preliminary planning, a few principles have been formulated. The return period of 10 000 years remains the basis for the establishment of the design water levels along the Dutch coast. Also the refinement of the economic reduction of the design water level for less important polders is kept.

However, a sea-level rise of approximately 0.60 m per century replaces the Delta value of 0.20 m. It is assumed that the extreme value distribution of storm-surge levels shifts by 0.60 m over its entire range. Consequently all design water levels increase by the same amount as the MSL.

Further assumptions are that the statistical distributions of the deep-water wave heights and the river discharges stay unaltered. These are heroic assumptions because it is by no means certain that global warming will not influence climatic variability. The importance of the assumption may be clarified from Table 10.1, which gives a breakdown of the level of wave attack on an average sea dike in the various components.

A 10% increase in the last two wind-driven components, which is quite possible if climatic variability increases, adds 0.90 m to the level of attack. This uncertainty puts refined discussions of the exact amount of sea-level rise in an interesting perspective.

The same reasoning holds for the river discharges. This has less dangerous consequences, since the incidence of storm surges and high river discharges is independent in the Netherlands. However, in countries with a monsoon-type climate the opposite might be true.

Table 10.1   Breakdown of the level of wave attack

Variable	(m)	Symbol
MSL in 1991	0.0	MSL
Sea-level rise	0.60	R
Astronomical HW	1.0	HW
Wind set-up	4.0	S
Wave run-up	5.0	
Level of attack in 2091	MSL + 10.6	Z

10.4.2   Effects of climatic change

A constant sea-level rise of 0.60 m per century shortens the planning period of a renewed sea dike from the original 50 to 16 years or longer, as will be shown later. So, when an increased rise is observed, society has still a decade or more to plan and execute the measures needed.

If the sea defence is attacked by depth-limited waves, a sea-level rise of 0.60 m will induce an increase of approximately 0.30 m in wave height. This effect may add an extra 0.50 m to the wave run-up because, in principle,

$$Z = \text{MSL} + R + \text{HW} + S + 8H_s \tan \alpha \qquad (10.1)$$

where H_s is the significant wave height (in m), $\tan \alpha$ the outer slope $(\frac{1}{4}-\frac{1}{6})$ $H_s = 0.5(R + \text{HW} + S + d)$ and d is the water depth relative to MSL. So the extra height needed is one to two times the sea-level rise. Besides this, the revetment at the outer slope should be extended to a higher level and, in the case of depth-limited wave attack, improved.

The planning period for well-maintained river levees will also be reduced to 16 years. The extra height then needed will be equal to the sea-level rise (neglecting discharge increases)—certainly in the lower reaches, where the backwater effects are strongest. The time scale of the process is so long that morphological adaptations of the river bed are likely. For rivers with slopes of the order of 10^{-4} and fine sediment, the morphological time scale (de Vries, 1975) is the order of a century. So the improvement of the levees may have to be extended further inland than the backwater curve suggests.

The basic erosion of sandy dune coasts will increase to 15–30 m in 50 years (Figure 10.12), depending on the beach slope, but the increased sand hunger of the deepening estuaries and the decreasing sediment output of the rivers will reinforce the recession process. An increased frequency of beach nourishments or a set of other defensive measures must be chosen to maintain the status quo.

With regard to structures, the sea-level rise will reduce the planning period from 100 to 32 years. After that period, the height must be increased, but

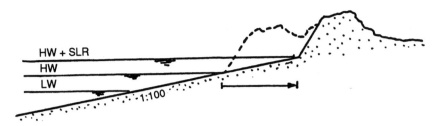

Figure 10.12 Dune erosion due to sea-level rise (SLR).

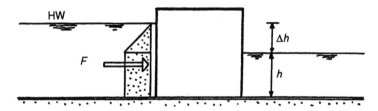

Figure 10.13 The static horizontal load on a structure caused by a head difference.

strengthening may be not necessary in many cases. The relative increase in the horizontal load F on the structure is given by

$$\frac{dF}{F} = \frac{h + \Delta h}{h + \frac{1}{2}\Delta h}\frac{R}{\Delta H} \tag{10.2}$$

where h is the water depth behind the structure, Δh is the head difference and R is the rise in HW.

The load (Figure 10.13) increases in principle with the same percentage as the head difference, although for shallow structures a considerable multiplication factor arises as shown above. It should be noted that the moments on the structure increase slightly more. For heavily loaded deep-water structures, the rise may be accommodated by a small reduction in safety factor or by a change in management procedures. As an example, the Eastern Scheldt storm surge barrier, with a water depth of 20 m and a head difference of 6 m, would see a static load increase of 11%, *ceteris paribus*. However, owing to ecological optimisation, the management procedures of the barrier have already been adapted, leading to a reduced loading level (Hendriks, 1989). The older shipping locks in IJmuiden, however, would be unable to cope with a 0.60 m rise in sea level.

10.4.3 Effect of sea-level rise on water management

Water management will see salt-intrusion problems and drainage difficulties grow in conjunction with sea-level rise. The intrusion of sea water in coastal aquifers will increase, and an augmentation of the infiltration of fresh water may be needed to avoid damage to agriculture or the environment. In principle, sea water will also penetrate further upstream in rivers and estuaries, inducing complex ecological changes. Here, however, a lot depends on the morphological reaction of the systems. If the time scales of the two processes are approximately equal, salt intrusion may not change to a great extent.

The draingae of low-lying areas by gravity or by fossil energy depends on the ratio between the sea-level rise and the head difference over the drainage structure (discharge sluice or pumping station). If the relative change is small, the drainage will not be seriously hampered. A larger relative change may necessitate constructive measures. The lake IJssel (former Zuiderzee) provides an illustrative example. The lake receives fresh water from a branch of the Rhine, the IJssel. The water is drained into the North Sea via discharge sluices in the Enclosure dike. If the sea-level rise is limited (less than 0.60 m), the present sluices can provide sufficient drainage capacity. A sea-level rise considerably exceeding 0.60 m makes constructive measures unavoidable. Four preliminary alternative solutions have been developed:

  (i) raising the water level of the lake;
 (ii) an additional discharge sluice and a pumping station;
(iii) constructing a discharge canal through the lake;
(iv) diverting the IJssel via the Rhine.

The second solution proved to be the most economical and the most flexible. It allow a wait-and-see policy, whereby measures are implemented quickly step by step, when the actual sea-level rise makes them necessary.

10.4.4   Effect of sea-level rise on ecology

The coastal and estuarine areas contain diverse habitats that sustain important and productive ecosystems. So a loss of these wetlands may be detrimental to ecological diversity and marine food production. If the sea level rises faster than the morphology can adapt, and coastal erosion, which would shift the tidal wetlands inland, is prevented by human intervention, a considerable loss of intertidal area may occur. As the adaptation velocity of morphological systems and ecosystems under the influence of sea-level rise is not well known, predictions are very uncertain.

When coastal erosion is not prevented, valuable dune area will be lost to the sea.

Along the rivers, a favourable development may be expected. When the sea-level rise leads to a rise in river level and after some time to an increase in bed level, the flood plain will be flooded more frequently. Thus the riverine forests, which were lost in the previous century by normalisation, might return (Figure 10.14), if the right policies are adopted.

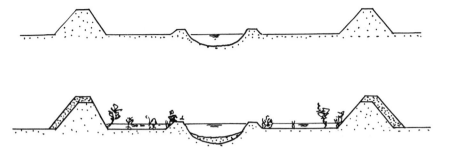

Figure 10.14   The possible natural development of flood plains.

10.4.5   Cost of protection and the optimal decision

All estimates of the costs of the extra protection needed if a sea-level rise of the order of 1 m occurs show huge amounts of money (IPPC, 1990). These amounts, however, seem surprisingly moderate, if they are related to the gross national product and the customary 100 year planning period in this field. The annual extra protection costs expressed as a percentage of GNP lie between 1 and 0.01%. The estimates for Japan and The Netherlands are presented in Table 10.2. The assumption underlying the table, namely that all present coastal defence needs are met, explains the low cost estimates for Bangladesh.

An important policy question is when and to what extent the sea-defences should be heightened in case of a rise is sea-level. To answer this question, the sea level rise can be incorporated in the economic model (10.1) in a relatively simple manner (Vrijling, 1990).

An expression is found for the first increase is the height of the sea-defence needed to counter the rise. However, for a consistent solution of the optimal strategy to battle sea-level rise, not only the first heightening of the sea defence but also all subsequent future adaptations have to be accounted for. From the minimal-cost requirement there follow an optimal design height of the sea defence and an optimal planning period (cycle) of strengthening (Figure 10.15). The crest of the sea defence should be kept in a band under the optimal design level calculated with sea-level rise. The width of the band is equal to the amount of sea-level rise during the planning period.

Table 10.2   Protection costs

	Total costs (US$M)	Annual cost as % of GNP
Japan	22 378	0.02%
Netherlands	4 211	0.03%
Bangladesh	1 200	0.10%

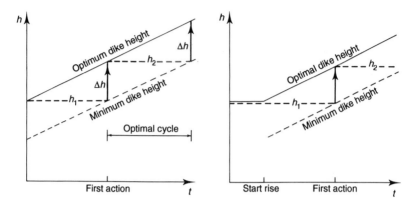

Figure 10.15   The economic management strategy in planning period and dike height increase.

If a sudden increase in the sea-level rise occurs, the band widens symmetrically, and the economic model shows that no action should be taken to strengthen the sea defence until the lower bound of the new regime is reached (Figure 10.15). So, from an economic point of view, there is even more time available than calculated in Section 3 if a temporary reduction in safety is acceptable. There are, however, other views on the acceptable risk of inundation that do not take the increasing risks so calmly (Vrijling, 1992).

10.5   SOME FINAL REMARKS

Global warming may increase the sea level compared with recent history. The acceleration predicted now is put into perspective if it is realised that the sea level has risen since 15 000 BC or earlier. Civilisations living along the sea shore have thus been involved in a age-long battle with the sea. And in the future they will still be fighting that battle with or without a man-made component in the sea-level rise. It should be noted that the relative sea-level rise has always contained such a component in the form of the sinking of the polders due to dewatering and mining, which mostly exceeded the actual sea-level rise.

In addition to the change caused by these physical processes, considerable change is forced on the sea defence by the political process. The inclusion of environmental values in the design of sea defences is a recent example.

Nevertheless, the most important task at this moment is to detect changes in the mean sea level, the tidal range, the wave climate and the climate itself that might be related to global warming. Recent observations do not yet show the predicted acceleration or a change in storm intensity or frequency, but an early discovery of decisive evidence is of the utmost importance.

If the threat becomes a reality, responses are required of the populations of low-lying countries. It should be stressed that protective measures and solutions have to be developed locally, since the diversity of conditions does not permit a general remedy. The influence of the local tidal, wave, surge and current conditions on the optimal solution is clear, but also the annual precipitation (which for instance differs considerably between temperate and monsoon regions), the soil type and the elevation vary to such extent that only locally optimised measures will be effective. Beside the natural boundary conditions, the social and economic situations in the threatened regions are of decisive importance. The options for a subsistence farming or fishing community differ completely from that for an 'information-based' society.

The most important point, however, is the political and social organisation in the threatened regions. If there is a network of ancient governing bodies, like water boards, that have for centuries combatted the relative rise of the sea successfully, an acceleration of the rise will fall completely within the scope of these institutions. In view of its low velocity ($6 \, \mathrm{mm} \, \mathrm{y}^{-1}$), the sea-level rise will seem a minor disturbance among the many changes, technical as well as social and political, with which these bodies have to cope. Regions that lack, for some reason, the network and the social experience to combat the sea in an organised way will face grave problems. These problems can only be alleviated by a combination of outside help and local initiative.

Help should be offered by international institutions and developed countries with the appropriate technological know-how. But assistance should only be provided on request, in view of the great social and economic effort required from the threatened region itself.

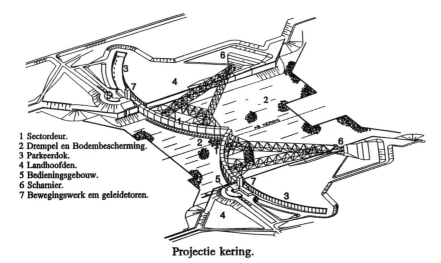

1 Sectordeur.
2 Drempel en Bodembescherming.
3 Parkeerdok.
4 Landhoofden.
5 Bedieningsgebouw.
6 Scharnier.
7 Bewegingswerk em geleidetoren.

Projectie kering.

Figure 10.16   Bird's-eye view of the storm-surge barrier Nieuwe Waterweg.

The surprising cheapness of sea defence measures, indicated by several studies and by historical experience, will be a good factor in the battle against this effect of global warming.

Finally, it seems wise to wait, see and think before acting against the predicted acceleration of the sea-level rise. No-regret measures and thorough 'what if" planning must at this moment be preferred over the adaptation of sea defences to unknown changes. If the sea defences of a country are well maintained immediate action is not advised in case of an increase in sea-level rise.

But if for some other reason sea defence measures are taken, it may be wise to include a provision for the increased sea-level rise. Therefore 90.60 m sea-level rise was included in the design of the storm-surge barrier in the Nieuwe Waterweg that is currently under construction (Figure 10.16).

11

Assessing the Wave Energy Resource

D. Mollison

Heriot-Watt University, Edinburgh, Scotland

11.1   INTRODUCTION

Wave energy, meaning the energy of ocean surface waves, is created by the drag of winds blowing over the sea. While the input density is low, seldom as much as $1\,\mathrm{W\,m^{-2}}$, the energy travels with little loss across oceans, so that the produce of a whole ocean can be 'harvested' at its boundary, where the mean power flux can average more than $40\,\mathrm{kW\,m^{-1}}$ ($40\,\mathrm{MW\,km^{-1}}$).

Locally, on a scale of tens of kilometers and minutes, the state of the sea is accurately described by a stationary Gaussian process. However, there are theoretical and practical limitations on the accuracy with which the spectrum of the sea state can be measured or estimated, and the sea state itself varies over any longer time scale, from occasional flat calms to storms with power levels of over $1000\,\mathrm{kW\,m^{-1}}$.

This chapter reviews some of the statistical problems that arise, particularly those of interest to the designer of wave power plants, both for productivity and survival. A brief account of the origin and nature of ocean waves is given in Section 11.2 (for more see, e.g., Mollison, 1986). The simulation of ocean waves in wave tanks is described in Section 11.3. Variability, over the range of time scales from individual waves to climatic change is discussed in Section 11.4. The evaluation of meteorological model estimates ('hindcasts') against more direct measurements of waves is described in Section 11.5. The selection of representative sets of spectra, and their use in tank tests and productivity estimates, are described in Section 11.6. The transformation of waves in shallow water and the efficiencies of 'greater than 100%' possible for a small-scale device are described in Section 11.7.

Statistics for the Environment 2: Water Related Issues. Edited by Vic Barnett and K. Feridun Turkman
© 1994 John Wiley & Sons Ltd.

We conclude in Section 12.8 with a brief discussion of some of areas of interest outwith the scope of this chapter, and an invitation to environmental statisticians to attack some of the fascinating problems associated with wave energy.

11.2   ORIGIN AND NATURE

11.2.1   Waves and the Earth's energy balance

Ocean waves, impressive as they can be, form only a small and inessential component of the Earth's energy balance. They are a side-effect of the movement of the major air masses that redistribute heat energy from the equatorial regions towards the poles. In doing so, the air masses lose some energy through drag to the sea surface, thus setting up 'deep-water waves', that is, oscillations of the sea surface under gravity. As a proportion of the global flux the power in these waves is only about 1 part in 10^5, with the input power flux into the ocean usually just a few $mW\,m^{-2}$, as against the original solar input which averages $350\,W\,m^{-2}$.

However, wave energy can travel thousands of kilometres with very little loss, so that a considerable proportion of the input to an oceanic area reaches its boundary. The year-round average, for a coast with good oceanic exposure (notably if facing west in the temperate zones) can reach 40–$50\,kW\,m^{-1}$ (*net*—see below, and Figure 11.3).

Another advantage of wave energy, which it shares with the wind, is that it is mechanical energy; moreover, unlike wind, its energy is in the oscillations rather than movement of its medium, so that the theoretical limit on efficiency of extraction is 100% (for wind energy, the limit is $16/27 \approx 57\%$).

A disadvantage of ocean wave energy is that its typical frequency is around $0.1\,Hz$, which is not ideal from the engineering point of view. Worse—a disadvantage it shares with wind, though with different details—it is highly variable on all time-scales.

For an excellent review of the wide variety of solutions that engineers have proposed to make wave power exploitation practical, see Salter (1989).

11.2.2   Description of sea states

The creation of waves is a complex nonlinear process, in which energy is slowly exchanged between different components. However, on a scale of tens of kilometres and minutes, the local state of the sea surface in deep water is accurately described by a stationary Gaussian random process.

Thus the local behaviour of the waves is fully determined by the spectrum of the sea state $S(f, \theta)$, which specifies how the wave energy, which is proportional to the variance of the surface elevation, is distributed in terms of frequency and direction. From this can be deduced many properties of the local sea state, ranging from the simple fact that the distribution of the instantaneous sea level at any point has a normal distribution to elegant results on the joint distribution of parameters describing the sea surface (Longuet–Higgins, 1957).

The spectrum in turn can be summarised quite accurately by a small number of basic statistics. The most important of these are

(i) the root mean square wave height H_{rms} (i.e. the standard deviation of the sea level)—it is common, especially in the engineering literature, to use $H_S \equiv 4H_{rms}$: H_S is approximately equal to the highest one-third of trough-to-crest wave heights, and thus matches reasonably well one's visual impression of wave height; and

(ii) the energy period T_e, the mean wave period with respect to the spectral distribution of energy, i.e. m_{-1}/m_0, where $m_n \equiv \int f^n \, dS(f)$.

The mean power flux in a sea state is $P = kH_{rms}^2 T_e$ (in $kW\,m^{-1}$), where $k \approx 7.87\,kW\,m^{-3}\,s^{-1}$; thus P is approximately $\frac{1}{2}H_S^2 T_e$ (in $kW\,m^{-1}$). Typical oceanic values of T_e are in the range 5–15 s; H_S varies from 0 (flat calm) to around 15 metres (severe Atlantic storm), with median values of about 2 m in summer and 4 m in winter (Mollison *et al.*, 1976).

The third crucial parameter is the principal direction of the power flux. Often an oceanic sea state will include both locally generated wind sea, whose principal direction should be that of the local wind, and swell generated up to several days earlier by distant weather patterns, which may have a quite different principal direction. In this case an adequate summary of the sea state will require separate heights, periods and principal directions of wind sea and (occasionally more than one) swell components. For a more precise description, one can add standard deviations of period and direction for each component, or a numerical summary of the complete directional spectrum.

Note that for resource estimation the relevant quantity is usually the power flux in a given direction, for instance the power flux crossing a line of prospective wave power devices. Even for the optimal direction, this net power flux in deep water will on average be at best about 75% of the gross power flux; though in shallow water, where wave components line up perpendicular to the depth contours, the two may be virtually equal (see e.g. Mollison, 1983). (The complexities of waves in shallow water will be discussed briefly in Section 11.7.)

The complete directional spectrum, or a good approximation to it, is usually more than is needed for studies of any one site, but is essential if we are to use data from one site to estimate the wave climate elsewhere (see Section 11.5).

11.3   SIMULATION

As a test facility for wave energy devices, the Edinburgh Wave Power Project led by Stephen Salter designed and built a computer-controlled wave tank capable of making simultaneously of the order of 100 wave components, each a simple wave of specified amplitude, frequency, direction and phase (Jeffrey *et al.*, 1978). The additivity of components implied here is possible because of the linearity of the Navier–Stokes equations. The wave tank can also absorb any incident waves: this feature, which is essential for the energy accounting required in tests of model device efficiencies, relies on the linearity and time reversibility of the Navier–Stokes equations.

Time reversibility also means that there is a close theoretical connection between the characteristics of wave makers and of wave energy absorbing devices. For both, good performance over a wide frequency range is important; however, an important practical difference is that for economic optimisation a wave energy device should be small relative to the waves. For instance, for the Salter Duck it was easy to obtain very high efficiencies at wavelength-to-diameter ratios of about 6:1; it took several years of research to obtain similar performance at ratios of around 20:1 (implying a roughly tenfold improvement in the cross-sectional area required for a given output). The implications of improved efficiencies $\eta(f)$ for overall output in a given wave climate is clearly shown when η is plotted against a 'stretchy' frequency axis, that is, where the scale of the frequency axis represents its distribution function, in this case with respect to power; the area under the efficiency curve then represents the overall mean output of the device in this wave climate (Figure 11.1).

Interesting statistical questions arise when a relatively small number of simple waves needs to be chosen to simulate a given directional spectrum. The target spectrum may be based on data or on one of a number of theoretical forms, and typically consists of a mixture of one to three smooth unimodal distributions.

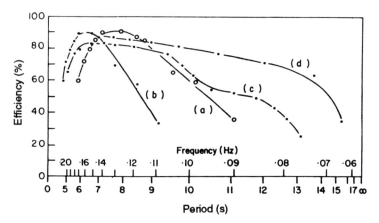

Figure 11.1   Efficiency curves for successive designs of the Salter Duck, plotted against a frequency/period axis 'stretched' to represent the empirical distribution of wave power at South Uist: (a) design of April 1975, scaled to 15 metres diameter full size; (b) the same, scaled to 10 m diameter; (c) design of September 1976, scaled to 10 m diameter; (d) design of December 1979, scaled to 10 m diameter. (From Mollison (1980).)

The simulation of the frequency spectrum on its own is relatively straightforward: dividing the spectrum, or one of its unimodal components, into successive intervals of equal energy generally gives satisfactory results. It is convenient to choose frequencies that are multiples of some (very small) base frequency Δf; a fast Foruier transform can then be used to calculate the wave record, giving a sea state with repeat period $1/\Delta f$. One potential problem, which needs to be checked for though it does not often arise, is that too many of the frequencies chosen might be multiples of some multiple of the base frequency, thus reducing the repeat period of the sea state by the latter multiple.

Another problem, whose solution depends on the purpose of our simulations, is whether to constrain the energy of each frequency interval to its long-term mean or whether to allow natural variation (which is approximately chi-squared with parameter twice the frequency interval divided by Δf); the latter is correct (Tucker *et al.*, 1984) if we want a random sample (of length $1/\Delta f$) from our sea state, but the former is arguably more appropriate for performance testing of devices.

A deterministic choice of directions for each chosen frequency works well for realistic spectra in representing the observed fairly slow change in the conditional directional distribution with frequency (Mitsuyasu, 1975)—Figure 11.2 shows scatter plots of the chosen components for a representative selection of Atlantic Sea States (this is the subset of 46 spectra from Crabb's (1980) stratified sample of synthesised directional spectra referred to in Section 11.6).

For ordinary behavioural and productivity tests, it is appropriate to choose the phases of components randomly (i.e. uniformly). However, control of the phase of components allows us also to simulate waves that occur in real seas only rarely or not at all. An example of the former is where many (or all) of the components are brought into phase with each other at one particular space–time point within the simulation; it is easy to create an extreme wave such as would occur only once in a hundred years within the sea state. Examples of the latter include a circular wave, which is of interest for the study of breaking waves, as well as party pieces such as the ideal surfing wave or the Scottish flag.

11.4   VARIABILITY

In order to describe fully the wave climate at a site, that is, the long-term distribution of waves, we need to consider their variability over the whole range of time scales, from an appropriate sampling interval short compared with the wave period up to year-to-year variability and the even slower scales of climatic change.

As already mentioned, the short-term variability of waves, over a few hours and a few tens of kilometres (in the open deep sea), is well described as a Gaussian random process.

Thus the wave-to-wave and group-to-group variation, which are crucial for modelling the power take-off of devices, can be calculated with sufficient accuracy

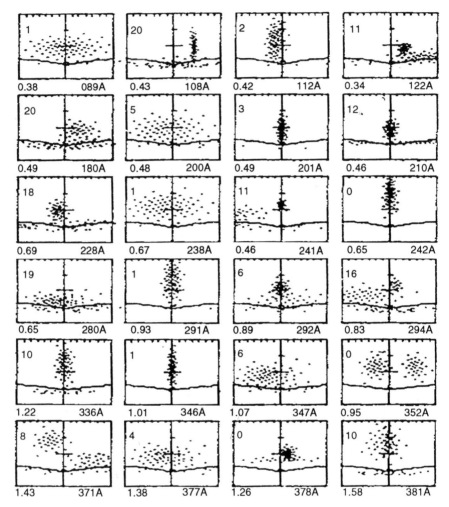

Figure 11.2 Scatter plots showing combinations of period and direction used in tank simulations of the wave climate off South Uist ('direction 0' here represents 260°, i.e. 10° S of W). Each spectrum is represented by about 75 'wave fronts' (simple sine waves), of equal amplitude within each component; over 50% of these sea states have two components, a concentrated long-period swell and a more scattered, shorter period wind sea. (Note that the curved line, and the number at the top of each scatter plot, refer to the number of short-period fronts that cannot be represented unambiguously in the tank because their wavelength is less than twice the wave-maker spacing; but these are all of very low power.) (From Taylor (1984).)

from knowledge of basic sea state parameters and the shape of the spectrum. For instance, 'groupiness' is associated with a spectrum that has only a narrow range of periods, such as arises in swell from distant storms. (See Longuet–Higgins

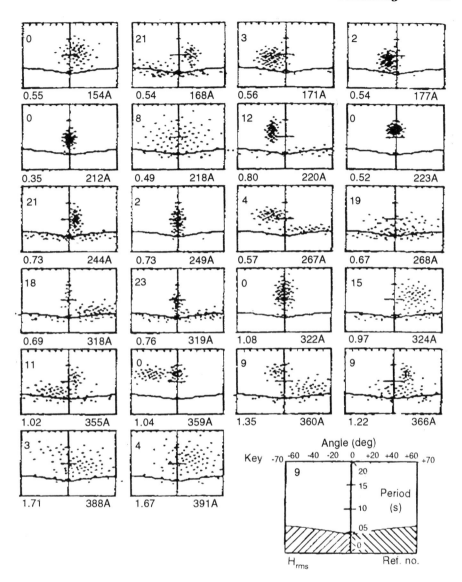

Figure 11.2   (*Continued*)

(1984) for a discussion of the modelling of groupiness, including Markov chain models; also Athanassoulis (1992).)

The duration of a sea state (typically a few hours) is important for estimating extreme waves within that state. The duration of weather systems (about 1–5 days) is important in determining the limits on forecasting, in particular forecasting calms when no power is available.

Although calms can occur at any time of year, power levels are generally much higher and more persistent in winter. In the Atlantic mean power levels are typically around five times as high between October and March as between May and August.

Year-to-year variability is also considerable—changes of 20–50% from one year to the next are common—making it difficult to estimate long-term means and trends when good measurement series of more than 2–3 years are rare (see Section 11.5.2). Year-to-year and longer-term climatic variability are especially important for estimating the lifetime extremes that a structure will experience.

The analysis of such long time series as do exist suggests considerable non-stationarity; in the North Atlantic wave power levels seem to have increased by 50–100% since the 1960s (Draper, 1988).

11.5   WAVE CLIMATE ESTIMATION

11.5.1   Requirements for device testing

The detail in which the designer of wave power devices requires knowledge of wave climate advances hand in hand with the development of device design. Basic information, such as the approximate overall mean power level, and the distribution of power over time and by frequency, is a prerequisite for matching any kind of device to its wave climate.

Details of spectral shape that will be more important for some devices than others include frequency bandwidth and the high-frequency tail of the spectrum. Narrow spectra will favour resonant features of device response, and will have relatively long runs of large waves. Wave breaking, which may pose a serious problem for some devices, depends on the high-frequency tail of the spectrum (Greenhow, 1989).

Fortunately, the need for a methodology of wave climate estimation that will allow evaluation of the resource at any site of interest necessitates full directional spectra (see Section 11.5.4) from which all the necessary details can be calculated (though some work may be needed to study whether certain details, such as the high-frequency tail, are adequately estimated).

11.5.2   Measurements

In the open sea the most widely used measuring devices are wave-recording buoys (scalar or directional), which integrate information from acceleration sensors to yield time series of buoy motions. To obtain spectral estimates, measurements are recorded (using a sampling interval of a second or less) over a time period, typically about 30 min, chosen as a compromise between the Scylla of non-stationarity and the Charybdis of sampling variability; spectra can then be calculated by Fourier transform. Such sample spectra are typically calculated at 3 or 6 h intervals; from them time series of spectral parameters, such a H_{rms} and T_e and the mean power level, can be derived.

Wave measurements close to the coast can be made by a number of devices besides buoys, including submerged pressure and ultrasonic probes, and wave-staffs and ultrasonic probes suspended above the sea.

Wave buoy measurements are expensive, and have therefore been made mainly for specific engineering purposes, often for durations of a year or less; some of the most useful open ocean measurements and climate studies have been made for the oil industry, and are subject to commercial confidentiality.

Satellites provide another source of wave information, but have a number of disadvantages. They do not, at present, give such detailed or accurate spectral estimates as wave buoys, and their data are only intermittent: in order to cover the widest possible area, a satellite is usually set in a shifted periodic orbit, which implies that any one location is covered only at fairly long intervals, varying from a week up to a month. However, this does give them the advantage of wide coverage, allowing a rough general assessment of the wave resource over very large areas.

11.5.3 Hindcasts from numerical wave models

Numerical wave models, developed over the last 30 years, provide the largest amount of wave information. Several centres around the world make routine runs of a wave model, driven by the output of a meteorological model. Their accuracy depends on the sophistication of the model itself and on the accuracy of the input wind field.

Such models have gone through three 'generations' in recent years, corresponding to increasing sophistication in the way that their equations represent numerically the physics of events. In practice, for extensive studies of very large areas only second- and third-generation models are sufficiently reliable.

The longest operating of these with wide coverage is that of the UK Meteorological Office (Golding, 1980, 1983). Data from this model for 1983–86 were compared with measurements from directional buoys for two offshore sites near the British Isles (in the SW Approaches and West of Shetland) by Mollison (1991), who found generally good agreement, though the numerical model gave slightly higher mean power levels. Since 1986 the model has undergone further refinement, and Pontes *et al.* (1993) found no evidence of bias in a comparison using post-1986 measurements from Portugal.

11.5.4 Methodology for resource evaluation

As described above, data of the quality required for a reliable estimate of wave climate are time-consuming and expensive to collect, and exist only for a few oceanic sites, whereas good quality indirect estimates for wave conditions in the open sea are now

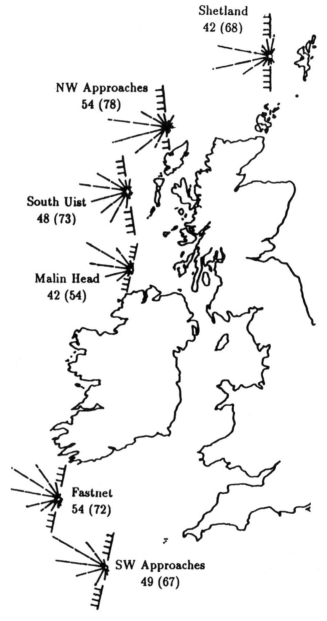

Figure 11.3 Wave power estimates for UK offshore sites, based on the UK Meterological Office's wind-wave hindcast model. Wave roses show mean power from each 22.5° sector, with marks at $5\,\mathrm{kW\,m^{-1}}$ intervals. Mean net (gross) power levels are shown in $\mathrm{kW\,m^{-1}}$; the net figure is for power P_θ crossing the line (——————) whose direction θ maximises P_θ for the particular site. (From Mollison (1991).)

routinely available from numerical wind wave models. Wave conditions at nearshore and coastal sites can be highly dependent on local topography.

This supports the following methodology, which was adopted for the evaluation of the nearshore wave energy resource in a recent study for the UK Department of Trade and Industry (Mollison, 1991), and which is the basis of a current EC project to compile an European 'Wave Energy Resource Atlas'.

A network of offshore reference sites is chosen, with a spacing of at most a few hundred kilometres, for which we obtain data from a numerical wave model; these data sets should ideally consist of full directional wave spectra for a time period of at least five years. This network of reference sites should include some for which direct measurements are available, as a check on the accuracy of the model. Figure 11.3 shows the main reference sites adopted for the UK study, which used hindcast data from the UK Met Office for 1983–86, calibrated against directional wave buoy measurements for 1984–86 for Shetland and the SW Approaches (Mollison, 1991).

An 'Atlas' of the offshore wave energy resource can then be compiled by filling in between reference sites, either with data from the numerical model if available, or by interpolation.

From such an Atlas the nearshore and shoreline resource can be calculated using one of the variety of hydrodynamic computer models available (see Section 11.7), taking the wave climate at one of the offshore reference sites, together with bathymetry from detailed charts, as input. (Again, quality checks should be carried out for the computer models, for a variety of situations where wave measurements exist, though this is at present problematic, because of the commercial confidentiality of many of the models.)

The level of detail required for these inshore calculations in such that it is not, at least at present, practical to carry them out for all of the European coastline of wave power interest. This suggests a two-level methodological approach, combining an 'Atlas' database for offshore reference sites with computer tools for calculating the resource at specific locations. This approach has major advantages of flexibility and long life, in that the components (both database and computer tools) can be updated individually as relative technological advances are made.

For Europe's Atlantic coasts and the North Sea, the quality of existing data and estimates from computer models is generally adequate to allow this methodology to be implemented now, but for parts of the Mediterranean better data and estimates are needed.

11.6 SAMPLES

Useful wave data sets are almost inevitably large, because of the complexity of spectral samples, the wide range of time scales of interest, and because to take small numbers of samples with long intervals between them does not make economic sense, whether for measurements or calculations. The question therefore arises as to how to choose a 'representative' subsample of our data. This might be for use as input to a numerical model to calculate nearshore wave conditions (see Section 11.5.3) or as input to a wave tank for testing models of wave energy devices or other structures (see Section 11.3).

Our criteria for selection will clearly vary, depending on whether we are interested in the full range of sea states or only in extremes (whether of wave height or some other aspect such as steepness). For the full range of sea states, we are faced with the problem of representing a high-dimensional parameter space; we can perhaps make do with three of these dimensions—height, period and principal direction—but, even then, for adequate coverage of the range of each we require a minimum of the order of 50 samples (e.g. taking $4 \times 4 \times 4$ samples, with a few missing cells).

At a time when long-term directional data were not yet available, Crabb (1980) applied a method for reconstructing directional spectra to 399 samples from a year's measurements off South Uist, chosen by a simple stratification by height, period and season. From these, a subset of 46 was subsequently chosen by the UK Department of Energy's consultants as a standard set for productivity tests of wave power device designs. This selection gave good coverage of the core of the distribution of Crabb's original sample, but discounted about 30% on grounds that they were extreme in one parameter or another. Given that we are interested in estimating average values of functions of sea state (such as device efficiency), this seems a poor procedure: missing values in the core of the distribution could be estimated relatively accurately by interpolation; the extrapolation required to estimate values in the unrepresented 'outlying' sea states cannot be done accurately.

More recently, for the UK Department of Industry's resource evaluation (Thorpe, 1993), Mollison (1991) chose samples of about 50 spectra each from UK Meteorological Office hindcast data for 10 sites around the British Isles (including the six shown in Figure 11.3). These samples were chosen to cover the full ranges of height, period and principal direction, and within those with a view to allowing as accurate estimates of device productivity as possible. Thus, for instance, sea states were chosen, within each period range, with heights successively representing low seas in which devices may be expected to operate at their maximum efficiency, two medium seas ranging up to a value at which devices could be expected to have reached their limit, and finally a sea of extreme steepness (see Figure 11.4).

An 'objective-oriented' method was also used to assign weights W_J to the chosen sea states J: the weight of each member of the full data set was shared between its (at most four) nearest representatives in a way that would give exactly correct productivity estimates if a device had constant efficiency η up to the third of the four represented heights, and constant output above that height; thus the mean power in the sea is $\sum_J W_J$, and the mean output for such a device is $\sum_J W_J \eta_J$ (for details see Mollison, 1991).

As mentioned in Section 11.4, even annual wave power averages can vary considerably. Long-term time series of wind speed and direction exist, and can be used to throw some light on this aspect, and to weight wave climate samples taken from relatively short time series so as to make them more representative of the long term distribution. (Though we should note, as a caution, that Draper (1988) did not find

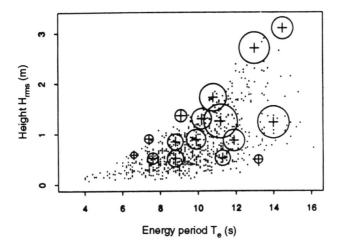

Figure 11.4   Scatter plot of H_{rms} and T_e for one directional sector (the 50–80th percentiles of the power distribution, 279–296°) for the SW Approaches, showing selected representatives J (+), with circles whose area is proportional to their weight W_J. (From Mollison (1991).)

an increase in wind speeds corresponding to the trend he described in wave data for the North Atlantic.)

Mollison (1980) fitted a log–linear model for the dependence of monthly averages of wave power P_i on monthly averages of wind input W_i, where the wind input is defined as the fifth power of the wind speed U (the physical justification for this is that $P \propto U^5$ for fully developed sea states; Pierson and Moskowitz, 1964). He found that a model using only the ordering of the W_i gave a lower standard error for the long-term wave power average, and had the advantage that it could be used directly to weight the existing data. The 95% confidence interval for the long-term average was $50.3 \pm 5\,\mathrm{kW\,m^{-1}}$, as compared with the crude average for two years' data of 41.0. This estimate agrees well with the average of $47.8\,\mathrm{kW\,m^{-1}}$ from Crabb's (1980) stratified sample (see above). This is particularly impressive in that that sample came only from the first year's data, for March 1976 to February 1977; and the wind input model suggests that this 12 month period had a lower power average ($35.7\,\mathrm{kW\,m^{-1}}$) than would have been obtained if measurements had started in any of the preceding 132 months—a confirmation of the Law of Bad Luck (putting it politely) significant at the 1% level!

11.7   THE SMALL-SCALE AND SHORELINE RESOURCE

In the early years of modern wave energy research, the mid to late 1970s, the emphasis was on the search for solutions to a large-scale energy crisis. A key result in persuading designers to think small was the discovery by Budal, Evans and Newman (see e.g. Evans, 1988) of the 'point-absorber' effect, which says that a device can absorb power from a 'capture width' wider than its own physical width by up to L/π (where L is the wavelength).

This has led to a wide variety of designs of small-scale devices, with widths of the order of 10 m but exploiting the resource over a capture width of the order of 50 m, and thus with potential mean output in the range 100–1000 kW: for instance the pioneering Norwegian prototypes of Kvae:ner–Brug (Malmo and Reitan, 1986) and TAPCHAN (Mehlum, 1986). The devices built to date have been at the shoreline, because of advantages such as the availability of specific accessible sites, and the greater potential for using established technology, but the point-absorber effect applies equally to offshore designs, as is illustrated in

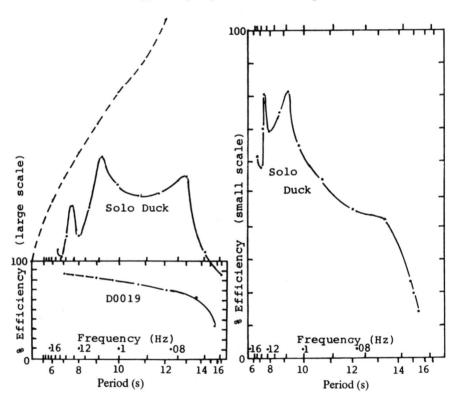

Figure 11.5 *Left.* Dependence of output on frequency for a solo Duck model at 1 : 100 scale, compared with efficiency curve for a spine-based unit, curve (d) of Figure 11.1; as in that figure, the frequency/period axis is 'stretched' to represent the empirical distribution of wave power at South Uist. The dashed curve shows the theoretical output limit for the solo device. *Right.* The same output curve for the solo Duck, replotted as a percentage of the theoretical output limit. The frequency/period axis has been modified correspondingly, so that the area under the efficiency curve still represents the overall mean efficiency in this wave climate. (From Mollison (1986).)

Figure 11.5, showing how efficiencies nominally of well over 100% can be achieved.

The estimation of the shoreline resource poses considerable problems. The offshore wave climate varies slowly over space, being approximately steady over distances of tens of kilometres (Mediterranean, European continental shelf) to a few hundred kilometres (North Atlantic). But in the nearshore region (water depth 15–25 m) or at the shoreline the wave climate can vary significantly over distances of tens of metres (see e.g. Pontes and Pires, 1992), the resource generally being lower compared with offshore conditions.

As the waves travel towards a coast through waters of decreasing depth, interaction with the seabed (and currents) may lead to major changes. These include energy-conserving effects such as shoaling and refraction, diffraction and certain types of reflection; these, especially refraction, can be a positive factor for wave energy utilisation, concentrating wave energy into specific areas ('hot spots'); however, such focusing will normally only apply to part of the directional spectrum, so that there may also be an undesirable increase in variability in the wave climate.

The principal energy-dissipating mechanism is wave breaking, though over wide continental platforms, such as off the Hebrides and in the North Sea, energy loss by bottom friction can have a major effect (Mollison, 1983).

Computational models have been developed (see e.g. Southgate, 1987) which describe the individual shallow-water phenomena satisfactorily (except perhaps wave breaking), but the interaction of all the phenomena is too complex to be fully modelled at present.

Thus a 'Wave Energy Atlas' as described in Section 11.5 can be extended to give estimates of the nearshore and shoreline resource, but this requires judgement in choosing an appropriate computational model, and the accuracy of the result will depend on the complexity of the local topography.

It should be noted that long stretches of coastline may be ruled out for wave power exploitation on practical grounds such as access and shoreline structure (Mollison and Pontes, 1992; Whittaker *et al.*, 1992).

11.8 DISCUSSION

This chapter has ranged over a variety of topics, principally simulation techniques, methodology for the estimation of wave climates, and the choice of representative samples of wave data.

Current problems in the estimation of wave climates are mainly practical: how to bring together, calibrating or verifying where necessary, a range of existing sources of data and numerical calculation programmes. Both data gathering and programming are expensive and specialist activities, and their results are consequently commercially sensitive, which hampers progress both in utilising and comparing them.

One particular problem requiring further study is that of extremes, where our knowledge is necessarily imperfect because of inadequately long time series of reliable measurements and estimates, and because of the possibility of significant climate change over periods of a decade or more. In estimating extremes, and the correlation between sites, we could learn from work on other environmental series, for instance of sea levels (Tawn, 1993) and of wind speed and direction (Haslett and Raftery, 1989). It must also be borne in mind that what constitute extreme conditions for a wave power plant will depend on the design: for one device it might be the highest wave or largest surge, for another the steepest wave, or a combination of large waves with extreme crest length. In calculating extreme individual waves from spectra we need to take account of nonlinear effects: steep waves have more pointed crests and flatter troughs than linear theory predicts.

It is important for engineering optimisation that wave power device designs should be tested in realistic and representative sea states, with careful attention to their effect on the successive stages of the power conversion chain (see e.g. Mollison, 1980). The methods described here are somewhat heuristic, and there must be scope for a more general approach to the problem of choosing 'representative samples'. Another problem that also could be generalised is the weighting of samples from a shorter time series of an environmental variable using a long term series of a related variable.

The fascination and reward of studying wave energy statistics lies in the wide range of problems that arises, not just within statistics—though these range from exploratory data analysis to theoretical stochastic processes—but across a range of other subjects, principally physics and engineering.

One could also mention economics. Earlier UK assessments of wave energy were marred by statistics abuse, or at least inexcusably poor statistical practice (see Mollison, 1984). While the recent assessment of Thorpe (1993) shows a very welcome improvement, there remain problems of great social importance in evaluating costs and benefits in a field such as renewable energy. Such problems are often ignored, being regarded as either too simple or too intractable, but statisticians should take an interest, if only to press the need for clear thinking and proper recognition of uncertainty.

ACKNOWLEDGEMENTS

I should like to acknowledge the help of Stephen Salter and colleagues at the Edinburgh Wave Power Project (1973–), Trevor Whittaker and colleagues in the Department of Trade and Industry supported review of the UK's shoreline and nearshore wave energy resource (1990–92), and Teresa Pontes and colleagues in the European Community supported review of wave resource assessment methodology (1992–93).

12

Extreme Sea Levels: Modelling Interaction between Tide and Surge

M. J. Dixon and J. A. Tawn
Lancaster University, UK

12.1 INTRODUCTION

The design of coastal flood defences requires accurate estimates of the probability of extreme sea levels. Recent developments in the statistical modelling of these, which enable temporal and spatial properties of the meteorological and astronomical components of the sea level to be exploited, give considerably improved estimates of their distribution (Pugh and Vassie, 1980; Tawn, 1992; Coles and Tawn, 1990). In this chapter we describe one component of a current Ministry of Agriculture Fisheries and Food funded study to further develop these models and apply them in a full-scale analysis of all UK coastal sites.

The sea-level process X_j observed for a specific site at time j consists of two separate physical processes: a deterministic tidal component T_j and a meteorologically induced stochastic surge component S_j, i.e.

$$X_j = T_j + S_j. \tag{12.1}$$

Standard tidal analysis techniques can be used to separate these constituent processes from a series of observations of X_j. Pugh and Vassie (1980) proposed that the distribution of extreme sea levels can be obtained by estimating the distribution of surges and combining this distribution with the deterministic tidal series. Tawn and Vassie (1989) and Tawn (1992) proposed successive refinements to this approach that reduced the analysis of the surge distribution to the estimation

Statistics for the Environment 2: Water Related Issues. Edited by Vic Barnett and K. Feridun Turkman
© 1994 John Wiley & Sons Ltd.

of the upper tail of the conditional distribution of the surges given the concomitant tidal level.

If the tides and surges are independent processes, this conditional distribution of surges is the same for all tidal levels, so standard extreme value techniques (Smith, 1989) can be used to estimate it. However, in shallow-water areas dynamic processes, such as bottom friction, cause the tidal and surge components of the sealevel process to interact; in particular, surge values that occur at the time of a high tide tend to be damped, whereas surge values at rising tide typically are amplified. Accounting for this form of interaction in the modelling of extreme surges is important, since ignoring this feature and proceeding as if the processes were independent is liable to result in significant overestimation of extreme sea levels.

The methodology for combining the upper tail of the conditional surge distribution with the tidal series to obtain the distribution of extreme sea levels is well defined (Tawn, 1992, 1994). The current methods for estimating this conditional tail are restricted to site-by-site analyses, and thus fail to exploit available physical knowledge about the spatial coherence of the processes involved (Flather, 1987), and of the interaction in particular (Wolf, 1978). A consequence of such approaches is that little understanding of the interaction process is obtained, and spatial interpolation of the joint tail is likely to be poor. In this chapter we develop spatial methods for modelling the conditional surge tail.

The standard oceanographic approach used to assess the level of tide–surge interaction is the examination of the standard deviation of the conditional surge process (Prandle and Wolf, 1978; Pugh, 1987; Walden *et al.*, 1982). Since our interest here is in the behaviour of the extreme surges, interaction will be characterised using non-parametrically estimated high quantiles of the distribu- tion of surges conditional on tidal level. In Section 12.2 our methods for separate site estimation of the conditional surge tail and of the interaction process are outlined. In Section 12.3 these methods are extended to spatial procedures by use of a combination of non-parametric spatial smoothing and parametric covariate models. Finally, in Section 12.4 these methods are applied using extensive hourly surge and tide data from eight sites along the UK east coast. The results provide increased understanding of the interaction process over time and space, and confirm hydrodynamical model forms for the spatial form of interaction along this coastline.

12.2 MODELLING EXTREME SURGES IN THE PRESENCE OF TIDES

12.2.1 The independence case

In this section we briefly outline how, through the use of standard extreme value techniques, the distribution of extreme surges is obtained in the case of tide–surge independence. First consider the idealized case of a sequence of i.i.d. variables

$Y_1, \ldots, Y_n$. On the basis of a limiting point process result, Smith (1989) argues that, provided a high threshold u is taken, the process of exceedances of u can be approximated by a non-homogeneous Poisson process with intensity $\lambda(y)$ given by

$$\lambda(y) = \sigma^{-1} \left[1 + \frac{\xi(y - \mu)}{\sigma} \right]_+^{-(\xi + 1)/\xi} \qquad \text{for } y > u, \qquad (12.2)$$

where μ, $\sigma (\sigma > 0)$ and ξ are location, scale and shape parameters respectively, and $y_+ = \max(y, 0)$. Consequences of this approximation are that

(i) all exceedances of the threshold are used in the estimation of the extremal parameters μ, σ and ξ;
(ii) the upper tail of the distribution of Y is given by a generalised Pareto model, i.e.

$$\Pr\{Y_i > u + y \,|\, Y_i > u\} \approx \left[1 + \frac{\xi y}{\sigma + \xi(u - \mu)} \right]_+^{-1/\xi}.$$

Clearly, if the surge process is i.i.d., this model provides all the information we require. In practice, the hourly surge series exhibits temporal dependence, so some form of declustering of the exceedances of the threshold is required in order to apply this model. Similarly, seasonality of the surge process complicates the analysis. In general, seasonality and dependence can easily be incorporated into the analysis (Smith, 1989; Tawn, 1992), but since our problem of tide–surge interaction is more fundamental, we do not dwell on these issues here.

12.2.2   The interaction case

When tides and surges do interact, this has the effect of making the surge series non-stationary, with distribution depending on the concomitant tidal level. Correspondingly, the distribution of the surges that exceed a high threshold also depends on the concomitant tidal level. One way of handling this dependence is to model the upper tail of the conditional surge distribution—in other words, to extend the non-homogeneous Poisson process approximation so that the intensity function depends on the concomitant tidal level, i.e. by taking $\lambda(y, T_j)$ in (12.2). The simplest such approach is to replace the parameters μ, σ, and ξ by functions $\mu(T_j)$, $\sigma(T_j)$ and $\xi(T_j)$ of the concomitant tide (Tawn, 1992). Tawn (1988) applied this approach to data from some east coast sites, and found, for each site, that the location and scale parameters could be suitably modelled as functions of tidal level using low-order polynomial regression functions, and that there was no evidence for dependence of the shape parameter on the tide. The disadvantage with such an approach, however, is that simple parametric polynomial regression functions are not flexible enough to model the complex tide–surge interaction process observed over the coastal network. A more flexible procedure is to use covariate forms obtained from the joint distribution of the tide and surge. Tawn (1994) suggested that if $a(T)$ and $b(T)$ are functions of tidal level such that the location–scale normalization of the surge series

$$[S_j - a(T_j)]/b(T_j)$$

produces a series that is stationary with respect to the tide then the methods of Section 12.2.1 could be directly applied to the transformed series. The functions $a(T)$ and $b(T)$ were taken as suitably selected aspects of the empirical joint tide–surge distribution. Viewed more generally, this proposal amounts to applying intensity (12.2) in a regression setting with

$$\mu(T) = \mu b(T) + a(T), \quad \sigma(T) = \sigma b(T), \quad \xi(T) = \xi, \tag{12.3}$$

i.e. $a(T)$ and $b(T)$ are the required covariates, so the structure of this model is in line with findings of Tawn (1988). The selection of suitable $a(T)$ and $b(T)$ requires considerable care, but, on the basis of extensive analysis, large quantiles of the surge distribution conditional on the tidal level T are chosen. Specifically, letting $q_i(T)$ $(i = 1, 2)$ denote two high quantiles of this distribution, with $q_1(T) < q_2(T)$, we take

$$a(T) = q_1(T), \quad b(T) = q_2(T) - q_1(T). \tag{12.4}$$

The quantiles $q_i(T)$ $(i = 1, 2)$, are chosen to be high enough to contain sufficient information about the extremal tail, but low enough to be estimated empirically with sufficient accuracy. Although the covariate choice is to some extent *ad hoc*, the actual covariate form is natural in an extreme value setting and gives rise to invariance of the extremal parameters μ, σ and ξ with respect to location and scale changes of the surge series. This invariance is of key importance when these extremal parameters are examined spatially in Section 12.3.

An apparent limitation of the model with regression functions (12.3) is that the shape parameter ξ is homogeneous over the tidal range. In principle, the regression function $\xi(T)$ could easily be extended to allow the shape parameter to depend on some aspect of the empirical conditional distribution of surge given tide. However, since the covariate forms (12.3) adequately describe all the properties of the observed tide–surge interaction associated with extreme surges, and the selection of a suitable covariate based on the joint distribution is not simple, here we use models of the form (12.3) with covariates (12.4).

12.3   SPATIAL ANALYSIS

Many sites have short data series consisting of only a few years of hourly observations, and consequently if the procedure of Section 12.2 is applied on a site-by-site basis, estimates of the covariates $a(T)$ and $b(T)$ and the extremal parameters μ, σ and ξ are likely to have large sampling variability for each site. However, as the parameters and the covariates all correspond to physical features of the process, and it is known that the sea-level processes are spatially coherent, these aspects of our model should reflect this. Here we exploit this spatial coherence by separately spatially modelling the covariates and the extremal parameters.

First consider the covariates. As defined in Section 12.2, for each site, $a(T)$ and $b(T)$ each depend upon a combination of two physically distinct aspects of the process; the interaction between tides and extreme surges and the marginal distribution of extreme surges. Thus, although these covariates should be spatially reasonably smooth, they provide little physical insight into the separate mechanisms of interaction and marginal heterogeneity along the coastline. As, by definition, the surge has zero mean, the principal change in marginal distribution from site to site is a scale change, so one way of distinguishing the two physical processes is to remove the amplification effect on the extremes by standardising the marginal extreme surge distribution at each site. This standardisation could be achieved by dividing the surge series by the sample standard deviation, but since our interest is in the upper tail, an alternative measure of variation, which is of more relevance, is a high quantile, q_s say. Here the choice of which quantile to use is to some extent *ad hoc*, so sensitivity of results needs to be examined. Applying the method described in Section 12.2.2 to this standardised surge series gives covariates $a^*(T)$ and $b^*(T)$ that measure only the interaction of the extreme surges and tide. These functions can then be mapped against coastal distance, d, and non-parametric spatial models $a^*(T, d)$ and $b^*(T, d)$ fitted to provide a model of the interaction process at any tidal level and coastal position. Similarly, q_s can be mapped against coastal distance, and a non-parametric spatial model of marginal amplification $q_s(d)$ obtained for any coastal site. Consequently, if we are interested in the covariates $a(T, d)$ and $b(T, d)$, at any coastal distance, these may be obtained from our separate spatial estimates of amplification and interaction by the property

$$a(T, d) = q_s(d)a^*(T, d), \qquad b(T, d) = q_s(d)b^*(T, d). \qquad (12.5)$$

Now consider the tail parameters μ, σ and ξ. As noted at the end of Section 12.2.2, these parameters are invariant under scale transformations of the surge, so that the scheme proposed above leaves these parameters unchanged. At each site these parameters are characteristics of the extreme tail of the conditional surge variable once the effects of site position, i.e. amplification and interaction, are accounted for. Since the extreme surge process is strongly spatially coherent over entire coastlines, the extremal parameters μ, σ and ξ will also exhibit a high degree of spatial smoothness. This property can be incorporated into the model by use of simple covariate models of coastal distance. The resulting likelihood is complex owing to spatial dependence, but since similar procedures were proposed by Coles and Tawn (1990, 1991), details are omitted. However, it is worth noting that by spatially estimating these parameters a highly significant gain in precision is obtained relative to site-by-site analyses.

12.4   APPLICATION TO THE EAST COAST OF THE UK

The data are hourly surge levels with corresponding tidal levels for eight sites along the east coast of the UK, from Wick in northern Scotland to Sheerness at the mouth of the Thames estuary. In order of both coastal distance from Wick, and the propagation of the surge, the sites are Wick, Aberdeen, North Shields, Whitby, Immingham, Lowestoft, Southend and Sheerness.

First we construct our interaction covariates $a^*(T)$ and $b^*(T)$ for each site before considering the spatial versions $a^*(T, d)$ and $b^*(T, d)$. To aid comparison between sites, for each site the tidal levels are transformed to be uniform on $[0, 1]$. Consequently we refer to $a^*(t)$ and $b^*(t)$, i.e. transformed tide is denoted by t but the functional notation unchanged. Defining $f(s, t)$ to be the joint density of the normalised surge and transformed tide, we have that the surge distribution conditional on tide t is $f(s, t)$, since the transformed tide is uniform. The q_ith quantile $q_i(t_0)$ of this conditional distribution at $t = t_0$ satisfies

$$q_i = \int_{-\infty}^{q_i(t_0)} f(s, t_0)ds, \qquad (12.6)$$

where $q_1 = 0.98$ and $q_2 = 0.99$ were found to be suitable quantiles upon which to base $a(t)$ and $b(t)$ for the east coast data. For a given q_i, $q_i(t_0)$ can be estimated from (12.6) if f is replaced by a bivariate kernel density estimate. The sampling variability of this estimator of $q_i(t_0)$ can be obtained by applying this estimation procedure repeatedly to blocks (years) of data, which produces a sample of i.i.d estimates. Similarly, for these east coast data, the 98% surge quantile was found to be a suitable choice for q_s.

For each of the eight sites, Figures 12.1 and 12.2 show the estimated functions $a^*(t)$ and $b^*(t)$ that are obtained from estimates of $q_i(t_0)$, $t_0 \in [0, 1]$, using (12.4). Note that if the tide and surge were independent then for all $t, a^*(t) = 1$ and $b^*(t)$ would be constant. A clear interpretation can be given to $a^*(t)$ in particular, since the cases $a^*(t_0) = 1$, <1 and >1 correspond to the surge tail being unchanged, reduced and amplified respectively, at tidal state t_0, relative to the marginal surge distribution.

As a result of the transformation of the tide and the surge normalisation in Section 12.3, both $a^*(t)$ and $b^*(t)$ are not site-specific, and consequently can be compared over sites. The eight plots in Figures 12.1 and 12.2 show a gradual change in the profile of the functions over the sites, suggesting that interaction is a coherently changing process along this coast.

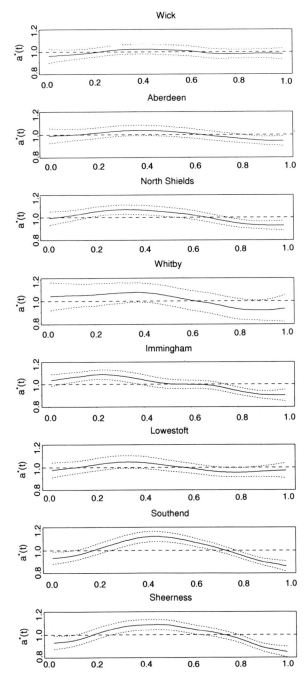

Figure 12.1 Estimates of the function $a^*(t)$ for each of the eight sites in increasing order of distance from Wick. The abscissa is standardised tidal level and the dotted lines are 95% pointwise confidence intervals. The dashed line $a^*(t) = 1$ is shown as a guide.

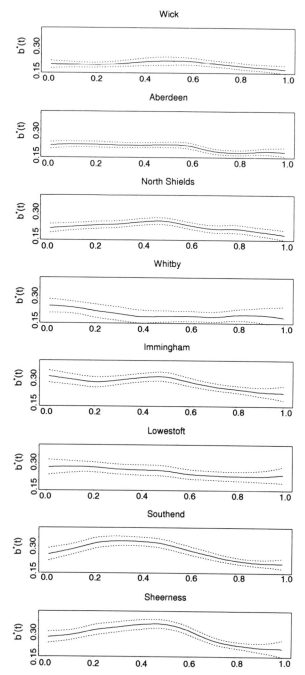

Figure 12.2 Estimates of the function $b^*(t)$ for each of the eight sites in increasing order of distance from Wick. The abscissa is standardised tidal level and the dotted lines are 95% pointwise confidence intervals.

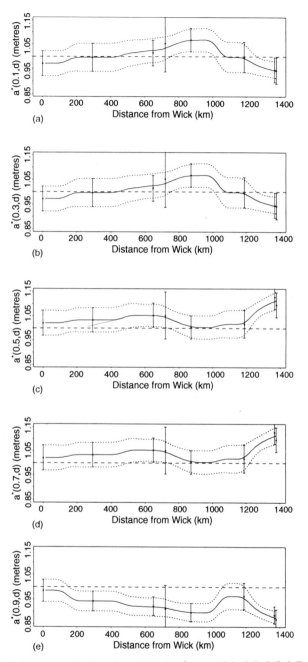

Figure 12.3 Estimates of the function $a^*(t_0)$ at tides $t_0 = 0.1, 0.3, 0.5, 0.7$ and 0.9 with 95% confidence intervals, against distance (in km) from Wick, obtained from the estimates in Figure 13.1. (a)–(e) correspond to $t_0 = 0.1, 0.3, 0.5, 0.7$ and 0.9 respectively. The dotted lines are 95% pointwise confidence intervals. The dashed line $a^*(t_0, d) = 1$ is shown for each figure.

In Figure 12.3 estimates of $a^*(t_0)$ for each site, together with a kernel regression estimate, weighted by the standard deviation of the corresponding site-based estimates, are plotted against coastal distance for values of $t_0 = 0.1, 0.3, 0.5, 0.7$ and 0.9. These transformed tidal levels t_0 correspond to low tide through to high tide in steps of equal tidal state. Also shown on Figure 12.3 are the weighted spatially smoothed estimates of $a^*(t_0, d)$. It is clear from Figure 12.3(a) that in the lowest tidal state there is a gradual change along the coastline from dampening of the surge at Wick to amplification at Immingham, returning to a dampened form towards Sheerness. At mid-tidal levels, Figures 12.3(b–d), there is slight amplification at Wick, which decreases to the Immingham–Lowestoft region before significantly increasing to Southend and Sheerness. The interaction process at high tidal states is the most important for extreme sea-level studies since the largest sea levels typically result from large surges occuring at high tides. Figure 12.3(e) shows that a dampening effect of various magnitudes occurs for all the coastline at high tides. Furthermore, the spatially estimated profile is remarkably similar to findings in Wolf (1978), where purely interaction terms were extracted from a hydrodynamical model of the sea-level process in the North Sea.

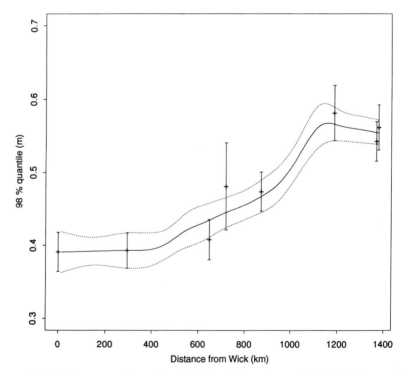

Figure 12.4   Estimates and associated 95% confidence intervals of the 98% quantile of the marginal surge distribution for each of the eight sites along the coast, against coastal distance. The solid (and dotted) lines represent the spatial smoothing (and associated 95% confidence intervals).

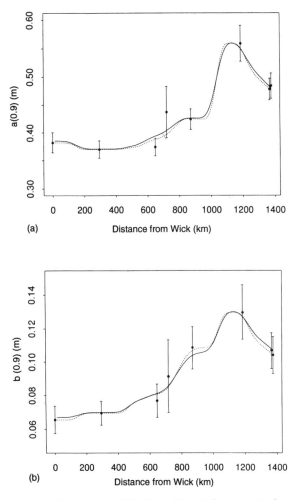

Figure 12.5 Estimates for $a(0.9)$ and $b(0.9)$, in (a) and (b) respectively, at each site with 95% confidence intervals. The lines are spatial estimates for $a(0.9, d)$ and $b(0.9, d)$ obtained by the two methods described in Section 12.4: the broken line is the spatial estimate obtained by direct smoothing of the $a(0.9)$ estimates, and the solid line is the estimate obtained using (12.5).

As described so far, these models provide information only about the interaction process. To obtain estimates of quantities more relevant to extreme sea-level analyses, these estimates need to be recombined with the marginal specific component of the model, $q_s(d)$. Figure 12.4 shows estimates of q_s together with a smooth spatial estimate obtained by applying weighted kernel regression. What this figure shows is a gradual amplification of surges down the east coast with a significant increase around the Wash region between Immingham and

Lowestoft. Viewed without either the smoothing or the confidence intervals, the estimate for Whitby looks inconsistent. One reason for this is that the sampling period for Whitby is the 1980s, whereas most other sites also include data for the 1970s, which is generally recognised as being a milder decade in terms of storm surges; hence the smoothing process corrects for this sampling bias.

We have now obtained separately estimated smoothed spatial estimates for the two fundamental physical components of the conditional surge process: amplification $q_s(d)$ and interaction $a^*(t, d)$ and $b^*(t, d)$. Consequently, spatial estimates of the covariates $a(t, d)$ and $b(t, d)$ for the tail of the conditional surge distribution can be obtained from (12.5). Alternatively, $a(t, d)$ and $b(t, d)$ can be estimated directly and smoothed spatially. This latter approach has the disadvantage that, although the two physical processes of amplification and interaction are separately spatially coherent, there is no reason that when combined the resulting process should be as simple to interpolate. Figure 12.5 shows that the estimates obtained from both approaches when $t_0 = 0.9$ give very similar results. Clearly, for this coastline, the advantage of decomposition of the covariate functions is minimal. However, more generally, our approach in using spatial estimation of the tail parameters μ, σ and ξ is capable of leading to considerable improvements, and our analysis of the covariate functions has revealed significant new physical insight into the spatial process of tide–surge interaction.

ACKNOWLEDGEMENTS

We thank J. M. Vassie, of the Proudman Oceanographic Laboratory, and C. W. Anderson for helpful comments on an earlier version. Both authors were partly supported by a Ministry of Agriculture Fisheries and Food grant. The work used equipment provided by SERC under the Complex Stochastic Systems Initiative.

PART III

Water Quality, Supply and Management

13

Bayesian Methods in Asset Management

A. O'Hagan
University of Nottingham, UK

13.1 INTRODUCTION

As part of the regulation of the privatised English and Welsh water industry, water companies are required to formulate an Asset Management Plan, involving the production of detailed estimates of all areas of capital expenditure over a 20 year period. Many companies are using a Bayesian approach first developed at privatisation in 1989. This paper outlines that method with particular reference to recent enhancements.

One benefit of a Bayesian approach is that, instead of examining a random sample of assets, it is possible to target detailed studies to areas and assets that will be most informative, in order to minimise posterior variances for key quantities. The paper describes the computation of a preposterior variance criterion, an approximation suitable for large models and an algorithm to search for optimal study programmes.

13.2 BAYESIAN STATISTICS IN THE WATER INDUSTRY

The water industry in England and Wales was privatised in 1989. As part of the regulatory framework for privatisation, the water companies were required to produce an Asset Management Plan (AMP), which essentially consisted of estimates of the company's capital investment need over a 20 year period. Companies are now required to submit a second AMP in 1994, and it is expected that the exercise will be repeated at five-yearly intervals. Two of the ten major companies that were privatised in 1989 used a Bayesian approach, which is

Statistics for the Environment 2: Water Related Issues. Edited by Vic Barnett and K. Feridun Turkman
© 1994 John Wiley & Sons Ltd.

described in O'Hagan *et al.* (1992) and more fully in O'Hagan and Wells (1993) for their first AMP (known as AMP1). For the latest exercise (AMP2), that approach has been substantially developed and is being used by five of the 'big ten' water and sewage companies and by two of the larger number of 'water-only' companies that were already privately run in 1989. It is also being used by the Northern Ireland water service, which is still publicly owned and is preparing its first AMP. This work, including development of software that has been jointly sponsored by nine companies, is being carried out by the Water Statistics Unit at Nottingham University. In a separate development based on the same AMP1 methods, Warwick University are applying Bayesian methods in preparation of AMP2 for another of the 'big ten' companies.

This chapter describes some of the principal features of the methodology being applied in AMP2, taking it beyond the AMP1 approach of O'Hagan and Wells (1993) and dealing with some of the limitations described there. Section 13.3 considers models for investment need—the more complex and flexible models of the costs themselves that are now available, the greater range of data that can be utilised, and the underlying statistical theory. Section 13.4 gives a detailed analysis of the problem of selecting a sample of engineering studies.

It is worth noting that many water companies are also using a Bayesian analysis for the problem of estimating the condition of assets. This needs to be done as one aspect of considering investment need for the AMP, but is required separately. Whereas the simplifying approach of the Bayes linear estimator is used to handle the large and complex task of overall investment estimation, it is possible to develop a more conventional, fully specified Bayesian model for the asset condition problem. A solution making use of the powerful computational tool of Gibbs sampling has been programmed by WRc (formerly the Water Research Centre), and is being widely used in the British water industry.

13.3 INVESTMENT MODELLING

13.3.1 AMP1 approach

A water company is a complex business, and there are a variety of ways in which investment needs arise. The Bayesian approach used for AMP1 has the following features (for further details see O'Hagan *et al.*, 1992; O'Hagan and Wells, 1993). The company's region is divided into zones, the number of zones being typically from as few as 50 to several hundreds, depending on the size and organisational structure of the company. The total investment need in zone i is a variable C_i to be estimated. In particular, the total cost $T = \sum_i C_i$ is of primary interest. Prior information about the C_i is constructed by first expressing them in terms of simpler quantities by means of one or more cost equations, then obtaining prior knowledge about each of these simpler quantities from relevant company personnel.

For instance, O'Hagan and Wells (1993) give a model used at AMP1 for the investment costs on sewerage for Anglian Water. The total cost in a zone is first expressed as a sum of costs to relay or renovate critical sewers and rising mains in poor condition, to remove the risk of flooding to properties, to lay new sewers to deal with growth, and to effect remedial work on unsatisfactory overflows. Each component of this sum is then further broken down into other variables. As a single example, the cost to remove the flooding risk is expressed as the number of properties at risk of flooding times an average cost per property. The model thereby expressed C_i in terms of 25 other quantities, which we call base quantities. Anglian Water supplied prior information about the base quantities, and the model was used to derive prior means, variances and covariances for the total zone costs.

After construction of prior information, a number of zones are selected for detailed engineering investigation, yielding improved estimates of the C_i in the studied zones. These data are to be combined with the prior information to give posterior estimates of the C_i and of the total cost T. The method used for this purpose is the Bayes linear estimator (BLE). Whereas a full Bayesian analysis would require the specification of the complete joint prior distribution of the C_i (in possibly hundreds of dimensions), the BLE requires only means, variances and covariances. As already described, these are in turn obtained from prior information about the simpler base quantities. The BLE can in general be thought of as an approximation to a full Bayesian analysis, which is exact if joint distributions are all normal. In this application approximate normality of the C_i is reasonable in view of their construction as a sum of other costs.

13.3.2   Updating base quantities

The simple methodology has been elaborated in several ways for AMP2. The principal step is to operate at the level of individual base quantities rather than the whole zone costs C_i. In AMP1 the prior information about each base quantity was used only to construct prior information about the C_i. Data in the form of improved estimates of the C_i was obtained, and the BLE used to give posterior information about the C_i. If the posterior estimate of cost C_i in a particular zone was 10% higher than its prior estimate, the company could not deduce in what ways the extra spending would arise. An early decision for AMP2 was to apply the BLE to update information on individual base quantities. Posterior estimates of the C_i are therefore obtained from posterior means, variances and covariances of base quantities in the same way as prior information about the C_i was derived in AMP1.

The change brings not only greater freedom and better use of information but also greater responsibilities. One difficulty in AMP1 was that even very intensive and costly engineering study in a zone could not identify the true value of its cost C_i. So the data used in the BLE are not observed values of the C_i, in a sample of zones, but simply better estimates of those C_i than the prior information had given. It was acknowledged in O'Hagan and Wells (1993) that a modification of the BLE used there to deal with this complication was not theoretically adequate. This difficulty has now been resolved by a theoretical model for systems of expert estimates of varying degrees of accuracy. This model is developed in Goldstein and O'Hagan (1993), and provides a sound basis for the use of the BLE. This now forms part of the methodology and is built into the software.

A second responsibility is to elicit prior information about base quantities more carefully. When this was used just to build up prior information about the C_i, there was a high degree of robustness to errors of specification in a few base quantities. However, such errors can produce very unpredictable posterior estimates if the prior information conflicts with the information coming from detailed engineering studies on individual base quantities. Companies have found the careful and accurate specification of prior information the most difficult part of the exercise, and this is an area where further research would be very beneficial.

13.3.3 Mappings

In order to compute prior or posterior information about T, the C_i or other derived quantities of interest, means, variances and covariances of these quantities must be expressible in terms of means, variances and covariances of base quantities. This is done recursively through the cost equations. For instance, if a quantity C is the sum of two other quantities A and B, then

$$E(C) = E(A) + E(B), \tag{13.1}$$

$$\text{Var}\,(C) = \text{Var}\,(A) + \text{Var}\,(B) + 2\,\text{Cov}\,(A, B), \tag{13.2}$$

or if C is the product of A and B, and if A and B are independent, then

$$E(C) = E(A)E(B), \tag{13.3}$$

$$\text{Var}\,(C) = \text{Var}\,(A)\,E(B)^2 + E(A)^2\,\text{Var}\,(B) + \text{Var}\,(A)\,\text{Var}\,(B). \tag{13.4}$$

Operations of addition and subtraction of any quantities, and multiplication of independent quantities, are allowed in the cost equations.

In AMP1 all base quantities were defined at the level of an individual zone. If, for instance, the zone cost included a cost for improving unsatisfactory overflows then that cost was a total for improving all such overflows in that zone. For AMP2 the overflow improvement cost could be defined at the level of an individual overflow. If the company wished to report on costs at the zone level then the total cost in the zone could be separately defined as a sum of individual overflow costs.

Expressions that relate quantities defined on different units are called mappings. Quite complex structures can be modelled using mappings, as the following example shows.

13.3.4 An example of cost modelling

Consider the total investment cost over the company's region for expansion of sewage treatment facilities to deal with growth in sewage over the 20 year AMP period. This is first written as a sum over the various sewage treatment catchments:

$$[\text{ST growth cost}] = \sum_i [\text{catchment growth cost}]_i. \qquad (13.5)$$

A catchment is the area over which sewage is collected and sent to a single sewage treatment works. [catchment growth cost]$_i$ is the expansion cost for sewage treatment in a given catchment. The summation $\sum_i$ is formally a mapping relating these quantities defined on the unit of a catchment to the total [ST growth cost] defined on the whole region.

There are two aspects to sewage treatment. First the sewage is treated at a sewage treatment works, and surplus water and treated waste is sent out into a river or to sea, but there remains a residue of sludge, which must be disposed of differently. This is sent to a sludge works, and one sludge works will usually deal with sludge from several sewage treatment works. Here the sludge may go through further stages of treatment or processing before final disposal. We therefore write

$$[\text{catchment growth cost}]_i = [\text{STW growth cost}]_i + [\text{sludge growth cost}]_i, \qquad (13.6)$$

$$[\text{sludge growth cost}]_i = \sum_{k(i)} [\text{SW growth cost}]_k. \qquad (13.7)$$

[STW growth cost]$_i$ is the cost for expansion of the sewage treatment works in catchment i, and [sludge growth cost]$_i$ is the cost for expansion of sludge facilities in catchment i. This is further expressed in terms of [SW growth cost]$_k$ which is the cost for expansion of the kth sludge works. The summation $\sum_{k(i)}$ sums over all sludge works in catchment i. For many catchments there will be no sludge works, and this sum will be empty. For others there will be a single sludge works (usually close to the sewage treatment works). Since for any given i there will be at most one term being 'summed' in $\sum_{k(i)}$, the term 'mapping' is really more appropriate than 'summation'.

The next step is to express the cost for expansion at either kind of works as a product of the amount of growth and a unit cost:

$$[\text{STW growth cost}]_i = [\text{STW growth}]_i\,[\text{STW expansion cost}]_i, \qquad (13.8)$$

$$[\text{SW growth cost}]_k = [\text{Sludge growth}]_k\,[\text{SW expansion cost}]_k. \qquad (13.9)$$

[STW growth]$_i$ is the amount of growth of sewage (in terms of population equivalent) to be experienced in catchment i, while [STW expansion cost]$_i$ is the unit cost per population equivalent to expand the treatment facilities at the ith sewage treatment works.

Further equations are needed to express the relationship between [STW growth]$_i$ and [sludge growth]$_k$, since growth of sludge to be processed is a consequence of growth in sewage for treatment:

$$[\text{STW growth}]_i = [\text{catchment growth}]_i + [\text{sludge returns growth}]_i, \tag{13.10}$$

$$[\text{catchment growth}]_i = \sum_{j(i)} [\text{DA growth}]_j. \tag{13.11}$$

The growth in sewage for processing at a sewage treatment works is made up mainly of [catchment growth]$_i$, which is the growth in sewage arriving at the works through the sewers. Equation (13.11) expresses this as a sum of [DA growth]$_j$, which is the growth in sewage in the jth drainage area. A drainage area is a more convenient unit in the case of large catchments. It comprises a reasonably compact area of sewers, all draining to a single point. For a small catchment there is just one drainage area, and the 'single point' is the treatment works. A large catchment may be divided into several drainage areas. Equation (13.11) is introduced first because the company will actually estimate growth in sewage at the level of an individual drainage area, and secondly because [DA growth]$_j$ will also feature in another part of the cost model dealing with costs for expanding the sewer system. The other component of sewage growth for treatment works i is [sludge returns growth]$_i$, which is occasioned by one form of sludge treatment known as dewatering. Dewatering extracts extra water from sludge to reduce its volume, but this water must be sent back to a sewage treatment works and processed as sewage. (Technically, this can then come back to the sludge works as more sludge, and back to the sewage treatment works after dewatering and so on. However, the volumes at this point are miniscule, and are ignored in the model.) Sludge returns are then written as

$$[\text{sludge returns growth}]_i = \sum_{k(i)} [\text{sludge return factor}]_k [\text{sludge growth}]_k. \tag{13.12}$$

The mapping $\sum_{k(i)}$ is used again to relate sludge works k to the catchment i in which it is located, and hence to the sewage treatment works that receives its dewatering returns. [sludge return factor]$_k$ is the proportion of sludge arriving at sludge works k that is returned after dewatering (and is zero if sludge works k does not dewater). Although the volume returned may be large, quantities in this model are expressed in population equivalent, and the sewage content of dewatering returns is small. A more realistic model would take account of different growth costs for a sewage treatment works dependent on volume increases and on population-equivalent increases. Volume increases can arise, for example, from a large supermarket car park being built, resulting in an increase in rainwater arriving at the works. This will necessitate an expansion of inlet and outfall facilities at the works, but not of processing facilities.

The final equation relates sludge growth to catchment growth:

$$[\text{sludge growth}]_k = \sum_{i(k)} [\text{catchment growth}]_i. \qquad (13.13)$$

The mapping $\sum_{i(k)}$ is the inverse of $\sum_{k(i)}$ and is a genuine summation. It sums over the sewage treatment works that supply sludge works k.

This example illustrates how complex some aspects of models used for AMP2 are. The total [ST growth cost] could no doubt be written more compactly, with fewer or simpler equations, but (13.5)–(13.13) express it in terms meaningful to the company. They also allow the company to receive posterior estimates for intermediate quantities like [SW growth cost]$_k$ or [sludge returns growth]$_i$, which are of interest in their own right. The system of equations rests on four sets of base quantities: [DA growth]$_j$, [sludge return factor]$_k$, [STW expansion cost]$_i$ and [SW expansion cost]$_k$. The company must first supply prior information about all of these, and then may carry out investment studies to obtain improved estimates of some of them. They might for instance visit a sludge works and draw up a cost-effective scheme of work for enlargement of that specific works, allowing for the capacity and condition of the existing equipment, the availability of land on the site for building new facilities, the accessibility of the site, and so on. This would provide an improved estimate of the unit cost [SW expansion cost]$_k$.

13.4 THE DESIGN PROBLEM

13.4.1 Programmes of studies

Estimates of quantities of interest such as the total cost T based only on prior information will almost certainly be too inaccurate for use in the AMP. It is therefore necessary for a company to obtain further data, in the form of improved estimates of individual base quantities obtained from specially commissioned engineering studies. A single study involves visiting an individual unit (e.g. a sewage treatment works, an overflow, a drainage area or a public water supply zone) and obtaining improved estimates of one or more base quantities defined for that unit. For instance, a study might consist of obtaining an improved estimate of the [SW expansion cost]$_k$ in sludge works k, and perhaps also other base quantities associated with maintenance of equipment and buildings at sludge works k that may appear in other parts of the cost model.

Companies may undertake studies of many different kinds, but we consider here designing a programme of studies of a particular type. By a programme of studies, we mean a collection of studies of that type, visiting a specified selection of units. For example, a programme of sludge works studies might consist of studies in sludge works numbers 1, 5, 7 and 23. Suppose that there are N units that can be studied, and a programme of n studies is required. There are $\binom{N}{n}$ possible study programmes. A classical simple random sample would give each of these $\binom{N}{n}$ an equal chance of being the selected programme. However, randomisation is not necessary in Bayesian analysis, and we search instead for the *best* programme.

A criterion for the 'best' programme is needed, and we assume that the objective is to minimise the posterior variance of a specific quantity D of interest. However, posterior variance will generally depend on the data observed in the studies. The general Bayesian solution is to minimise the preposterior variance of D, which is the prior expectation of the posterior variance. Denoting the data in general by x, the optimality criterion is therefore to minimise $E[\text{Var}(D|x)]$.

13.4.2   Preposterior variance formulae

Posterior means, variances and covariances for base quantities are obtained by applying the BLE. Posterior variances and covariances are then actually independent of the data x (although of course they depend on the choice of programme). Therefore, if B is a base quantity then $\text{Var}(B|x)$ may be calculated from the prior information, and is equal to $E[\text{Var}(B|x)]$. However, if D is a derived quantity as opposed to a base quantity then $\text{Var}(D|x)$ may depend on x. This will happen particularly through the use of (13.4). If D is the product of two (independent) base quantities A and B then

$$\text{Var}(D|x) = \text{Var}(A|x)E(B|x)^2 + E(A|x)^2\text{Var}(B|x) + \text{Var}(A|x)\text{Var}(B|x) \tag{13.14}$$

depends on $E(A|x)$ and $E(B|x)$, which will typically depend on x.

Equations (13.1)–(13.4) refer to prior information, but convert immediately to equations for posterior means and variances exactly as in (13.14). $E[\text{Var}(D|x)]$ can be calculated recursively if we can also obtain versions for preposterior means and variances (and also covariances). For means, we immediately have

$$E[E(D|x)] = E(D),$$

so the preposterior expectation is the prior expectation. Equation (13.2) also carries over immediately to give

$$E[\text{Var}(D|x)] = E[\text{Var}(A|x)] + E[\text{Var}(B|x)] + 2E[\text{Cov}(A, B|x)] \tag{13.15}$$

when $D = A + B$. However, when D is the product of A and B, (13.3) gives

$$E[\text{Var}(D|x)] = E[\text{Var}(A|x)]E[E(B|x)^2] + E[E(A|x)^2]E[\text{Var}(B|x)]$$
$$+ E[\text{Var}(A|x)]E[\text{Var}(B|x)], \tag{13.16}$$

using the assumed independence of A and B. Now

$$E[E(A|x)^2] = \text{Var}\,[E(A|x)] + E[E(A|x)]^2$$

$$= \text{Var}\,[E(A|X)] + E(A)^2, \tag{13.17}$$

$$\text{Var}\,[E(A|x)] = \text{Var}\,(A) - E[\text{Var}\,(A|x)]. \tag{13.18}$$

Substituting (13.18) and (13.17) into (13.16) leads to

$$E[\text{Var}\,(D|x)] = E[\text{Var}\,(A|x)]E(B)^2 + E[\text{Var}\,(A|x)]\,\text{Var}\,(B) + E(A)^2E[\text{Var}\,(B|x)]$$

$$+ \text{Var}\,(A)E[\text{Var}\,(B|x)] - E[\text{Var}\,(A|x)]E[\text{Var}\,(B|x)]. \tag{13.19}$$

As before, similar equations apply for covariances.

Therefore computation of $E[\text{Var}\,(D|x)]$ for a given quantity D of interest and a given study programme may be done recursively, expressing it ultimately in terms of prior means, variances and covariances (which for derived quantities are themselves obtained recursively from the prior means, variances and covariances of base quantities) and preposterior variances and covariances of base quantities (which are obtained directly from BLE theory).

13.4.3   A simplifying approximation

Although the computations described are sound in principle, in practice they may not be feasible. A large model will typically have thousands of base quantities, and for a high-level derived quantity D the computations will involve a great many operations on enormous variance–covariance matrices. For any given study programme, it could easily require an hour or more of computation on a very powerful computer to calculate $E[\text{Var}\,(D|x)]$. It then becomes completely impractical to conduct any sensible search of the $\binom{N}{n}$ possible study programmes to try to find the best. It is essential to find an approximation that can be computed relatively quickly for a wide range of programmes.

The first step is to suppose that the quantity of interest is expressible as a sum of quantities $D = \sum_k E_i$, where E_i is defined on the ith unit. For example, suppose that a sludge works study in works k obtains an improved estimate of [SW expansion cost]$_k$ only. The quantity of interest might be defined as $D = \sum_k$ [SW growth cost]$_k$, so that $E_k =$ [SW growth cost]$_k$. D is obviously a part of [ST growth cost], and hence a part of the company's total investment cost. It is the part most obviously affected by the studies in question.

Let $\boldsymbol{V}_1$ be the prior variance–covariance matrix of the E_i, and let $\boldsymbol{V}_2$ be their preposterior variance–covariance matrix if studies were conducted in *all* N units. Therefore $\boldsymbol{V}_2$ is the expected result of studying 100% of the units. Let $\boldsymbol{V} = \boldsymbol{V}_1 - \boldsymbol{V}_2$.

Now consider the effect of a programme of n studies. For simplicity, and without loss of generality, suppose that these are in units $1, \ldots, n$, and write

$$V = \begin{pmatrix} V_{11} & V_{12} \\ V'_{12} & V_{22} \end{pmatrix}, \tag{13.20}$$

where V_{11} is $n \times n$. The approximation is to say that the pre-posterior variance–covariance matrix of the E_i after this study programme is given by $V_2 + W$, where

$$W = \begin{pmatrix} 0 & 0 \\ 0 & V_{22} - V_{12}V_{11}^{-1}V_{12} \end{pmatrix}. \tag{13.21}$$

An outline justification of this approximation is given in Section 13.4.4. For further detail see Appendix C of Golbey and O'Hagan (1993).

The preposterior variance of D is, by (13.15), the sum of the N^2 elements of the preposterior variance–covariance matrix of the E_i. This may be written alternatively as

$$V_2 + W = V_1 - \begin{pmatrix} V_{11} \\ V_{12} \end{pmatrix} V_{11}^{-1} (V_{11} \quad V_{12}). \tag{13.22}$$

The best programme will therefore maximise the sum of the N^2 elements of the second term of (13.22), which is $v'_1 V_{11}^{-1} v_1$, where the vector v_1 is obtained by summing each of the n rows of $(V_{11} \quad V_{12})$. For any given programme, this criterion may be very quickly computed once V has been found. Computing V may itself take an hour or more for a substantial model, but the benefit of this approximation is that only one such computation is needed. Thereafter, many programmes may be compared quickly.

13.4.4   Justifying the approximation

The key to the approximation lies in the assumption that the preposterior variance–covariance matrix of $E_1, E_2, \ldots, E_n$ when the programme consists of studying units 1 to n is the same as for a programme of 100% studies. That is, it is assumed that no information about E_i is gained from a study of unit j over and above that gained from a study of unit i.

To see the effect of this, consider the example of sludge works studies that yield improved estimates of [SW expansion cost]$_k$. The knowledge gained from studying [SW expansion cost]$_j$ in works j will clearly be relevant to estimating $E_k = $ [SW growth cost]$_k$ in each works $k \neq j$. This is clear from comparing (13.20) and (13.21), where there is a reduction of $V'_{12}V_{11}^{-1}V_{12}$ in the preposterior variance–covariance matrix in the units that have not been studied. But the essence of the approximation is that the study in works j will *not* improve the estimation of [SW growth cost]$_i$ in works i if i has also been studied. The assumption is reasonable in this case. The study in works j improves the estimate of [SW expansion cost]$_j$ by taking account of local conditions at works j. This improves knowledge about [SW expansion cost]$_k$ at other works, because if the study produces a higher/lower estimate than the prior in works j then it suggests that conditions at other works might typically lead also to higher/lower costs. But if works i has already been studied, the knowledge gained at works j does not affect the estimate at works i because its local conditions have already been accounted for.

This behaviour is fundamental to the model for all base quantities, and lies at the heart of the general theory of Goldstein and O'Hagan (1993). The force of the assumption that underlies the approximation is in carrying this behaviour from the base quantities actually studied to the quantities E_i that make up the quantity of interest D. In the example we must accept also that learning about [SW expansion cost]$_k$ through studies of those unit costs does not provide information about [SW growth cost]$_k$ in any other way. That is, those studies do not provide information about [sludge growth]$_k$. In general, E_i will depend on the base quantities studied during a study of unit i and also on other quantities. Appendix C of Golbey and O'Hagan (1993) gives specific instances of how the approximation may fail in practice through the studies yielding information directly about those other quantities.

This assumption justifies the zeros in (13.21), and it remains for us to justify the specific way in which the studies provide information about the E_i for units that are not studied. It is here that (13.21) really is an approximation. The justification is that it would be exact if the E_i were themselves base quantities and knowledge about them were being updated by using the BLE after obtaining improved estimates directly of $E_1, E_2, \ldots, E_n$. The approximation is to ignore the detailed way in which the E_i depend on the base quantities being studied. Some numerical experiments have suggested that it is a good approximation in practical applications.

13.4.5   Search algorithm

Even with the simplification afforded by this approximation, it is still impractical to compute the effect of every one of the $\binom{N}{n}$ possible programmes in order to find the best, when typically N is in the hundreds and n may be ten or twenty. A stepwise search strategy has been developed based on adding and subtracting units one at a time from the programme. The form of (13.21) makes adding extra studies particularly simple: first replace V by W then apply (13.21) again, and the new W will be the same as would have been obtained by applying (13.21) for the combined programme of studies in one step. Stepwise procedures have been found to be effective in choosing a best subset of regressor variables, which has some similarities to the present problem.

The algorithm begins by finding the best single unit for study. Then it finds the best single unit to add to the first, and continuing in this way finds an initial programme of n studies by stepwise addition. This need not of course be the best such programme. Even after the second unit is added, the resulting programme of two studies need not be the best. However, the algorithm then proceeds iteratively to improve the programme by considering removing one of the current n selections and replacing it by the best unit to add to the remaining $n - 1$. Having considered each of the $n - 1$ candidates for removal (ignoring the latest addition, which need not be considered in this way), the one whose substitution gives the greatest improvement is accepted and a fresh iteration begun. The process ends when no further improvements are possible using this one-for-one substitution method. The procedure is still not guaranteed to find the global optimum, but it is unlikely in practice that the true optimum will be substantially better than the chosen programme.

13.5 ONGOING WORK

The methodology outlined here has proved to be a flexible and effective tool to assist water companies in Asset Management Planning. Companies have substantial prior information about the condition and performance of their asset stock, and about the schemes of work and costs that would be needed to remedy deficiencies. A Bayesian approach that makes use of this information is seen to have distinct advantages over classical methods. As an unexpected bonus, companies find that the discipline of constructing a cost model and marshalling the available prior information about base quantities is valuable, and provides them with insights and an overall perspective of the operation of the business that they might otherwise have missed. The modelling approach is flexible and allows inter-relationships within the system to be represented in a way that would be very difficult to handle in a classical statistical framework.

An important benefit in future will be the ability of the system to continue to adapt to new data after completion of AMP2, thus providing a mechanism for maintaining an up-to-date view of future investment needs as well as monitoring progress against the AMP2 projections. Classical statistics seems to be incapable of such a role, because of its need to gather data by random sampling, which would be an unacceptable way of behaving in practice and would miss a great deal of useful information. Some further discussion of the benefits of the Bayesian approach, from the perspective of the company, are given in O'Hagan and Wells (1993).

Nevertheless the methodology has various limitations that will need further research to overcome. Some are exposed below as a spur for the author to work for their solution in time for AMP3!

(i) *Data on derived quantities.* It is assumed that studies will produce improved estimates only of base quantities. One area of difficulty is that models often include products of amounts of work and unit costs for doing the work. Prior information about these is essentially independent, and the theory requires independence between quantities that are multiplied. However, a study will typically provide improved information about the amount of work and the total cost of the work. An estimate of the unit cost can be obtained as the ratio of these, but it is not now realistic to assume that it is then independent of the improved estimate of the amount of work. It would be more accurate to express the information in its original form, as improved estimates of a base quantity (amount of work) and a derived quantity (amount times unit cost). Goldstein and Wooff (1993) provide some evidence that the BLE may be robust to departures from strict independence (see also O'Hagan, 1993).

(ii) *Full Bayesian modelling.* Use of the BLE to obtain posterior information about base quantities may be a poor approximation in some cases to a fully specified Bayesian analysis, and furthermore a fully specified analysis may be possible. An example is where a cost may be zero if no work is needed on an asset, but otherwise could be substantial. It may be reasonable to think of the cost if work is needed as approximately normal, but normality is a poor approximation when there is a point mass of probability at zero. It should be possible to model such a case more fully and to apply a full Bayesian analysis rather than the BLE.

(iii) *Elicitation.* Eliciting prior information has been found to be difficult for engineers, and research is needed on effective ways to proceed. Unfortunately, the Bayesian community seems to have collectively put its head in the sand and ignored this crucial issue over many years. Little has been published that might be usable in complex practical situations.

(iv) *Study programme design.* It is often unrealistic to try to find the best programme of n studies. Zones and assets vary greatly in size and complexity, and a study of one may cost very much more than a study of another. Of course, studying the largest units proves to be the most informative generally, so optimal programmes of a fixed size tend to be costly. It would be better to assign a cost to each study and to find the optimal programme within a given budget. Also, a company may plan a variety of different kinds of studies to improve their knowledge in various ways. It would be helpful to be able to find an optimal portfolio comprising programmes for each kind of study.

ACKNOWLEDGEMENT

I am grateful to Michael Goldstein for various helpful discussions, notably on the question of study programme design.

14

Measuring and Modelling Pollutant Transport through Soils and Beyond. A Study in Uncertainty

T. M. Addiscott

Rothamsted Experimental Station, UK

14.1 OVERVIEW

Pollutant transport in soils must be measured or modelled with an appreciation of the uncertainties involved. These arise, for both activities, from the small-scale soil heterogeneity that leads to mobile and immobile categories of water in most soils, from larger-scale variability in soil properties and from the degree to which the scale of the measurement or model simulation is appropriate to the scale of the system studied.

Small-scale measurements with porous ceramic cups can show very variable concentrations in the water extracted from soil, particularly under grazed grassland, where nitrate concentrations tend to be distributed log-normally. The true mean of such a distribution can be assessed most reliably with the estimator of Finney (1941) and Sichel (1952). A lysimeter usually provides a more certain measure of solute concentrations in water draining from the soil, but the air–water interface at the base can affect both the amount and the composition of the drainage. Large-scale plots with field drains over impermeable subsoils are a very effective but expensive option.

Unless a model is linear with respect to its parameters, it cannot be used safely without reference to variability. Small-scale heterogeneity can be treated by taking account of mobile and immobile water. Modellers have responded to the problem of the larger-scale variability in four ways: deterministic models with

Statistics for the Environment 2: Water Related Issues. Edited by Vic Barnett and K. Feridun Turkman
© 1994 John Wiley & Sons Ltd.

distributed parameters, stochastic transfer function models, capacity-type models whose parameters vary little, and the pedon approach of Bouma (1993). The distinction between parameter variability and process variability should be noted. Nonlinearity seems to become an increasingly important problem as the scale at which a model is used is increased, and the simple capacity-type models may be as useful as any for catchment-scale use.

14.2   INTRODUCTION

Economic growth in industrialised societies has led to increasing expectations. People expect food to be plentiful and cheap, and water suitable for drinking and cooking to be available in virtually unlimited quantities. At the same time, they both expect more of the environment and place greater strains on it. Conflicts between these various expectations are beginning to emerge (see e.g. House of Lords Select Committee, 1989, p. 7). One expression of such conflict lies in the increasing concern about nitrate and other pollutants in natural waters that are used as supplies of potable water or for recreational purposes. The soil and the vadose zone, defined here as the zone whose upper surface is the greatest depth at which roots extract water and whose lower surface is the water table, form the main buffer between the agricultural and other activities that satisfy one set of expectations and the natural waters that reflect another. Viewed on a broader scale, the soil and the vadose zone are also the buffer between the atmosphere and the hydrosphere. The study of pollutant transport through this buffer is clearly very important for reconciling conflicting public expectations and for managing water resources in the more general sense.

The proper study of transport in the soil and the vadose zone involves both measurement and modelling, which in an ideal world would interact as described in the hypothetico-deductive scheme of Popper (1959) or in the analogy given by Hillel (1991) that perceives experimenters to be like actors hoping for a good play in which to perform and modellers to be like playwrights yearning for good performers to redeem their plots. This chapter attempts to discuss both aspects of the study, not least because some of the problems involved are common to both measurement and modelling. These problems include the uncertainty that arises from the interaction between the heterogeneity and the physics of the soil and the uncertainty as to whether a measurement or a simulation has been made at a scale appropriate to the system studied. As will be seen, these uncertainties interact with each other. 'Uncertainty' can be understood here in terms both of the general usage of the world and of a mathematical expression deriving from likelihood functions. Mention will be made of the generalised likelihood uncertainty estimation (GLUE) procedure of Beven and Binley (1992), although whether the procedure can be extended from the study of water flows to that of pollutants carried by water remains to be seen.

14.3   MEASUREMENTS OF POLLUTANT TRANSPORT

Measuring the concentrations of pollutants in water is usually a matter of routine chemical analysis; the problem lies in collecting the appropriate water. The water in which we are interested is that lost from the soil initially to the vadose zone and subsequently to streams, rivers or aquifers. The pollutants that the water carries may be ameliorated in the vadose zone; nitrate, for example, may be destroyed by denitrification and certain other pollutants may be sorbed. Processes in this zone, however, are not readily amenable to control, and in practice pollutants probably need to be considered to have 'escaped' when they leave the rooting zone, that is, the zone in which the withdrawal of water by roots prevents further downwards flow. This is not an unequivocal definition, because different crops root to different depths, and depth of rooting changes during the year, but the depth of measurement needed is likely to be 1–2 m for arable crops and grassland but larger for forests, which lie beyond the scope of this study.

Most public concerns about water pollutants centre on their *concentrations*, as does EC legislation, so measurements of concentration are very important, but the concentration is not the only information needed. We need to know too the total *quantity* of pollutant leaving the rooting zone of a given area of soil and the volume of water in which it is carried. Water and nitrate from one area will eventually mingle with water and nitrate from other areas, and the resulting concentration can be estimated only if we know how much water and how much nitrate has come from each area. This point becomes particularly important when catchments that include a variety of land uses are investigated.

The remainder of this section describes the various methods by which solute concentrations in water leaving the soil are measured and discusses the uncertainties associated with each.

14.3.1   Porous ceramic cups

One approach to sampling water likely to drain from the soil involves withdrawing it under suction, using an assembly comprising a porous ceramic cup attached to a length of plastic or stainless steel tube of similar diameter (stainless steel only if the solute of interest is likely to be sorbed by the plastic) (Figure 1). Two fine tubes pass through a cap on the top of the assembly; one is used to create the partial vacuum that draws soil water into the porous cup, while the other reaches into the base of the cup and is used to bring out to a collector on the surface the water that has accumulated in the cup. Porous cup assemblies are very simple in principle, but need to be installed and used with considerable care. In particular, the installer must ensure good contact between the porous cup and the surounding soil and prevent any channelling of water from the soil surface down around the tube to the region of the cup (Webster *et al.*, 1993).

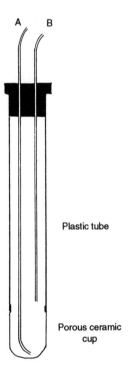

Figure 14.1   A porous ceramic cup assembly. Air is withdrawn through tube B to create a partial vacuum that draws soil solution into the porous cup. This solution is drawn out through tube A. Diagrams in this chapter are from Addiscott *et al.* (1991).

The main problem for users of porous cups tends to be the variability in the measured concentrations (see e.g. Wagner, 1962). Some of this variability is inherent in the cups themselves, some arises from the way they are installed, and the rest reflects the variability in the soil itself (Hansen and Harris, 1975). Within the soil there are different scales of variability, which depend both on the intrinsic nature of the soil and the way in which it is managed. The intrinsic variability arises from the structural heterogeneity of the soil, that is, from the existence of aggregates of greatly differing sizes and pores that differ to a similar extent, with the result that in many soils there exist, in effect, mobile and immobile categories of water. These categories were inferred first by Lawes *et al.* (1882) from observations of the concentrations of nitrate and other solutes draining from the soil at Rothamsted. The extrinsic, management-related variability arises from various factors, of which the most influential is probably the decision between arable and grassland in land use. Webster *et al.* (1993) found relatively modest variability in nitrate concentrations in water drawn from soil under arable cropping, but the concentrations found by Cuttle *et al.* (1992) in soil under grass grazed by sheep ranged over four orders of magnitude and were distributed log-normally. This extreme variability resulted from the non-uniform deposition of urine and faeces by the grazing animals. Although concentrations of nitrate-N exceeded $100 \, \text{mg} \, l^{-1}$ at some samplers on some occasions, at no time did the estimate of the mean concentration exceed the EC limit of $11.3 \, \text{mg} \, l^{-1}$ for potable water. Tillage is another factor likely to affect variability; it should lessen it in the plough layer, particularly in soils likely to crack.

The main uncertainty attached to the use of porous cups was defined nearly 20 years ago by England (1974), who commented that 'in the present state of the art one cannot be sure from what macroscopic volume of soil the sample was extracted nor from which pores it was drained'. A further source of uncertainty, indentified by van der Ploeg and Beese (1977), lies in the possibility that the suction applied to the porous cups could distort flow patterns in the soil. Thus measuring the flux of solute might change what was being measured—a variant of Heisenberg's uncertainty principle (which originally applied to electrons). Hansen and Harris (1975), for example, found that the nitrate concentration they measured seemed to change with the initial rate of water uptake.

Problems of this nature, which reflect the structural heterogeneity of the soil, should differ in importance between soils of various types. Sandy soils have least structural heterogeneity and ought to give the best results with this technique, and as the proportions of silt and clay increase and soils become heavier, with geater structural heterogeneity, the technique would be expected to become less useful. This is what was found in field experiments made by Barbee and Brown (1986), who applied choride to fallow soil and allowed it to leach under natural rainfall. Samples of the water flowing from the soil were taken weekly using either porous ceramic cups or simple lysimeters that comprised pan collectors inserted horizontally into the soil profile from trenches. In a sandy soil there were no significant differences between the chloride concentrations measured by the two sampling approaches, but there were marked differences in a silty soil, and in a clay soil the porous cups produced a large enough sample of water for analysis on only one occasion and seemd to have been by-passed by any water flow on all other occasions. Other authors (e.g. Shaffer *et al.*, 1979; Russell and Ewel, 1985) have also found evidence that porous cups can be by-passed by the water flow in the soil, particularly during heavy rainfall.

Although porous cups can give disappointing results when used inappropriately, the satisfactory agreement between this approach and a simple lysimeter technique in a sandy soil found by Barbee and Brown (1986) is supported by measurements made by Webster *et al.* (1993). These authors reported good agreement between solute concentrations measured in a sandy soil using porous cups and those obtained in simultaneous studies using a monolith lysimeter (defined below) after an initial year in which results were affected by soil disturbance. However, what porous ceramic cups do not measure is the flux of solute from the soil. Estimating this necessitates the estimation of the flux of water from the soil—another very variable quantity. Measuring water fluxes in the soil is difficult, and the simplest approach is to estimate the water flux from measurements of rain, evaporation and water storage. Some degree of uncertainty attaches to the resulting estimate of water flux, because no account is taken of the spatial variability of the water flux and because the evaporative loss from the soil and crop cannot be measured directly, but it is unlikely that the solute fluxes estimated this way err very seriously, provided that an appropriate statistical estimator has been used for the distribution of solute concentrations.

14.3.2   Statistical estimators for distributed solute concentrations

Porous ceramic cups are usually replicated in field experiments, and therefore yield a distribution of values, which may be log-normal, as in the data of Cuttle *et al.* (1992). This means that an appropriate statistical estimator is needed to represent the distribution, but, before choosing one, it is necessary to define carefully what information is wanted from the distribution, having in mind that we are concerned with both the concentration of solute leaving the soil and the corresponding flux.

If the concentrations are normally distributed, there is no problem; the arithmetic mean is an appropriate estimator for the concentration, and if a unique value for water flux is assumed then it can be multiplied safely by the concentration to give the solute flux. Where, however, concentrations are log-normally distributed, as found by Cuttle *et al.* (1992), more care is needed. If we want the best estimator of the concentration *per se*, one that is not unduly influenced by a few very large values, the mean of the log distribution is probably the most reliable; this is equivalent to using the geometric mean. If, however, we are concerned with the overall flux of solute, we need to give the few large concentrations their full weight because they will contribute substantially to the flux. This suggests that the arithmetic mean of the concentrations is needed, but this is not an efficient estimator for skewed distributions because of the large error attached to the estimate. Is it perhaps better to obtain an estimate of the arithmetic mean by transformaing back from the log-normal distribution, bearing in mind that this is not necessarily a straightforward procedure? There seem to be two main approaches, discussed by Parkin *et al.* (1988), one of which is much simpler to implement than the other. The choice between them is influenced by the number, variance and skewness of the data as well as by convenience.

The first method, which Parkin *et al.* (1988) describe as the 'maximum-likelihood method', depends on equations given by Aitchison and Brown (1957) for estimating the mean μ and variance σ^2 of a population represented by a log-normal distribution using the sample mean m and variance s^2:

$$\mu = \exp{(m + \tfrac{1}{2}s^2)}, \tag{14.1}$$

$$\sigma^2 = \mu^2[\exp{(s^2)} - 1]. \tag{14.2}$$

The second method involves 'uniformly minimum-variance unbiased estimators (UMVUE)', developed independently by Finney (1941) and Sichel (1952). It is

Table 14.1   The distributions examined by Parkin *et al.* (1988) (as abbreviated form of their Table 14.2); each population has 10 000 values.

Population	Mean	Median	Mode	Variance	Skew
Slight skew	10.0	8.944	7.155	25.0	1.625
Moderate skew	10.0	7.071	3.536	100.0	4.000
Marked skew	10.0	4.472	0.894	400.0	14.000

more difficult to use than the previous method because it involves a power series with which the population mean and variance are estimated by

$$\mu = \exp(m)\psi(\tfrac{1}{2}s^2), \tag{14.3}$$

$$\sigma^2 = \exp(2m)\left\{\psi(2s^2) - \psi\left[\frac{(n-2)}{(n-1)}s^2\right]\right\}, \tag{14.4}$$

where the power series ψ is given by

$$\psi(t) = 1 + \frac{t(n-1)}{n} + \frac{t^2(n-1)^3}{n^2(n+1)2!} + \frac{t^3(n-1)^5}{n^3(n+1)(n+3)3!}$$

$$+ \frac{t^4(n-1)^7}{n^4(n+1)(n+3)(n+5)4!} + \cdots, \tag{14.5}$$

where n is the sample size. Parkin *et al.* (1988) found in their evaluation of this method that, for the final term to be less than 1% of the sum of the preceding terms, six to ten of the terms in the series had to be calculated, but this is not a problem with modern computers. They wished to determine how much more efficient were the estimators given in (14.1)–(14.4) than the straightforward computation of the mean and variance of the untransformed data. To do so, they generated three truly log-normal populations, each of 10 000 values but with differing skews (Table 14.1) and sampled from each of them using sample sizes ranging from 4 to 100. They evaluated the estimators by comparing the estimated and true values of the mean, variance and coefficient of variation (CV) in terms of the mean square errors (Barnett, 1973); this was done for each population for a range of sample sizes. The results for the mean are summarised in Table 14.2, which shows the degree of preference for each estimator.

For estimating the mean of the slightly skewed population (Table 14.2), the estimator of Finney (1941) and Sichel (1952) gave results that were virtually identical to those from the standard untransformed estimator, so the latter is preferred because it is much simpler. The estimator of Aitchison and Brown (1957) gave a larger mean square error than either, and was not therefore acceptable, for sample sizes less than 40. For the moderately and markedly skewed populations, the Finney–Sichel estimator was preferred for all ranges of

Table 14.2 Preferability of the estimators for assessing the mean

		Preferability		
Population	Sample size	Standard untransformed	Aitchison–Brown	Finney–Sichel
Slight skew	4–20	***		**
	20–40	***		**
	40–100	***	*	**
Moderate skew	4–20	*		***
	20–40	*		***
	40–100	**	*	***
Marked skew	4–20			***
	20–40		**	***
	40–100		**	***

*** most appropriate.

sample size, but the Aitchison–Brown estimator was quite acceptable for the larger samples from the markedly skewed population. The standard untransformed estimator was not appropriate to the latter population.

For estimating the variance, the Finney–Sichel estimator was preferred for all the combinations of skewness and sample size shown in Table 14.2 except when samples of sizes 4–20 were taken from the slightly skewed population, when the standard untransformed estimator was preferred for its simplicity. The Aitchison–Brown estimator was acceptable for sample sizes greater than 40 from the slightly skewed population. The Finney–Sichel estimator was also preferred for estimating the CV for all the combinations of skew and sample size, but the Aitchison–Brown estimator was acceptable when the sample size exceeded 40, or when it exceeded 20 and the skew was slight.

The study by Parkin *et al.* (1988) summarised above is very interesting to soil scientists because a number of soil properties in addition to nitrate concentrations tend to show log-normal distributions. The study, however, enjoyed the twin 'luxuries' of a truly log-normal distribution and a very large population, and it is interesting in the present context to compare its outcome with the estimates of the mean nitrate concentration in the soil solution obtained from the three estimators in the study by Cuttle *et al.* (1992). These authors measured nitrate concentrations during three years in which sheep grazed either a grass–clover ley or grass given N fertilizer. Each year/system combination included 60 observations which were markedly skewed, so Table 14.2 suggests that the Finney–Sichel estimator would have been the most effective, that the Aitchison–Brown one would have also been satisfactory, but that the standard untransformed estimator would not. Table 14.3, taken from the results of Cuttle *et al.* (1992), shows that the estimates given by the first two estimators were generally similar to each other, and that they were larger than that given by the standard untransformed estimator and

Table 14.3 Estimates of mean loss of nitrate-N (kg ha^{-1}) based on measurements with porous ceramic cups by Cuttle *et al.* (1992) (based on their Table 6); 60 observations

Year/system	Standard untransformed	Aitchison–Brown	Finney–Sichel	Geometric-mean
1/Grass–clover,	17	38	33	2.5
N-fertilized grass	3	8	7	0.7
2/Grass–clover,	9	8	8	1.5
N-fertilized grass	2	2	2	0.6
3/Grass–clover,	10	6	6	1.0
N-fertilized grass	23	27	25	2.9

very much larger than the geometric mean. White *et al.* (1987), who studied nitrate concentrations in soil samples taken from grassland grazed by cattle, found log-normally distributed populations of values similar to those observed by Cuttle *et al.* (1992) in samples from porous cups. These authors concluded that the Aitchison–Brown estimator was satisfactory for sample sizes greater than 50 and when the variance of the log-normal distribution was less than 0.75 μg N cm^{-3} but that the Finney–Sichel estimator was more efficient when the samples were fewer or the variance larger. This again accords with Table 14.2.

White *et al.* (1987) were also concerned that the skewness of the population of concentrations that they measured might be so great that not even the log-normal distribution described its properties satisfactorily. They therefore compared the fit of their data to this distribution with its fit to the gamma distribution, but χ^2 tests showed that the log-normal distribution gave the better fit in the majority of cases. These authors also found, as might be expected from the origin of the variability, that the skewness of the population increased with the stocking density (the number of animals per hectare). The skewness was least where the animals had been excluded from the grassland for 14 months, but the population of nitrate concentrations was still described best by a log-normal distribution.

Variability in nitrate concentrations caused by excretion has most impact in grazed grassland, but this is because of its scale; a cow can deposit 2 l of urine on about 0.5 m^2 of grassland, giving a localised application of possibly 500 kg ha^{-1} of N. Other organisms excrete nitrate and cause intense variability in its concentration in the soil, but these are bacteria and other micro-organisms, and the scale of our measurements in the soil do not detect this variability. This is one reason for the relatively smaller variability in nitrate concentrations found by Webster *et al.* (1993) in arable land, but tillage must also have contributed.

14.3.3 Lysimeters

A lysimeter comprises a block of soil held in position either by a retaining vessel or by the soil around it. Its upper surface is subjected to rain and evaporation, usually naturally occurring, and may grow plants. Water passing through the

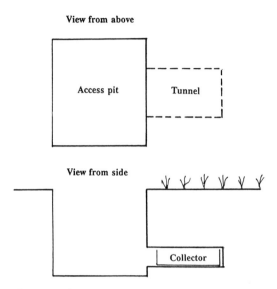

Figure 14.2   An Ebermeyer lysimeter.

soil is collected at the base and sampled for analysis. Some lysimeters can be weighed, usually to keep a check on the quantity of water held in them. There are two basic forms of lysimeter. One involves digging a trench or pit for access and inserting a collecting vessel horizontally beneath the block of soil of interest (Figure 14.2). The other is made by isolating the block of soil by encasing it in a suitable container, often a piece of fibreglass pipe, severing the block at the base, and attaching a suitable collector; it is then installed either *in situ* or in a bank of lysimeters elsewhere. The procedure is illustrated in Figure 14.3.

The first type of lysimeter (Figure 14.2) is often known as an Ebermeyer lysimeter, in honour of a Bavarian meteorologist who, towards the end of the nineteenth century, planned to undermine 0.25 ha of forest, complete with trees, to a depth of 1.5 m and collect drainage from it. He was, unfortunately, overtaken by technical and financial problems. The second type of lysimeter is usually known, more prosaically, as a monolith lysimeter. Both types have uncertainties associated with them.

One key question is whether the making of the lysimeter has altered the flow of water and solutes within the block of soil, which could happen either within the main body of the lysimeter or at the base. Problems identified by an experienced collector of monoliths (Belford, 1979) included the compression of the soil when it is encased—both vertically, as the casing is driven into the soil, and laterally next to the casing wall—and the collapse of the structure of unstable soils. Another possible problem, which could occur when clay soils are encased, is the shrinkage of the clay away from the wall of the casing during dry weather. This could leave preferential pathways for water at the wall that were not characteristic of the

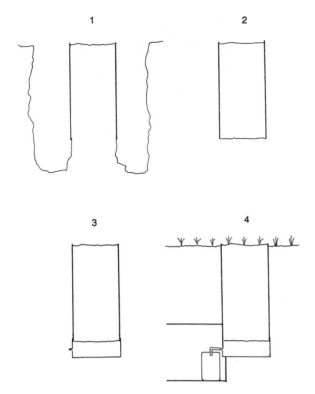

Figure 14.3 Stages in the preparation of the monolith lysimeter. (1) A metal or fibreglass pipe is driven into the soil and the soil cut away. (2) The soil is severed as cleanly as possible at the base. (3) A collector for water is attached. (4) The lysimeter is installed.

undisturbed soil. The absence of walls in an Ebermayer lysimeter avoids these problems but introduces an element of uncertainty about the origin of water draining from it. The water may have been deposited on the soil above the collector and percolated vertically downwards, but it could have been carried to the collector in part by lateral flow after deposition elsewhere. There is no way of knowing.

One potentially important problem is common to both types of lysimeters; this is the inevitable air–water interface that occurs at the base of the soil above the collector. The pores that conduct water through the soil are usually fairly continuous, so there is a 'hanging column' of water that extends to a reasonable depth without a horizontal interface between water and air. At such an interface the surface tension holds back water from draining from the soil until the soil becomes saturated, even in a freely draining soil (Richards *et al.*, 1939). This problem can be avoided by applying suction to the base of the soil, and Coleman (1946) showed that such suction controlled both the rate at which water drained from the soil and the amount of water held in the base of the soil. Pores of different sizes drain at different suctions, so the amount of suction applied determines which particular suite of pores is releasing water. This could influence the quantity and concentration of solute leaving the soil, as was shown by Haines *et al.* (1982). They compared two Ebermayer lysimeters, each with the same depth of soil; one had the collector subjected to a 1 m hanging column of water while the other had no suction. The suction doubled the average flow of water but decreased by a factor of three its nitrate concentration. The key point here is that the concentration *changed* to such an extent; the fact that it decreased probably just reflected the distribution of the organic material that was mineralised to produce the nitrate. It is clear that the suction applied needs to be as close as possible to the suction that would arise from the hanging column of water if it was there. Morton *et al.* (1988) adjusted the suction they applied to lysimeters by referring to tensiometers inserted in nearby soil to the depth of the lysimeter base.

14.3.4   Application of statistics in lysimeter studies

A porous ceramic cup has a diameter of the order of 2–4 cm, but draws water from a rather larger but unknown diameter. A monolith lysimeter typically has a diameter of 70–80 cm. The differing scales of measurement are offset to some extent by replication, porous cups practically almost always being used in larger numbers than lysimeters. In the experiments of Webster *et al.* (1993) the standard errors of measurement reported did not differ greatly between the two techniques, probably mainly because of the greater replication of the porous cups.

There is, however, one question that arises with lysimeters but not with porous cups, and that is whether the scale of the measurement is likely to be large enough to encompass the variability that is likely to occur. Here we have to consider the spatial structure of the variability. Webster and Burgess (1980) concluded that the spatial analysis of soil data was likely to show large nugget variances. Several studies have shown nitrate concentrations to be characterised by large nugget variance, with much of the variability occurring within a small range; see e.g. White *et al.* (1987) for grazed grassland and Whitmore *et al.* (1983) for fallow arable land, although the latter authors noted that the range increased between autumn and spring. The autumn measurements of Whitmore *et al.* (1983) suggested that 95% of the variability in soil nitrate concentrations occurred within $4 m^2$, so a fair proportion would have been encompassed by a monolith lysimeter of 80 cm diameter ($0.5 m^2$), particularly as much of the variance occurred at the first lag, 0.5 m. Unfortunately, their spring measurements showed that $64 m^2$ was needed to accommodate 95% of the variability in nitrate concentrations. Because we are concerned with nitrate losses in water, we also need to consider the rates of flow of water at the site. These were very variable, and showed a large nugget variance, but fitting a satisfactory variogram was not easy (Webster and Addiscott, 1990).

14.3.5   Large-scale drainage experiments

One form of experimentation takes advantage of heavy, impervious clay subsoils, which need artificial drainage to make them useful for farming. An example of this approach is the Brimstone experiment (Cannell *et al.*, 1984). Here very little water seeps through the subsoil, and the largest proportion of the water leaving the soil does so through the field drains, a combination of mole and pipe drains. Some water also leaves the soil by surface run-off and by interflow, that is, lateral flow at the base of the plough-layer. All these flows can be sampled and analysed. The plots in this experiment are isolated laterally by barriers of polythene sheet, and are large enough to permit normal agricultural operations. They also have the advantage that the flows of water in them are similar to the flows in soil drained by the same standard approach elsewhere. They also present few statistical problems, because they are large enough to encompass at least 95% of the variability in both solute concentrations and rates of water flow. The main uncertainty is probably whether the would-be experimenter can meet the cost of such an experiment, of the order of £250 000.

14.4   MODELLING POLLUTANT TRANSPORT

Most forms of uncertainty in modelling can be traced back to the problem of non-linearity. One aspect can be seen by considering a function $f(x, y)$ of x and y. The way in which models are sometimes used seems to presuppose that

$$\mu_{f(x,y)} = f(\mu_x \mu_y) \tag{14.6}$$

Equation (14.6) holds, however, only if $f(x, y)$ is linear in x and y or if the variances of x and y are zero.

A substantial proportion of the models for solute transport in soils combine the Richards equation, which describes water flow, with the convection–dispersion equation, which relates the flow of solute to that of water (Wagenet, 1983). These models are deterministic and non-linear, and their parameters usually show large variability and often a substantial degree of skewness. The resulting problem has been recognised and widely discussed (see e.g. Addiscott and Wagenet, 1985), and various responses to it have evolved. We need to consider the problem and the responses at two scales: within a small volume of soil and at field level. At the smaller scale we are in effect considering one aspect of the heterogeneity of the soil: the soil consists of aggregates and structural units that contain very fine pores that do not contribute appreciably to the downward flow of water and between which there are larger pores in which such flow does occur. At this scale we are not usually concerned with parameter variability, because the scale is too small for it to be measured. This scale is of the same order of magnitude as the problems discussed earlier concerning the precise location of a porous ceramic cup within a given soil volume. At field scale we are concerned with the spatial variability of the soil properties, such as hydraulic conductivity, that are used as parameters in transport models, although this variability may occur over relatively short ranges (Webster and Addiscott, 1990). This scale corresponds approximately to that at which differences between porous cups are discussed.

14.4.1   Responding to the problem of heterogeneity

The main response to the problem at the small scale has been the evolution of transport models that assume the soil to contain both mobile and immobile water. Some are more mechanistic than others; that is, they seek to define the system at a more fundamental level. The model of van Genuchten and Wierenga (1976) provides an example. This model was developed and applied initially at the laboratory column scale, but Barraclough (1989) applied a similar model at the field scale. Non-mechanistic mobile/immobile models appropriate to the field plot scale have been developed at Rothamsted (Addiscott, 1977; Addiscott and Whitmore, 1991).

14.4.2   Responding to the problem of field-scale variability

Modellers have evolved a variety of responses to the variability problem, of which four are described here.

The most fundamental approach is to use a mechanistic model with randomly distributed parameter values to produce a distribution of output values. The parameter values are normally drawn by a Monte Carlo technique from the corresponding measured distributions. This approach has been applied both to saturated soil with steady-state flow and to unsaturated soil. Steady-state flow makes both parameter estimation and the modelling procedure easier to manage, but the phenomenon is usually found only in experiments designed to accommodate analytical solutions to models (see e.g. Nielsen *et al.*, 1973). Models appropriate to unsaturated soil are complex and they also depend on the relationship between the hydraulic conductivity, the volumetric moisture content and the hydraulic potential. This relationship can be defined fairly satisfactorily for soils in laboratory columns, but defining it for a soil in the field is not so easy. Defining it in such a way as to encompass the field-scale variability of the relationship is very difficult (Bresler *et al.*, 1979; Wagenet and Rao, 1983). It also leads to the uncertainty identified by Wagenet and Addiscott (1987). These authors assessed the different mathematical methods used to estimate the distribution of values of the hydraulic conductivity K from measurements of changes in the volumetric moisture content θ in the soil. They found that all the first four moments of the distribution of K differed considerably according to the mathematical method used to extract them from the field data. Clearly, this approach, though satisfying from a theoretical standpoint, is fraught with practical difficulties and carries, despite its comprehensiveness, a fair amount of residual uncertainty. It is very much a research tool.

An alternative to seeking to accommodate stochastic variation in a deterministic model is to assume that the variability rather than the mechanism dominates the process. It is therefore irrelevant to use other than the simplest model for the process, and what becomes necessary is a good description of the variability. This concept led Jury (1982) to develop the stochastic transfer function model, which considers the distribution of solute travel times from the surface of the soil to some reference depth. The model takes the form of a distribution function

$$P_L(I) = \int_0^1 f_L(I)\, dI,$$

in which the probability density function $f_L(I)$ summarises the probability P_L that a solute applied at the soil surface will arrive at depth L as the quantity of water added at the surface increases from I to $I + dI$. The model regards the soil as a bundle of twisted capillaries of differing lengths through which water moves by piston flow. Biggar and Nielsen (1976) and van de Pol *et al.* (1977) found $f_L(I)$ to be log-normally distributed. This is an interesting approach to the problem, but the main difficulty involved is that the model requires calibration. Calibration at one depth enables predictions to be made for other depths or other applications of water. How well it works in soils that change with depth remains to be seen. The model was initially developed for solutes applied at the soil surface, but White (1989) has applied it to solutes, such as nitrate, that are produced within the soil.

Another simpler approach takes advantage of the fact that the volumetric moisture content of the soil is very much less variable than the hydraulic conductivity. Models with capacity parameters, that is, parameters that reflect the capacity to hold water, largely avoid the variability problem and have been shown to give satisfactory simulations of solute movement in field plots (see e.g. Burns, 1974, 1975; Addiscott, 1977, Addiscott *et al.* 1978). One model, the SLIM model of Addiscott and Whitmore (1991), has both a capacity parameter and a simplified rate parameter. The latter is the more variable (in terms of the CV), and is also the parameter to which the model shows the greater sensitivity (Addiscott and Bland, 1988).

The last of the four responses to the variability problem differs greatly from the rest. The 'pedon' approach of Bouma (1993) seeks to characterise identifiable soil units whose behaviour in processes such as leaching is assessed on the basis of experimental experience rather than on the basis of measuring specific parameters at particular sites.

14.4.3   Parameter variability and process variability

The previous section was concerned with the impact of parameter variability on the proper functioning of models, a topic that has attracted great interest from soil physicists in particular. Hillel (1991), however, has correctly pointed out that parameter variability and process variability, though clearly overlapping, are not necessarily the same entity. The properties measured as parameters, for example, may not be truly representative of the process, and problems can certainly arise when parameters have to be derived mathematically from measurements of other properties that are quite clearly not directly representative of the process. Wagenet and Addiscott (1987) showed the consequences of deriving the distribution of a flow parameter from changes in distributed values of a property that is a measure of capacity.

Further problems could obviously arise from failing to consider the variability of all the parameters of a model. It was suggested (Addiscott, 1993) that models may suppress or exaggerate the variability in their parameters and that taking account of the variability of one parameter may amend this suppression or exaggeration. Wagenet and Rao (1983) simulated spatially variable nitrate concentrations. They found that the variability in the flow parameters did not produce as much variability in the nitrate concentrations when crop uptake was included as would have been expected from a simulation for bare soil. If, as seems likely, the model presumed the crop to take up nitrate in proportion to its concentration, it would have shown the resulting variable crop uptake to have lessened the variability in nitrate concentration. It seems clear that the practice of sensitivity analysis needs to be extended to assessing the sensitivity of the model to changes in the variances of its parameters as well as the means. This is discussed further elsewhere (Addiscott, 1993).

14.4.4   Parameter variability with spatial structure

Many soil properties are not only randomly variable but also spatially correlated. These include several used as parameters in models. Obtaining an unbiased estimate of such a property involves *kriging*, a process of weighted averaging in which the weights are chosen to avoid bias in the estimate and to minimise the estimation variance (see e.g. Webster and Oliver, 1990). Parameter estimation for models should therefore ideally involve kriging, but there are two problems.

One is that the theory underlying kriging was developed for properties such as concentrations of minerals in rocks that are additive in nature. Not all soil properties used as parameters are additive. Hydraulic properties, in particular, are both non-additive and skewed.

Another question arises by analogy with (14.6). Is the kriged estimate of $\mu_{f(x,y)}$ the same as the estimate of $f(x, y)$ obtained by inserting the kriged estimates of x and y into $f(x, y)$? The analogy suggests that this will only be so if $f(x, y)$ is linear with respect to x and y, and two studies lend a certain amount of support to this conclusion. Both asked more or less the same question, but of rather different models. If a model has parameters that show spatial structure and there is a need to produce spatially averaged output that can be used for interpolative purposes, there are two possibilities:

(i) run the model using the measured but uninterpolated values of the parameters, determine the variogram of the output from it and interpolate from it;
(ii) determine the variogram of each parameter and produce interpolated values that are used in the model to give 'pre-interpolated' output.

The question is, do these procedures give the same outcome?

De Jong *et al.* (1992) asked this question with respect to the versatile soil water budget model of Baier *et al.* (1979). They applied a spatial averaging procedure (the Thiessen polygon weighting technique) to either the weather inputs to the model or to the output from the model. They found no difference between procedures (i) and (ii) above for temperature-related outputs, and only very slight differences, not likely to be of practical importance, for outputs related to moisture. It seems that the model is concerned with gains and losses of water, and is therefore effectively linear with respect to the moisture inputs. Presumably the same is true of the temperature inputs.

Addiscott and Bailey (1990) asked the same question with respect to the SLIM leaching model (Addiscott and Whitmore, 1991). The spatial averaging procedure used was that in the commercial UNIMAP package, which does not use kriging, and a follow-up study is in progress using a kriging procedure. This study showed major differences between procedures (i) and (ii). The SLIM model has an element of nonlinearity with respect to one of its parameters, evidenced by the observation that excluding the variance of this parameter in simulations of field data on solute leaching made the fit between simulation and measurements poorer (Addiscott and Bland, 1988). This nonlinearity is not very marked, but it was sufficient to cause highly significant ($p < 0.001$) differences between the simulations obtained by the two procedures.

The results of these studies suggest that the spatial averaging of $f(x, y)$ does not give the same result as inserting spatially averaged values of x and y in the function unless the function is linear with respect to x and y. This conclusion raises an uncertainty in large-scale simulations in which nonlinear models are used, particularly if the way in which the model is used carries the implicit assumption that using spatially averaged parameters in a model is equivalent to applying the spatial averaging to the model output.

14.4.5 The problem of scale

Most models for soil processes carry an inherent assumption, which may or may not be made explicit, about the scale at which the model is to be used. This can often be inferred from the scale at which the model is developed and validated. We need to ask—but do not always do so—whether the model can properly be applied at other scales. Is, for example, a model that is developed and validated with respect to one kilogram of soil in a laboratory column appropriate to the many millions of kilograms of soil in a farmer's field or on a moorland hillslope? This question was raised recently by Beven (1989) in a paper concerned with catchment-scale modelling of water flows. He asked whether the small-scale physics of homogeneous systems could be applied to large-scale grids in catchments. Could any credence be attached to an 'effective' parameter value for a 'physically based' model (based on the Richards equation) used in a grid of $250\,m \times 250\,m$, that is, $62\,500\,m^2$? The use of an 'effective' parameter value implies a degree of uniformity that everybody knows does not exist. Beven's conclusion was that a 'physically based' model used in this way became in effect a 'lumped-parameter' model, by which he probably meant a simplified model whose parameters cannot be defined in mechanistic terms. Beven also discussed whether the 'physically based' models could be redeemed by increasing the scale of averaging to decrease the variance of the model's parameters. He concluded that the necessary spatial averaging carried inherent problems, and the discussion in the previous section, taken with the nonlinearity of the Richards equation, strongly suggests that he was right and that the solution does not lie in spatial averaging, whether implicit or explicit.

Beven was modelling water flows at catchment scale, but his conclusions seem fairly closely applicable to modelling pollutant transport, even at somewhat smaller scales. The 'physically based', or mechanistic, models for pollutant transport involve the Richards equation, used in conjunction with the convection–dispersion equation, and are therefore subject to the problem of non-linearity and all that this entails.

14.4.6   Solute transport models for large-scale use

If we wish to make valid use of models at large scales on a grid basis or for areas defined in other ways, certain characteristics of the model and its parameters are desirable, though not necessarily achievable:

(i) the model should be linear with respect to its parameters;
(ii) the parameters should not be excessively variable;
(iii) parameters that are spatially correlated should be additive in nature.

The models that fulfil these conditions most readily are the simple capacity models, such as those of Burns (1974) and Addiscott (1977). Neither of these fairly early models seem to have been tested for linearity. The more recent SLIM model (Addiscott and Whitmore, 1991) has a simplified rate parameter in addition to the capacity parameter, and this introduces an element of non- linearity, as discussed earlier. A straight capacity model may well be the best option for modelling solute movement in sandy and other relatively unstructured soils at large scales, but it may not be suitable for soils with more clearly defined structure, especially those that crack. Such soils ideally need a capacity-type model that takes account of mobile and immobile water. Possibilities include the original model of Addiscott (1977), the TETrans model of Corwin *et al.* (1991) and the SLM model of Hall (1993).

Defining the type of model needed goes only part of the way to determining how to simulate losses of pollutants and other solutes at catchment and other scales greater than that of the individual field. We need to decide too how the variability of the system will be treated in the model. This will depend to a large extent on the level of information available. The simplest option will be to consider parameter variation without reference to spatial correlation, and to treat variation in inputs that are not strictly parameters as if they were parameters. The next, and higher, option will be to consider land use units in some form of geographic information system, perhaps overlaid with soil survey information, but considering the unstructured variability within each unit. The highest option will be to consider the intrinsic spatial variability of the parameters, using the stratification procedure (Stein *et al.*, 1988) to take account of the (extrinsic) land use pattern.

One of the ways in which studies of this kind can be made more useful is to use a disjunctive kriging approach (Webster and Oliver, 1989). This enables the results of the study to be presented as maps showing the probability of, for example, exceeding a given pollutant concentration in water draining from the soil. In this way, the uncertainty associated with simulations of this nature is treated realistically and presented in a manner that facilitates the use of the information for practical purposes.

14.4.7 Generalised likelihood uncertainty estimation

Much of the discussion on parameters in this paper has presumed that the parameters have been derived from measured soil properties. This is certainly not always so; many modellers estimate parameters by fitting procedures. Some even proceed to demonstrate that the resulting parameters enable the model to simulate well the data from which the parameters were obtained! Most, however, restrict the use of fitted parameters to the simulation of other sets of data. This may not be a great problem when not too many parameters have to be optimised, but, as Beven (1989) pointed out, there must be interaction between parameters obtained in this way. He noted, for example, that some 2400 parameter values were needed to implement the SHE model in a study on the Wye catchment, only some of which could be measured in the field or estimated a *priori*. Many others had to be obtained by optimization. Suppose that only 100 of these parameters needed to be optimised, that each was allowed to take 10 trial values and that all combinations of parameter values were tested. In all, 10^{20} combinations would be tested. If 40 trial values were allowed, the number of combinations would be about the same as the number of fundamental particles in the universe. There is therefore a very large probability that at least two combinations of parameter values would give equally satisfactory results in comparisons of simulation and observation, and quite a strong probability that 1000 or more combinations would be equally useful. This being so, identifying the truly representative set of parameter values becomes impossible. Beven and Binley (1992) stated that they wished 'to avoid the idea that there is one set of parameter values that can, with any given model structure, represent a catchment area when it is known that the model structure must be in error'.

Beven and Binley have accepted the uncertainty identified above and sought to frame it in a structure that enables its implications to be pursued. Their generalised likelihood uncertainty estimation (GLUE) procedure works with multiple sets of parameter values and allows that (within the limitations of a given model structure and errors in boundary conditions and field observations) different sets of values may be equally likely as simulators of a catchment. There are procedures for incorporating different types of observations into the calibration process and evaluating new observations. The procedure operates with likelihood distributions of parameter sets, which are updated according to Bayes' equation. It is very demanding of computer power. Beven and Binley (1992) applied GLUE to a model for water flows in a catchment. With three parameters included in the uncertainty estimation, GLUE required very heavy use of parallel processing facilities. If the system was extended to the simulation of pollutant transport in the water flow, two further parameters would probably need to be included for the modelling to remain at the same level of complexity, and the demand for computer power would be prohibitive. Beven is currently using GLUE with a simplified solute transport model (K. J. Beven, personal communication).

14.5 CONCLUSIONS

People prefer certainties. This may be part of the appeal of the more deterministic branches of science, such as physics. The soil is subject to the influence of its parent material, weather, microbes and other soil organisms, plant growth, and in many areas cultivation, irrigation and other human activities. These multiple influences make the soil a heterogeneous medium, which, according to the dictionary, means that it is diverse in character, composed of diverse elements and, in mathematical terms, 'incommensurable because of different kinds'—in short, not a medium to encourage certainties or determinism. 'Incommensurable', though not a widely used word, is appropriate in this context because it reflects the problem of dealing with entities that are not comparable in magnitude, that is, with the problem of scale. Measuring and modelling have both been shown to be influenced by the heterogeneity of the soil as defined above, with parallels between the problems inherent in the two activities. Both have to take account of the structural heterogeneity of the soil that results in the existence of mobile and immobile categories of water in nearly all soils. Both also have to take account of the larger-scale variability to be found in any soil, and therefore to accept the element of uncertainty that results from this variability. Indeed, there is a need for more emphasis to be placed on this uncertainty when measurements or modelling are presented to decision-makers. Probability-based risk analysis seems likely to become more and more important.

The nonlinearity of models has emerged as one of the key topics in this chapter, and is probably the main cause of uncertainty in modelling the flows of water and pollutants. It is a problem that has to be faced if we are to produce models, or select existing ones, that can be used realistically at large scales. The nonlinearity discussed here is that which occurs because the second partial derivative of $f(x, y)$ with respect to x and/or y is not zero (here $f(x, y)$ represents any function or model). Another form of nonlinearity, that resulting from a feedback loop in the model, can lead to chaotic behaviour. No example of this has been found in transport modelling yet, probably because the scale of current studies is not large enough to accommodate such behaviour. Models of soil moisture at continental scale can incorporate a feedback loop in which moisture evaporated from the soil is returned as precipitation, and a model of this kind has been shown to exhibit chaotic behaviour (Rodriguez–Iturbe *et al.*, 1991).

Although people prefer certainties, they cannot always enjoy them, and they do not necessarily make progress by sticking to them. This is as true of science as it is of most aspects of life, and it seems particularly true of the study of pollutant transport in soils. As Sir Karl Popper observed, we have to live with a certain uncertainty.

ACKNOWLEDGEMENTS

This chapter has benefitted considerably from discussions with a number of other researchers. I am particularly grateful to A. J. Gold, R. J. Wagenet and R. Webster.

15

Mathematical Modelling Techniques to Evaluate the Impact of Phosphorus Reduction on Phosphate Loads to Lough Neagh

R. V. Smith and R. H. Foy
Aquatic Sciences Research Division

S. D. Lennox
Biometrics Division
Department of Agriculture for Northern Ireland

15.1 INTRODUCTION

Lough Neagh is a large shallow lake covering $387\,km^2$ with a mean depth of 8.9 m. In addition to supporting a valuable commercial fishery, the lake is the largest single source of potable water in Northern Ireland, but has become highly eutrophic, producing large crops of blue-green algae (Gibson and Fitzsimons, 1982). On occasions, water treatment works have been unable to maintain their optimum output from the Lough because of blocking of sand filters by algae. Parr and Smith (1976) had earlier shown that phosphorus (P) was the only critical nutrient for algal growth in Lough Neagh that was amenable to a reduction in loading. In 1981, drawing on these and earlier findings (Smith, 1977, 1979), the Department of Environment (NI) installed phosphorus reduction at 10 of the largest sewage treatment works (STWs) in the Lough Neagh catchment area.

Statistics for the Environment 2: Water Related Issues. Edited by Vic Barnett and K. Feridun Turkman
© 1994 John Wiley & Sons Ltd.

Based on data collected prior to 1980, a budget of soluble reactive P (SRP) inputs to Lough Neagh estimated that these STWs contributed about 40% of the annual SRP input of 256 tonnes (Foy *et al.*, 1982). Phosphorus reduction was achieved by the use of ferrous aluminium sulphate, ferrous sulphate or alum dosing. In this chapter we analyse the impact of P reduction at STWs on river SRP loadings rather than on total phosphorus (TP) loadings, because the SRP fraction is readily available for algal growth (Jordan and Dinsmore, 1985) and SRP loadings have been shown to have a strong correlation with algal levels in the Lough (Smith, 1986).

The present study was prompted by the problems noted by Gibson (1986) in evaluating the impact of phosphorus reduction at STWs on the overall phosphorus loading to Lough Neagh in the context of the substantial year-on-year variability of phosphorus loading associated with flow variation, and of the possibility that P inputs from other sources within the Lough Neagh catchment have been increasing. In order to identify underlying trends in river SRP loadings, we have compared two methods of removing the influence of flow on annual SRP loads. The first employs regression analyses of river loads with flow and time variables. In the second method river flows were simulated on a daily basis, to give synthesised annual hydrological cycles. These, in turn, were used to generate river SRP loads from SRP load versus flow equations derived from the results of river samples taken over three distinct time periods: 1974–76, 1981–83 and 1988–90. The two methods have employed SRP concentrations and flows from two of the principal rivers in the Lough Neagh catchment: the River Main and the Ballinderry River. The SRP inputs to the former river from STWs were substantially reduced by the introduction of P reduction, but STWs serving towns in the Ballinderry River catchment have not been subjected to P reduction. The study was part of an assessment of the overall impact of the reduction of P from point sources on SRP loadings to Lough Neagh (Foy *et al.*, 1993).

15.2   THE STUDY AREA

Located in the north-east of Ireland, the Main and Ballinderry rivers flow into Lough Neagh and have catchment areas of 709 and 430 km^2 respectively (Figure 15.1). Over the period of comparison from 1974 to 1991, the urban population in the Main catchment increased from 36 100 to 43 200. However, approximately 73% of the Main urban population is located in the Ballymena area, where P reduction at two STWs commenced in 1981 (Figure 15.1). Excluding the Ballymena area, the urban population of the Main increased by 2100, from 9800 in 1974 to 11 900 in 1991. This population increase, when combined with changes in the human P per capita values given by Foy *et al.* (1993), is estimated to have increased the River Main P loadings by a maximum of 1.2 tonnes P between 1974 and 1991. For the Ballinderry catchment, the urban population increased from 10 800 to 12 500 over the same period, giving a maximum

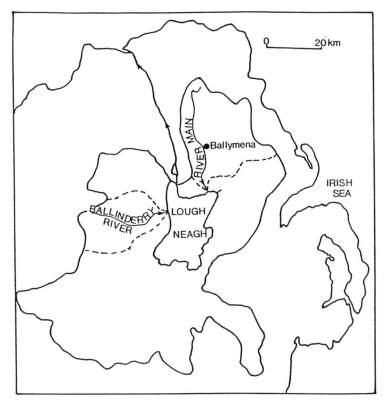

Figure 15.1   Lough Neagh catchment area, showing River Main and Balliderry River catchments Reduction of P commenced at Ballymena sewage treatment works in 1981.

P loading increase from urban sources of 1.0 tonnes P. It should be noted that these potential increases in river loadings refer to TP. In terms of an increase in SRP loadings, they may be considered the upper limits for increase, since not all P leaving an STW is in the SRP form, and a small retention of P within a conventional STW can also be assumed. The major land use of each catchment was pasture and rough grazing, which together covered 85–87% of the area with only an additional 7–9% under cultivation. Further details of land use are given by Smith (1977) and Foy *et al.* (1982).

15.3   METHODS

15.3.1   River loadings

Each river was sampled weekly from 1974 to 1991 at a location close to the river entry point into Lough Neagh (Smith, 1977). SRP was determined in 0.45 μm filtered samples by the method of Murphy and Riley (1962). Annual SRP loadings were estimated from the following regression equation, considered by Smith and Stewart (1977) as the most reliable for predicting SRP loadings:

$$\log L = a + b \log F, \qquad (15.1)$$

where L is the daily SRP load and F is the daily flow.

Regression equations were derived annually for each river from days on which flow and chemical concentration were available, and, by using daily flow observations, daily loads were predicted.

15.3.2   P Removal at STWs

Loadings of TP from the Ballymena STWs were calculated by multiplying daily composite concentrations by daily flows. Net P removed at these STWs was estimated by difference from P loads measured before P-reduction treatment by Gray (1984). Based on a survey undertaken by Storey (1990) the reduction in SRP loading was estimated at 72.8% of the reduction in TP loading.

15.3.3   Estimation of loadings using a simulated flow year

In order to investigate trends in loadings of SRP for winter and summer periods, log load versus log flow equations for the rivers Ballinderry and Main were prepared as in (15.1) for the time periods 1974–76, 1981–83 and 1988–90. Each time period was based on the hydrological year beginning in October and ending the following September, since the calendar year splits the hydrological cycle, which has high winter flows and low summer flows (Smith and Stewart, 1989). The three different time periods were further classified as winter and summer periods, which included the months October–March and April–September respectively. As there is considerable variation in flows between the time periods 1974–76, 1981–83 and 1988–90, the equations cannot be employed using actual flow data for these time periods to assess changes in SRP loadings that are not attributable to changes in flow. To remove the effect of this year-to-year variation in flows, it was necessary to create simulated sets of flow data that were representative of flows measured over an annual cycle.

The simulated data sets were created by randomly selecting 365 daily flow observations from the 1974–91 data set. Sampling was stratified by employing day number for the calendar year, and the selection process was performed using the RANDOMISE directive in GENSTAT 5 to permute the numbers 1, 2, ..., 18 into a random order. The first number in the permuted list was then used to select the year from which to extract the flow value for the day number in question, where 1 corresponded to 1974, 2 corresponded to 1975, and so on, with 18 corresponding to 1991. This process was carried out for each day of the year using a new set of the permuted numbers for each selection. One hundred simulated flow years were created for both rivers. SRP loadings were then estimated using the regression equations described above, and the hundred estimated loadings averaged to give predicted loads for the summer and winter months of each time period.

It is accepted that the actual flow data in the various time periods are auto-correlated and that this autocorrelation was not maintained in the simulated data sets. Since the derived data sets were only to be used for substitution in regression equations, it was felt that the autocorrelation of daily flow observations was not a crucial factor in this context.

15.4 RESULTS

The P-reduction programme at the Ballymena STWs commenced in 1981 and was fully operational by 1984, removing an average of 13.7 tonnes P year between 1984 and 1991 (Table 15.1). This TP reduction was equivalent to a reduction in SRP loading of 10.0 tonnes P yr^{-1}, which can be compared to the mean River Main SRP loading of 42.5 tonnes between 1974 and 1980 when P reduction was not in operation (Table 15.1). However, the reduction in

Table 15.1 Annual river flows, river SRP loads and TP removed at STWs for the Main and Ballinderry River catchments 1974–91

	River Main			Ballinderry River	
Year	River flow $(10^6 \, m^3 \, yr^{-1})$	River SRP load (tonnes P yr^{-1})	STW TP reduction	River flow $(10^6 \, m^3 \, yr^{-1})$	River SRP load (tonnes P yr^{-1})
1974	472.2	37.6		290.7	17.1
1975	343.1	29.9		178.8	13.1
1976	464.6	36.9		271.8	21.7
1977	518.9	44.3		337.3	22.6
1978	596.0	50.9		323.1	23.7
1979	606.0	50.0		361.2	25.7
1980	554.0	47.8		320.8	26.6
1981	679.1	44.4	− 6.9	331.7	27.0
1982	588.7	36.0	− 10.9	322.3	25.5
1983	357.9	23.0	− 8.4	273.2	25.3
1984	520.1	31.4	− 15.9	291.7	25.1
1985	586.9	34.2	− 14.2	342.9	30.6
1986	596.0	38.5	− 16.0	359.9	32.5
1987	480.1	38.3	− 12.5	295.3	32.5
1988	650.1	53.0	− 10.3	368.3	40.7
1989	391.2	29.9	− 13.2	227.8	24.6
1990	629.8	57.2	− 11.2	312.3	35.3
1991	536.6	40.2	− 16.2	239.6	24.6

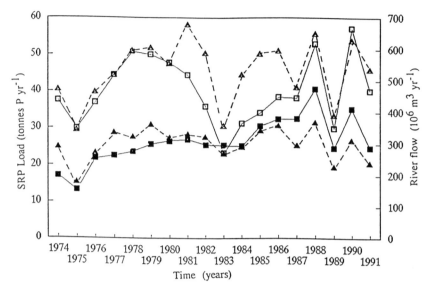

Figure 15.2   River Main annual flow (Δ) and river loads (□) and Ballinderry River annual flows (▲) and river loads (■) for 1974–91.

P loading from STWs was less than the range of 20.0 tonnes P yr^{-1} of SRP river loads between 1974 and 1980. In general, variation in river SRP loads tended to correspond to variation in river flows, with the low flows in 1974, 1983 and 1989 producing low river SRP loads and high loadings in 1979, 1980, 1988 and 1990 occurring when flows were high (Figure 15.2). The time plot of River Main SRP loadings shows some evidence of a decline in loads from 1981 to 1983, which was a very dry year (Figure 15.2). Despite comparatively high flows in 1984 to 1986, river SRP loads were substantially lower than those measured under similar flows from 1977 to 1980. However, the SRP loads in the high-flow years of 1988 and 1990 were higher than any previously recorded, despite the reduction of P from STWs (Figure 15.2 and Table 15.1).

The impact of P reduction on the seasonal variation of SRP loadings of the River Main was examined in the years following P reduction and in years prior to P reduction. Figure 15.3 shows monthly variation in Main SRP loadings for the years 1974–76 (pre P reduction) compared with those for 1981–83 (immediately following P reduction) and 1988–1990. It is apparent that the winter loadings of SRP from October to March show no consistent pattern of reduction with time. Indeed, the 1974–76 and 1981–83 winter loadings show a marked similarity with each other. The expected reduction in river loading due to P reduction at STWs of about 0.75 tonnes of SRP per month is only apparent in the summer months for 1981–83 and is absent for the years 1988–90. However, examination of flow data for the three periods shows the extent to which the flow regimes for each time period varied (Figure 15.4). It is apparent that any meaningful

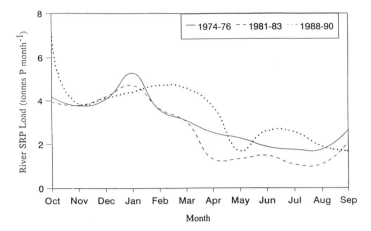

Figure 15.3 Mean monthly SRP loadings in the River main for hydrological years during 1974–76, 1981–83 and 1988–90 (1974 is the hydrological year October 1974 to September 1975).

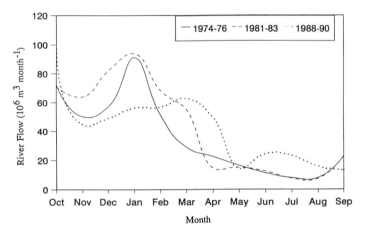

Figure 15.4 Mean monthly flows in the River Main for hydrological years during 1974–76, 1981–83 and 1988–90.

investigation into the effects of the P reduction programme on SRP loadings in the River Main must therefore take into account the influences of flow on loadings.

In contrast to the River Main, the River Ballinderry, which had no P reduction at STWs, shows a progressive increase in winter SRP loadings with time (Figure 15.5). The increase in winter loadings of SRP evident in the Ballinderry River for the Period 1988–90 occurred despite substantially higher winter flows in the years 1981–83 than from 1988 to 1990 (Figures 15.5 and 15.6).

In order to examine the impact of P reduction on the SRP loadings of the River Main, exclusive of the influence of changing flow regimes, a regression analysis

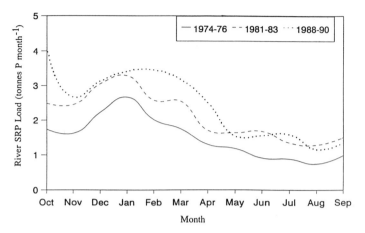

Figure 15.5 Mean monthly SRP loads in the River ballinderry for hydrological years during 1974–76, 1981–83 and 1988–90.

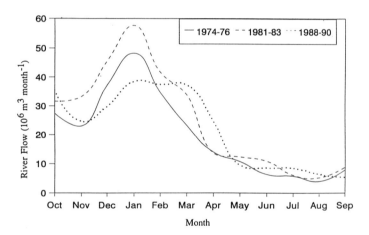

Figure 15.6 Mean monthly flows in the River Ballinderry for hydrological years during 1974–76, 1981–83 and 1988–90.

was undertaken on the SRP loadings for the years prior to P reduction (1974–80) versus the reciprocal of flow (1/flow). The SRP loadings show a good correlation with this variable:

$$y = \frac{77.37 - 17.14}{\text{flow}}, \quad n = 7, \quad r^2 = 0.89, \quad p < 0.01, \quad (15.2)$$

where y is the annual SRP loading (in tonnes P yr^{-1}) and 'flow' is the annual river flow (is 10^9 m^3 yr^{-1}). The time plot of SRP load residual values calculated as the difference between observed river loads and loads predicted from (15.2) is

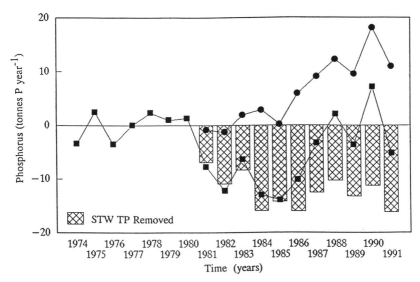

Figure 15.7 Time plot of residuals (■) from the 1974–80 regression equation (15.2) of River Main SRP loadings versus 1/flow. (●) denotes net residual values corrected for TP reduction at STWs. Bars denote TP loading reduction from Ballymena sewage treatment works.

plotted in Figure 15.7. Not surprisingly, given the high correlation coefficient of (15.2), the residuals for the period 1974–80 are small (< 5 tonnes P yr^{-1}) and show no particular trend with time. Their marked decline observed from 1981 to 1985/86 was comparable to the reduction in TP at the Ballymena STWs. However, in subsequent years the residual values increased with time despite the continuing reduction of P from STWs.

When the residual values are corrected for the reduction in P loadings from STWs, an upward trend in the resulting net residuals is apparent, particularly from 1985, implying that an increase in SRP loadings from other sources became significant after that date (Figure 15.7). However, this interpretation may not be justified, since the time plot of annual flows for 1974–80 shows that they tended to increase with time (Figure 15.2). The correlation coefficient r^2 between time and the reciprocal of flow for this period was 0.49, which is significant at the $p < 0.1$ level. This substantial correlation raises the possibility that a significant proportion of the correlation between time and flow in (15.2) was a consequence of a time-based increase in SRP loads independent of flow.

In an attempt to overcome this limitation in the data set, a multiple regression analysis was undertaken for the period 1974–91 in which the dependent variable was river SRP load and the independent variables were the reciprocal of flow, time and the quantity of TP removed at STWs. Over this extended period the r^2 value for the correlation between the time and flow variables was only 0.05. The slope coefficients of the resulting equation (15.3) were all significant at the

$p < 0.001$ level:

$$z = 6.76 - \frac{15.71}{\text{flow}} + 1.65\,\text{yr} + 0.92\text{STWP}, \quad n = 18, \quad r^2 = 0.92 \quad (15.3)$$

where z is the year number (with $1974 = 1$) and STWTP is the TP reduction at STWs (in tonnes P yr^{-1}). The STW TP variable was associated with 10% of the total variation. The slope coefficient of this variable of 0.92, with 95% confidence range of ± 0.45, indicates that the reduction of TP from STWs was matched by a similar reduction in river SRP loadings. The time variable was associated with 13% of the total variation, and estimates an upward trend in SRP loadings with time in the River Main of 1.65 ± 0.58 tonnes P yr^{-1} or, on a catchment area basis, 2.32 ± 0.82 kg P $\text{km}^2\,\text{yr}^{-1}$.

The Ballinderry river provides the opportunity to examine changes in river SRP loads over an extended period that are independent of major changes in urban P inputs and not influenced by P reduction. The loadings of SRP for this river for the years 1974–91 were regressed against the 1/flow variable:

$$y = 45.64 - \frac{5.65}{\text{flow}}, \quad n = 18, \quad r^2 = 0.39, \quad p < 0.01. \quad (15.4)$$

The residual values of observed Ballinderry River loads less loads predicted from (15.4) were plotted against time (Figure 15.8). They demonstrate an upward

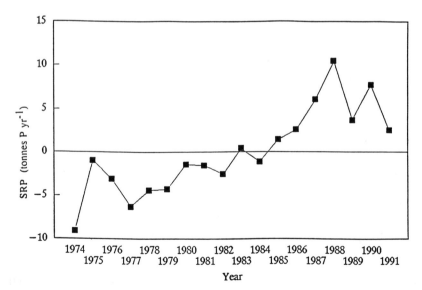

Figure 15.8 Time plot of residuals from the 1974–91 regression equation (15.4) of River Ballinderry SRP loadings versus 1/flow.

trend for the period 1974–91, which was quantified by introducing a time variable (with $1974 = 1$) as an additional independent variable, which gave the multiple regression equation

$$y = 35.77 - \frac{5.04}{\text{flow}} + 0.79\,\text{yr}, \quad n = 18, \quad r^2 = 0.83, \quad 15.5)$$

where $1/\text{flow}$ has a t statistic of -5.32 ($p < 0.001$) and yr (year number) has a t statistic of 6.28 ($p < 0.001$).

As in (15.4), time is positively correlated with Ballinderry River SRP load and is associated with 44% of its variation. The slope coefficient of 0.75 tonnes P year^{-1} gives a measure of the rate of increase of SRP loading over time, which, normalised on a catchment area basis, becomes 1.85 kg P km^{-2} yr^{-1}, with 95% confidence limits of 1.23–2.47 kg P km^{-2} yr^{-1}.

Our alternative approach of removing the influence of flow on SRP loading was to calculate SRP loadings for the simulated annual hydrological flow cycles from log load versus log flow equations for the periods 1974–76, 1981–83 and 1988–89. By combining the measured SRP concentration and sample day flows of three successive years, it was possible to obtain separate equations of log load versus log flow for the winter, October–March, and summer, April–September, periods (Table 15.2). Although separate winter and summer equations could be generated for individual years, in practice these individual six-month data sets were of limited use for predictive purposes because of the often limited range of flows and hence loads on the respective sampling days from which they were derived. Although autocorrelation exists in the original data, its absence in the

Table 15.2 Log load versus log flow regressions for summer (April–September) and winter (October–March) periods; loads measured as kg P day^{-1} and flows as 10^6 m^3 day^{-1}; $\pm$ denotes 95% confidence limits

River	Period[a]	Years	Equation	r^2
Main	Winter	1974–76	log load = $-1.08 \pm 0.03 + 0.74 \pm 0.08$ log flow	0.82
		1981–83	log load = $-1.14 \pm 0.04 + 0.71 \pm 0.09$ log flow	0.78
		1988–90	log load = $-1.09 \pm 0.02 + 0.93 \pm 0.07$ log flow	0.90
	Summer	1974–76	log load = $-0.92 \pm 0.08 + 0.47 \pm 0.13$ log flow	0.43
		1981–83	log load = $-1.06 \pm 0.10 + 0.59 \pm 0.15$ log flow	0.45
		1988–90	log load = $-1.00 \pm 0.07 + 0.65 \pm 0.16$ log flow	0.46
Ballinderry	Winter	1974–76	log load = $-1.19 \pm 0.04 + 0.85 \pm 0.11$ log flow	0.76
		1981–83	log load = $-1.12 \pm 0.03 + 0.75 \pm 0.10$ log flow	0.75
		1988–90	log load = $-1.00 \pm 0.03 + 0.83 \pm 0.10$ log flow	0.80
	Summer	1974–76	log load = $-1.15 \pm 0.08 + 0.48 \pm 0.11$ log flow	0.53
		1981–83	log load = $-1.09 \pm 0.09 + 0.32 \pm 0.14$ log flow	0.21
		1988–90	log load = $-0.99 \pm 0.07 + 0.51 \pm 0.10$ log flow	0.55

[a]number of observations for each period regression: 72–75.

Table 15.3 Comparison of winter and summer loads; values presented are the mean of 100 flow simulations; ± denotes 95% confidence limits[a]

River	Period	Summer load (tonnes P)	Winter load (tonnes P)
Main	1974–76	16.9 ± 0.1	24.8 ± 0.2
	1981–83	12.0 ± 0.1	20.9 ± 0.2
	1988–90	13.7 ± 0.2	29.2 ± 0.4
Ballinderry	1974–76	7.8 ± 0.0	13.3 ± 0.1
	1981–83	10.5 ± 0.0	15.4 ± 0.1
	1988–90	11.1 ± 0.1	20.8 ± 0.2

[a]Note that the variation in calculated loads is due to the variation in the simulated flows rather than to the variation in the coefficients of the log load versus log flow equations listed in Table 15.2.

simulated data has not caused a problem. The mean values of the 100 simulated flows for the Rivers Main and Ballinderry were 533 and $304 \times 10^6 \, \text{m}^3 \, \text{yr}^{-1}$, respectively, with standard errors of 0.6 and 0.5%.

The reduction in the simulated Main winter and summer SRP loadings for 1981–83 in comparison with those for 1974–76 were similar in magnitude at 4.9 and 3.9 tonnes P, which in combination, gives an annual reduction in SRP loading of 8.8 tonnes P year^{-1} (Table 15.3). This value can be compared with the measured reduction in TP of 8.7 tonnes P year^{-1} from the Ballymena STWs between 1981 and 1983 (Table 15.1). In contrast the 1988–90 data show a marked increase in the winter loading of 4.4 tonnes P with respect to the 1974–76 winter loads and only a 3.2 tonnes P reduction in summer SRP loads. Combined, the 1988–90 annual load of 42.9 tonnes P is 1.2 tonnes P higher than the value for 1974–76. Applied to the Ballinderry River, flow simulation exercises produced progressive increase in SRP loadings for both winter and summer months from 1974–76 to 1988–90 (Table 15.3). The overall increase from 1974–76 to 1988–90 was 3.3 tonnes of SRP in the summer and 7.5 tonnes in the winter. If this increase is normalised to the size of catchment, it gives an increase of $1.79 \, \text{kg P km}^{-2} \, \text{yr}^{-1}$, which in turn is close to the value estimated by (15.4) for the same period.

15.5 DISCUSSION

One of the major causes of annual variation in river SRP loadings is the variation in flows from year to year. In the present study this source of variation was removed by employing two independent methods. In the first annual loadings were regressed against 1/flow and time to yield a background increases of SRP loading close to $2.0 \, \text{kg P km}^{-2} \, \text{yr}^{-1}$ for each of the rivers. These increases in SRP loadings were in substantial agreement with results from the second method employing a simulation model of a typical flow year, which suggested that the background increase rate from the Ballinderry River was $1.79\,kg\,P\,km^{-2}\,yr^{-1}$. The magnitude of these increases accumulating over a period of 8–10 years would be sufficient to counterbalance the reduction in P from STWs discharging into the River Main. The multiple regression analysis of the River Main data indicate that the reductions in River Main SRP loadings were directly comparable to the reduced P loading from STWs.

We conclude that an increase in background loading of SRP is occurring in the rivers of the Lough Neagh system and that, although relatively small on an annual basis, it has, over the last decade, significantly reduced the impact of reductions in the P discharges from creameries and STWs (Foy *et al.*, 1993). Although the importance of long-term monitoring of nutrient inputs is an important requirement in the study of eutrophication of lakes (Sas, 1989), there appears to be a dearth of global long term P-loading data from rivers. Krug (1993) reported a similar phenomenon of a long-term increase in P from agricultural sources at a time of reduced loadings from urban and industrial sources in a Swedish catchment where the dominant farming practice was arable. It is interesting to note that the Swedish increase in P loadings from agricultural sources took place over a period when the use of P fertilisers remained constant. As the annual increase in background loading observed in the present study is relatively small, it can be readily obscured by annual variations in flow when only a few years' data are compared.

The underlying cause of the increase in river SRP loadings in the Lough Neagh catchment area must at this stage be a matter of speculation. However, the twofold greater increase in winter loadings calculated from the flow simulation exercise compared with those in summer indicates the absence of an increase in point source loadings, which would make equal contributions in both winter and summer. The potential increase in SRP inputs from urban population growth and changes in detergent P consumption in each river catchment are unlikely to have exceeded 1 tonne of SRP over the 18 years of study. Similarly, the gradual increase in SRP loading observed in the River Ballinderry with time, rather than a sudden increase, argues against the introduction of a new point source in the catchment. We conclude that the increase in background loading is probably associated with an increase in SRP loading from agricultural sources. It has taken place over a period of increased agricultural production, which, together with an absence of significant increases in known point sources and a trend towards increased loading under higher flows in winter, points to an increase in SRP loading from land drainage. In the Republic of Ireland an eightfold increase in soil P has been observed between 1950 and 1990 (Tunney, 1990). Although clay mineral soils have a high capacity for retaining P (Cooke, 1976) loss rate increases of about $2.0\,kg\,P\,km^{-2}\,yr^{-1}$ are sufficiently small not to be at variance with net increases in soil P of about $50–1000\,kg\,P\,km^{2}\,yr^{-1}$ observed in Ireland (Tunney, 1990).

ACKNOWLEDGEMENTS

We wish to thank the following sections of the Department of the Environment for Northern Ireland for their assistance in providing data employed in this paper. From the Water Service, Dr Bill Storey and Dr Tom Horridge for data relating to P reduction at STWs. From the Water Data Unit, Mr Phillip Holland and John Waterworth for river flows.

16

A Statistical Evaluation of Toxicity Tests of Waste Waters

J. A. Branco and A. M. Pires
Technical University of Lisbon (IST), Portugal

E. Mendonça and A. M. Picado
ITA, National Institute of Technology and Industrial Engineering (INETI), Lisbon, Portugal

The measurement of toxicity levels of industrial effluents has become crucial for the evaluation of the impact of these effluents on ecosystems.

The determination of chemical characteristics is a common procedure to evaluate toxicity of waste waters. Another possible way, which may complement this approach, is the use of toxicity tests based on bioeffects of the pollutants on selected species. The aim of this study is to assess and compare the performance of two such tests: Daphnia and Microtox. These tests may be used as a screening tool along the different phases of an industrial process, including the control of the final effluent, as well as in measuring the efficiency of waste-water treatment plants.

The available data were samples obtained at the outfall of a surface treatment process. Some of the samples were treated to reduce toxicity (the treatments included oxidation, floculation and tangential filtration). All of the samples—treated and untreated—were chemically analysed and submitted to the toxicity tests. After suitably transforming the original variables, an exploratory analysis was performed on the eight-dimensional space of the chemical variates in order to reduce the dimensionality and pick up the most relevant features of the data. The methods used were cluster analysis and principal components.

Statistics for the Environment 2: Water Related Issues. Edited by Vic Barnett and K. Feridun Turkman
© 1994 John Wiley & Sons Ltd.

Several linear models were fitted to the observations. The tests could then be compared and their relationships with the chemical characterisation could be determined.

16.1   INTRODUCTION

Several biological tests have been proposed in recent years in order to evaluate the toxicity of effluents. These toxicity tests are mainly performed on fish, microcrustaceans, molluscs or algae. The aim of this study is to assess and compare the performance of two such tests: The Daphnia test and the Microtox test.

The Daphnia test has been in use for quite a long time—the first reference dates back to 1933, according to Anderson (1980)—and has achieved a level of standardisation. The criterion used in this test to quantify the toxicity is the effective concentration (EC) of the effluent that reduces 50% of the mobility of the water flea *Daphnia magna* Straus.

The Microtox test is much more recent (Bulich, 1979) and is based on the luminescent properties of a marine bacteria (*Photobacterium phosphoreum*). The test quantifies the toxicity of an effluent by the effective concentration which reduces 50% of the bioluminescence. It is widely accepted that the Microtox test is quicker, simpler and cheaper than the Daphnia test, and is nowadays commonly used in Canada and the USA to monitor the toxicity of waters. In Europe the Microtox test is not currently used, and has not yet been the matter of legislation.

Some studies have been performed to compare the results of the two tests. For instance, Vasseur *et al.* (1984) have concluded that they are in good agreement but that one may be more sensitive than the other, depending on the kind of products that are synthesised and released by an industrial process. For these authors, effluents are considered toxic when EC_{50} is below 45% in terms of effluent dilution. Any comparison based on this definition is most likely to lead to good agreement between tests in cases where the effluents tested are either too toxic or insufficiently toxic. More sensitive methods are desirable, and in this chapter much stronger agreement between tests is considered.

It is also intended in this study to relate the results of the tests to the chemical characterisation of the effluent. The effluents analysed are all from the same type of industry: the surface treatment industry, which induces metallic pollution. This industry releases two kinds of effluents—acid and alkaline—and results are available for differently treated effluents as well as for the untreated effluents. So it is expected to have several levels of toxicity associated with a relatively wide range of chemical characteristics.

In Sections 16.2 to 16.4 a more detailed description of the data is given, together with the results of the statistical analysis. The discussion and conclusions of the results are presented in Section 16.5.

16.2 DESCRIPTION OF THE DATA

The sample data considered here have 21 multivariate observations. There are four observations from the acid effluent ($Ac_2, \ldots, Ac_5$) and five from the alkaline ($Al_1, \ldots, Al_5$). The remaining twelve correspond to treated effluents ($T_0, \ldots, T_{11}$) and may not constitute a homogeneous group, since the treatment processes varied (these processes included oxidation, floculation and tangential filtration in order to reduce the high level of cyanides and heavy metals present in the untreated effluent).

For each observation, 11 variables were recorded. The first eight ($x_1, \ldots, x_8$) are of a chemical nature:

x_1 is pH;
x_2 is the concentration of CN (in mg 1^{-1});
x_3 is the concentration of Cu (in ppm);
x_4 is the concentration of Cr (in ppm);
x_5 is the concentration of Fe (in ppm);
x_6 is the concentration of Ni (in ppm);
x_7 is the concentration of Ti (in ppm);
x_8 is the concentration of Zn (in ppm).

The remaining three (y_1, y_2, y_3) correspond to the results of the toxicity tests, y_1 and y_2 coming from the Microtox tests and y_3 from the Daphnia test:

y_1 is the effective concentration of the effluent that reduces 50% of the biolu-
    minescence of the bacteria after 5 min contact with the solution (EC_{50-5});
y_2 is the same, but after 15 min (EC_{50-15});
y_3 is the effective concentration that imobilises 50% of the Daphnia after 24 h in
    contact (EC_{50-24})

These effective concentrations are estimated by standard probit analysis after obtaining the results of bioluminescence or mobility reduction for different concentrations, as is described in Microbics Inc. (1988) and International Norm ISO 6341 (1989) respectively.

A careful look at the data revealed some difficulties that had to be eliminated before any statistical analysis was applied:

(i) Even in a small experiment like this, it was impossible to avoid the ocurrence of missing values. With regard to the chemical variables, two of the observations (T_0 and Al_1) were highly incomplete, so they had to be discarded from the computations and were only used for some descriptive purposes. There were three other missing values, one per observation, that could be estimated by group means since those observations belonged to the untreated effluent groups, which can be considered homogeneous as the opposite to the group of treated effluents. The results of the tests also had many missing values, 13 in all, that had to be dealt with.

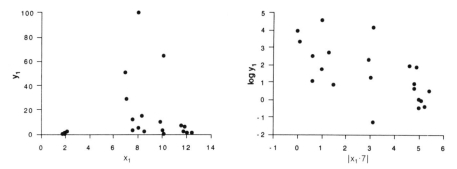

Figure 16.1   Plot of y_1 versus x_1, before and after transformation.

(ii) A small number of values had been recorded as 'undected', meaning that the true value was below the detection limit of the equipment. These were replaced by a number equal to half of that detection limit. This procedure was found to be reasonable, and may be seen as a consequence of assuming a uniform distribution in the small interval of nondetection and estimating the unknown value by the corresponding mean. The consequences of the inadequacy of this assumption are not expected to be very important.

(iii) High assymetry, nonlinearity and consequently non-normality are also present in this set of data. It was thus found convenient to transform the original data. The choice of transformation was informally based on graphical displays. For the variables $x_2, \ldots, x_8$ and y_1, y_2, y_3 the log transformation was chosen. We recall that these variables are concentrations, and it is known from experience that the log transformation works well with such variables. For x_1 a simple $|x_1 - 7|$, which may be called 'departure from neutrality', was found to be adequate. The effects are shown on Figures 16.1 and 16.2, which represent plots of two pairs of variables before and after the transformation.

(iv) Another difficulty is the small ratio between the number of observations and the number of independent variables (19/8). This is of course impossible to

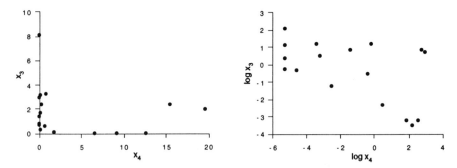

Figure 16.2   Plot of x_3 versus x_4, before and after transformation.

modify, but will condition the rest of the analysis in that it will be necessary to use techniques to reduce the dimensionality. The methods used are described in the Section 16.3.

16.3 STRUCTURE OF THE CHEMICAL VARIABLES

In multivariate analysis it is sometimes much more informative to use several complementary methods than to select a single 'most adequate' method (which in many cases does not exist). The results of some of the methods chosen in order to explore the most relevant features of the data and achieve a reduction of dimensionality are discussed next.

16.3.1 Cluster analysis

There is already a known group structure, but it is interesting to know whether this is confirmed by the data and if there are other groups. A hierarchical agglomerative clustering algorithm based on Euclidean distance and single linkage (Dixon, 1990) was chosen. (Other distances and amalgamation rules led to similar trees and the same grouping, thus showing the stability of these results.) The results showed that the separation between the three expected groups—acid, alkaline (both with four observations) and treated effluents—is good and that the group of treated observations may be split into two smaller groups, one with seven observations $(T_2, \ldots, T_8)$ and the other with four $(T_1, T_9, T_{10}, T_{11})$. This separation of the observations into four groups can be considered as a reduction of dimension, although a crude one, and will be used subsequently.

16.3.2 Principal components

Principal component analysis is perhaps the best known dimension-reduction tool of multivariate analysis.

As the classical principal components are known to be sensitive to outlying observations, both classical and robust principal components (by Campbell's (1980) method), were determined. The two methods led to similar results, showing that extreme observation points are not present and so the classical components are reliable. The first two principal components account for respectively 41.9% and 30.8% of the total variation of the chemical set of variables. Although none of these components has a straightforward interpretation, both reveal some kind of contrasts. It is possible to interpret the first as a general index of toxicity, for reasons that will become clear in Section 16.4.1. The plot in Figure 16.3 represents the observations on the plane of the first two principal components. The groups shown are those obtained by the cluster analysis. It is possible to conclude that the two different methods are in good agreement.

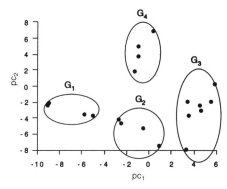

Figure 16.3   Plot of observations on the plane of the first two principal components.

16.4   COMPARISON OF THE TESTS

16.4.1   Preliminary analysis

Before trying to model the dependent variables, some very simple methods were used. First the tests were compared without making use of the chemical characteristics.

If the results of the tests are given by the classification into 'toxic' and 'non-toxic' using the criterion $EC < 45\%$ and $EC \geqslant 45\%$ then the agreement between the results of the different tests is almost perfect. This is not surprising if one takes into account the broad type of agreement required and the fact that in only one case is $EC_{50-15} < 45\%$ but EC_{50-5} and EC_{50-24} are both greater than 45%. This case may point to a better sensitivity of the Microtox at 15 min, but of course more observations would be necessary to reach a reliable conclusion.

The correlations between y_1 and y_3 (0.753) and y_2 and y_3 (0.877), which are significantly different from zero (under the usual assumptions, $p \leqslant 0.011$), also support the conclusion that the results of the Daphnia test and the Microtox test are related. The correlation between the two Microtox results (0.830) is also significant ($p = 0.000$).

If (y_1, y_2, y_3) are compared pairwise by means of a paired t-test, one must again conclude that there is no significant difference between the Microtox and Daphnia tests ($p = 0.36$ for both comparisons). However, a difference may exist between the two Microtox results ($p = 0.0028$). The apparent contradiction may be justified by the existence of some missing results for the Daphnia test.

It is also important to know how the results of the tests reflect the chemical variability. A relation between the result of each test and the chemical variables may be determined using the groups formed in Section 16.3.1 or the first principal component obtained in Section 16.3.2. One-way analysis of variance of each variable, y_1, y_2 and y_3, with the factor group showed the significance of this factor ($p \leqslant 0.009$). Also, the correlation between each y and the first principal component is significant (with y_1, we have $r = 0.542$, $p = 0.016$; with y_2 we have $r = 0.607$, $p = 0.007$; and with y_3 we have $r = 0.908$, $p = 0.005$). This is not surprising, since it was expected that both the groups and the first principal component reflect the chemical variation.

16.4.2 An explanatory model

It was possible to conclude that, on one hand, the results of the two tests are related and that, on the other, they reflect the effect of the chemical variability.

One first possible model for comparing the results of the tests is model I:

$$y_{ij} = \mu + \alpha_i + \beta_j + \varepsilon_{ij}, \quad i = 1, 2, 3, \quad j = 1, \dots, 19,$$

where i refers to the test and j to the effluent. Thus α_i represents the effect of the test (treatment), β_j the effect of the effluent (block), μ the grand mean and ε_{ij} the error term. Assuming the ε_{ij} are i.i.d. $N(0, \sigma^2)$, the usual analysis of variance may be applied. The conclusions are that the block effect is significant ($p = 0.000$) and the treatment effect is significant at levels greater than 2.8%.

However, with this model, we do not make use of the information given by the chemical variables, and it is impossible to consider the interaction between the tests. This is very important, because it may point to a better understanding of the relative sensitivity of the tests. A plot of the log EC_{50} values against the effluent number (Figure 16.4) shows that some interaction coefficients may indeed be present.

If instead we introduce the group structure determined in Section 16.3.1 as a factor with four levels, representing the chemical characteristics, and retain the

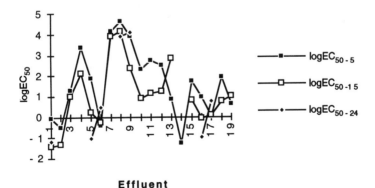

Figure 16.4 Plot of log EC50 against effluent number.

factor effluents, it is possible to consider an interaction term between groups and tests (model II):

$$y_{ij(k)k} = \mu + \alpha'_i + \beta'_{j(k)} + (\alpha'\beta')_{ij(k)} + \gamma_k + \varepsilon_{ij(k)k},$$

$$i = 1, 2, 3, \qquad j(k) = \begin{cases} 1 & (k = 1, 5, 16, 18), \\ 2 & (k = 2, 6, 17, 19), \\ 3 & (k = 4, 7, \ldots, 12), \\ 4 & (k = 3, 13, 14, 15), \end{cases} \qquad k = 1, \ldots, 19,$$

where j refers to the group and k to the effluent. Under the same assumptions, the conclusions are now different. The stronger effects are from groups and effluents (given group) with $p = 0.000$. But tests and test $\times$ groups are also significant with $p = 0.003$ and $p = 0.004$ respectively. The diagnostics (Q–Q plot of the residuals and plot of residuals versus fitted values) show that this model fits well (which is not surprising since it reproduces the cell means for tests $\times$ groups). An analysis of the coefficients seams to reveal a better sensitivity of the Microtox 15, especially for group 3 (the less toxic), and of the Daphnia for the alkaline group. (By a 'better sensitivity', we always mean a smaller value of the EC_{50}.)

16.5   CONCLUSIONS

The first part of our conclusions relates to the statistical analysis. The small number of observations compared with the number of variables posed some difficulties. The choice of the appropriate transformations turned out to be essential, because it improved linearity and stabilised the variance. Also essential was the grouping obtained with the cluster analysis. Finally, the use of different methods and models with the same aim not only helped the interpretation but also acted as a validation criterion for the results. With a better set of observations, it would be desirable to do further work. Some possibilities would be to define a proper global index of toxicity, to build up a calibration model based on the index for the tests, and to discuss the precision of all tests.

The second part of our conclusions relates to the toxicity tests themselves. It can be concluded that the two different organisms have comparable reactions in the presence of differently polluted effluents, although their sensitivities are related to the type of effluent, and that those reactions can be related to the chemical characteristics of the effluents by a causal model.

Although the analysis of a complete and fresh set of data would be important to reinforce the conclusions, some recommendations for those responsible for environmental legislation can emerge from the study.

The Microtox test should be the subject of norms in Europe as it is in America, because it presents economic advantages over the Daphnia test already mentioned.

Another aspect of great interest is the control of the effluents released by industry. At present laws impose limits only on each of the chemical components of such effluents. Laws imposing limits in terms of toxicity would be of great value because, with toxicity tests, it is possible to measure the quality of waste waters in a global way, and since these are single tests and cheaper than chemical analysis, they could be used more often—as a screening tool—without increasing costs.

ACKNOWLEDGEMENT

This study was partially supported by the contract BRITE/EURAM-0084-C(SMA).

PART IV

Hydrological Modelling

Evaluation of Predictive Uncertainty in Nonlinear Hydrological Models Using a Bayesian Approach

R. Romanowicz, K. Beven and J. A. Tawn

Lancaster University, UK

17.1 INTRODUCTION

Water flows are an important driving force for many environmental processes, particularly geochemical, ecological and geomorphological processes, over both long and short time scales. The prediction of water fluxes using hydrological models consequently has important practical applications in extending the knowledge of hydrological behaviour obtained from the few, generally point, measurement series that are available. Predictive equations have been used in hydrology since the 'rational formula' or Mulvaney (1850). All such models of hydrological systems have one thing in common: they are dependent on the specification of some empirical constants or parameters. This applies both to the simple areally lumped models that treat a catchment area as a black box system and to the complex distributed models based on physical principles that have been developed in recent years as computers have become more powerful.

Hydrological systems are a useful analogue to many terrestrial environmental systems. The task of the model is relatively modest—to predict the movement of water through the terrestrial system from when it arrives at the soil surface or

Statistics for the Environment 2: Water Related Issues. Edited by Vic Barnett and K. Feridun Turkman

vegetation as rainfall to when it leaves the area of interest, either as stream discharge or as vapour back to the atmosphere. This is, for the most part, a purely physical problem governed by principles that are well established at the 'point' scale. What makes this task so difficult is that the water flows take place in a highly complex geometry of coupled flow processes in a heterogeneous system that, at least for subsurface flows, is not directly observable. Water flows also show highly nonlinear responses to the inputs, very dependent on the antecedent pattern of water storage in the catchment. The difficulty of the task is also compounded by the measurements available, which are often restricted to local 'point' values sampled from a heterogeneous flow domain and may be of limited accuracy.

Thus it is that, despite a long history of modelling activity in hydrology and a wide variety of available models (O'Connell, 1991), there is little consensus among hydrologists as to the best methodology to use for different problems. There is also an expectation that models will not be particularly accurate in reproducing the observed data, especially catchment discharge responses under dry antecedent conditions when runoff production processes exhibit highly nonlinear responses. It is easy to demonstrate, in fact, that even the most 'physically based' models do not take proper account of the complexities of the system, so that their distributed predictions cannot be validated (see discussions in Beven, 1989, 1993; Grayson *et al.*, 1992; Konikow and Bredehoft, 1992).

Consequently, hydrological modelling relies heavily upon calibration of parameter values using some set of observables. Parameters often show interdependence and insensitivity in some parts of the parameter space used in the hydrological simulation, residuals are invariably correlated in time and often show changes in mean and variance over time. Optimisation techniques to overcome such problems have been the subject of considerable study in the past, from Ibbitt and O'Donnell (1971) to Duan *et al.* (1992), but the ranking of different parameter sets will depend on the chosen objective function used in the optimisation. Very few studies have recognised that such optimisation implies an associated uncertainty in the predictions and have attempted to estimate that uncertainty (see Garen and Burges, 1981; Bates and Townley, 1988; Hromadka and McCuen, 1988; Kuczera, 1988; Melching, 1992).

More recently, it has been argued that the concept of an optimum parameter set should be rejected in favour of a concept of equifinality of model structures and parameter sets (Beven and Binley, 1992; Beven, 1993). This recognises that, with such calibration of parameter values, nearly all models will provide some reasonable simulations of the data, especially allowing for the fact that the data themselves are not error-free. There may indeed, for a particular hydrological model, be many sets of parameter values (or model/parameter combinations) that might be almost equivalent in terms of reproducing the observations of discharge and perhaps other variables such as water table levels, sometimes from very different parts of the parameter space (see Duan *et al.*, 1992; Beven, 1993).

An uncertainty estimation procedure based on this concept of equifinality has been presented by Beven and Binley (1992). Their generalised likelihood uncertainty estimation (GLUE) technique was developed heuristically, but is, in essence, a Bayesian approach to uncertainty estimation for nonlinear hydrological models that recognises explicitly the equivalence, or near equivalence, of different parameter sets (or possibly model structures) in the representation of hydrological processes. A likelihood measure is used in evaluating the acceptability of any particular set of parameter values. These likelihood measures are used in evaluating the probability of the model predictions and so can be used to define the confidence limits of the predictions. Related approaches have been described by Hornberger and Spear (1981), Fedra (1980), van Straten and Keesman (1991) and Dilks *et al.* (1992).

In this chapter the relationship of the GLUE approach to more traditional likelihood and Bayesian estimation techniques is explored. It is worth re-emphasising the problems that this calibration methodology aims to overcome. Hydrological models are nonlinear, have parameters that may interact strongly, residual error series that are highly structured, and there may be non-unique parameter sets or even the whole regions of parameter space that are almost equally good simulators of the system.

17.2   BAYESIAN UNCERTAINTY ESTIMATION

The Bayesian approach is used here to estimate the uncertainty associated with the model predictions. The distribution of predicted flows can be calculated from a posterior measure associated with the particular model/set of parameter values; θ, that produces that prediction, given the observed data. The posterior distribution can be calculated from the product of some prior distribution, which reflects the knowledge of the modeller about the parameters, and a likelihood measure calculated from the model predictions and some observed set of data, z (Box and Tiao, 1992). In what follows, only one hydrological model will be considered, and the observed data will be the measured discharges for the catchment over a number T of time steps. The hydrological model also requires a time series of measured rainfall data and estimates of evapotranspiration as inputs, but it will be assumed that the results of any errors in this data are reflected in the likelihood measures associated with errors in the predicted discharges. As all our analysis is conditional on the rainfall inputs, we omit explicit reference to this in our notation, thus the data z are taken to be the observed flows only.

Thus, using Bayes' formula, the posterior distribution, conditional on the observations, can be calculated from

$$f(\theta|z) = f(\theta)L(z|\theta)/L(z) \qquad (17.1)$$

where $f(\theta)$ is the prior distribution of the parameters, and $L(z|\theta)$ is the likelihood measure for z given the parameter set θ. Here $L(z)$ is a scaling factor, which in the case that the posterior distribution is integrable over the admissible parameter space can be calculated from $L(z) = \int L(\theta|z)d\theta$.

The density function defined by (17.1) can be used in the calculation of the confidence limits for the predictions of the model. It is worth noting that (17.1) involves the entire vector of parameters, so that any parameter interactions or other difficulties will be implicitly reflected in the calculated posterior distributions. It is not necessary to consider each parameter separately, although the marginal distributions for individual parameters or groups of parameters can be calculated by an integration of the posterior distribution over the appropriate dimensions of the parameter space as necessary.

Equation (17.1) can be applied sequentially as new data or periods of data become available such that the existing posterior distribution is used as the prior for the new period, i.e.

$$f(\theta|z_1, \ldots, z_t) \propto f(z_1, \ldots, z_{t-1})L(\theta|z_t) \qquad (17.2)$$

where t is a calibration period index and $f(\theta|z_1) = f(\theta)L(\theta|z_1)/L(z_1)$, where $f(\theta)$ is the initial prior distribution of parameters defined by the modeller. In this study, as in the previous study by Beven and Binley (1992), the distributions are updated on the basis of likelihoods calculated for calibration periods of data containing one or more storms.

17.3   THE HYDROLOGICAL MODEL AND DATA

The model used in this study is a simplified version of the semi-distributed model TOPMODEL (Beven and Kirkby, 1979; Beven *et al.*, 1984; Quinn and Beven, 1993) in an application to the Institute of Hydrology River Severn experimental catchment at Plynlimon, mid-Wales. The model requires a sequence of rainfall and evapotranspiration data to run, and predicts the resulting stream discharges. Initialisation of the model makes use of an estimate of the catchment discharge at the start of a simulation period. The available data are hourly discharges and rainfall series for the year 1988. The hourly evaporation is estimated from the meteorological measurements using the Penman–Monteith formula. For calibration and validation pruposes, different time periods of 1000 h have been chosen from the observed records. This period is long enough to cover the length of rainfall events in the chosen catchment.

In the TOPMODEL formulation the predicted hydrological responses depend upon the distribution of an index of hydrological similarity that is derived from an analysis of the topography of the particular catchment being studied. Simple steady-state theory is used to develop a relationship between the topographic index and the local saturation of the soil as the catchment wets and dries. This relationship can be used to predict runoff contributing areas in a nonlinear way. The model also describes the subsurface drainage to the stream through

$$Q_b = T_0 e^{-\bar{\lambda}} e^{-S/m} \tag{17.3}$$

where Q_b is the subsurface drainage, T_0 is the average effective transmissivity of the soil when the profile is just saturated, $\bar{\lambda}$ is the catchment average of the topographic index, S is the catchment average storage deficit and m is a parameter that can be derived from an analysis of the recession curves for the catchment when it can be assumed that all the flow is derived from subsurface drainage. This recession parameter corresponds to a time constant for the subsurface drainage of the catchment. Equation (17.3) can be inverted, given an initial flow at the start of the prediction, to given an initial value of the storage deficit S. Then S is updated at each time step by a mass balance calculation involving vertical recharge to the unsaturated zone and the subsurface drainage Q_b.

The parameter m also enters as a scaling parameter in the calculation of the runoff production areas in the catchment through the relationship

$$\Delta S = m(\bar{\lambda} - \lambda) \tag{17.4}$$

where ΔS is the difference between local and catchment average storage deficits, and λ is the local value of the topographic index. Surface runoff contributing areas are predicted in locations where the local storage deficit is zero. This occurs first at points with high values of λ, and, as the catchment wets up, the contributing area will spread to lower values of the index. The index can be calculated using digital terrain maps of a catchment (Quinn *et al.*, 1991). Both the pattern and distribution of the index for the River Severn catchment are shown in Figure 17.1(a, b).

In the simplest version of the hydrological model (17.3) and (17.4) represent the most important functions in the model, with parameters m and T_0. Under relatively wet conditions, TOPMODEL has proven to be generally successful in predicting stream discharges in catchments of relatively shallow soils and moderate to steep slopes. Beven (1993) has used a five parameter version of TOPMODEL to illustrate the equifinality of different parameter sets in simulating the stream discharges of a small catchment in New Zealand. In this study parameters of the model that affect calculations of vertical recharge have been kept constant at values based on previous experience, since they are relatively unimportant to the wet period simulations of flows that will be reported.

In this application the model is used to simulate hourly flows using hourly rainfall inputs. The nature of the flows implies that they are positive and correlated in time. The 'best-fit' values of the parameters m and T_0 were obtained using a simple least-squares objective function (the widely used 'efficiency' criterion of Nash and Sutcliffe, 1970) and a Powell (1970) optimisation algorithm.

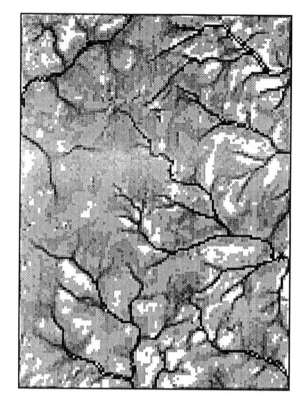

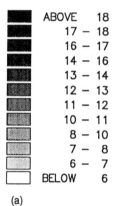

ABOVE		18
	17 −	18
	16 −	17
	14 −	16
	13 −	14
	12 −	13
	11 −	12
	10 −	11
	8 −	10
	7 −	8
	6 −	7
BELOW		6

(a)

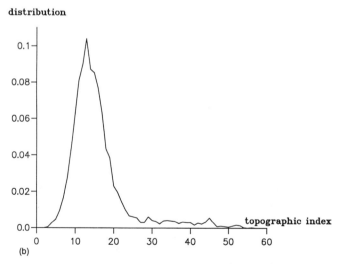

(b)

Figure 17.1 (a) Spatial pattern of the topographic index, over the upper Severn catchment and surrounding area, mid-Wales. The cross shows the location of the flow gauge. (b) Distribution of the topographic index for the upper Severn catchment.

The 'efficiency' is defined by

$$E = 1 - \frac{\displaystyle\sum_{t=1}^{T}(z_t - y_t)^2}{\displaystyle\sum_{t=1}^{T}(z_t - \bar{z})^2} \tag{17.5}$$

where z_t is the observed flow at time t, $\bar{z}$ denotes the average over the whole time period T of observed flows, y_t is the predicted flow at time t and the summations are made over the time steps $t = 1, \ldots, T$.

Some typical characteristics of error series of hydrological models can be seen in the results of simulation shown in Figure 17.4. The errors tend to be correlated in time, with large errors associated with the peak flows and smaller errors at low flows. This is a result of the effect of the dispersive catchment system driven by discrete periods of rainfall forcing. Previous studies of likelihood based methods to obtain 'optimal' parameter values take account of the correlation and hetero-scedascity of the error series (see e.g. Sorooshian and Dracup, 1980; Sorooshian *et al.*, 1983). In the latter work the comparison of two different maximum-likelihood criteria based on different noise model structures was considered. A discussion of the causes of the non-uniqueness of the parameter sets for conceptual rainfall-runoff models was presented by Sorooshian and Gupta (1983). All this work had a particular objective, which was the search for an 'optimal' parameter set. The GLUE approach, introduced by Beven and Binley (1992), has a very different standpoint. Instead of looking for optimal param-eters, GLUE, as well as the approach that follows in this chapter, searches for vectors of parameters in the parameter space that give a reasonable model performance for a wide range of rainfall events and different conditions in the catchment. Model performance is then analysed on the basis of the posterior probability surface, which can be updated when new observations are available and can be used for the derivation of the confidence intervals of the predicted flows.

17.4 FORMULATION OF THE ERROR MODEL

We take the statistical model describing the relation between the observed and predicted flows to have a multiplicative form as

$$Z_t = Y_t(r_T, \theta)\delta_t = Y_t\delta_t \tag{17.6}$$

where r_T denotes the observed rainfall series, $\theta = [m, T_0]$, δ_t is the model perform-ance error ($\delta_t > 0$), and the hydrological model TOPMODEL acts as a regression function. Taking the logarithm of (17.6) gives the observation equation in the logarithmic form:

$$Z_t^* = Y_t^* + \delta_t^* \qquad (17.7)$$

where the asterisk indicates the logarithm of the variable.

This type of logarithmic transform to stabilise the variance is standard in statistics (Box and Cox, 1964) and has been used before in hydrology by Dawdy and Lichty (1968). In order to take into account the correlated errors and a potential imbalance between rainfall and observed flows, we shall model δ_t^* by the Gaussian pth-order autoregressive model AR (p) with non-zero mean:

$$\delta_t^* - \mu = \sum_{i=1}^{p} \alpha_i(\delta_{t-i}^* - \mu) + \varepsilon_t \qquad (17.8)$$

where the mean μ of the δ_t process is constant in time, the α_i are the autoregressive parameters for the error series and the ε_t is assumed to be a normal white-noise error variable with variance σ^2. Hence the δ_t^* will also be normally distributed, and the likelihood function of the predicted flows can be expressed as the likelihood of the error variate δ_t^* with parameters $\phi = (\mu, \sigma, \alpha)$, depending on the hydrological model parameters θ. In what follows it will be assumed that the error series can be described adequately by a first-order autoregressive process ($p = 1$). This selection of p is purely illustrative, and in practice a higher-order process may be found to be a better description of the errors.

17.5   THE LIKELIHOOD FUNCTION AND THE POSTERIOR DISTRIBUTION

The likelihood function will consist of the T multiplications of the conditional probability function for each observation of the error ε_t given previous data. For the error model (17.8), with $p = 1$, it would take the form

$$L(\theta, \phi) = \prod_{t=1}^{T} f_{\varepsilon_t}(\varepsilon_t) = \prod_{t=1}^{T} f_{\varepsilon_t}((\delta_t^* - \mu) - \alpha(\delta_{t-1}^* - \mu))$$

$$= \prod_{t=1}^{T} f_{\varepsilon_t}((z_t^* - y_t^* - \mu) - \alpha(z_{t-1}^* - y_{t-1}^* - \mu)). \qquad (17.9)$$

Under the assumption of normally distributed errors, the joint probability density function of the simulation error series is then given by

$$f(\varepsilon_1^*, \ldots, \varepsilon_T^* | \theta, \phi) = (2\pi\sigma^2)^{-\frac{1}{2}T}(1 - \alpha^2)^{\frac{1}{2}} \exp\left[-\frac{1}{2\sigma^2} \Psi(\mu, \alpha, \theta) \right] \qquad (17.10)$$

where

$$\Psi(\mu, \alpha, \theta) = (1 - \alpha^2)(\delta_1^* - \mu)^2 + \sum_{t=2}^{T} [\delta_t^* - \mu - \alpha(\delta_{t-1}^* - \mu)]^2. \quad (17.11)$$

Hence the joint posterior distribution of parameters of the hydrological and statistical models is given by

$$f(\theta, \phi | z) = Cf(\theta, \phi)(2\pi\sigma^2)^{-\frac{1}{2}T}(1 - \alpha^2)^{\frac{1}{2}} \exp\left[-\frac{1}{2\sigma^2} \Psi(\mu, \alpha, \theta) \right] \quad (17.12)$$

where z denotes the observations (flows and rainfall/evaporation data) and C is a normalising constant. Here $f(\theta, \phi)$ denotes the prior distribution of the random parameters of the hydrological model TOPMODEL and the error model. It can be chosen in the form.

$$f(\theta, \phi) = f(\theta)f(\phi | \theta). \quad (17.13)$$

In the general case the distribution of the error model parameters may be expected to depend on the parameters θ, but, adopting a position of ignorance here, we take $f(\theta, \phi) = f(\theta)f(\phi)$.

It is possible to obtain some prior information about the parameter m from an analysis of recession curves, but it is also known that the simulations are quite sensitive to this parameter. Thus a non-informative uniform prior has been used, within a range defined from prior knowledge, in the expectation that this would be refined rapidly in the first period of simulation. Much less prior information is available for the T_0 parameter, but it is known to vary over a wide range. Accordingly a uniform prior distribution has been chosen for $\log T_0$.

Hydrological modellers have limited prior knowledge about ϕ, thus we take flat independent priors for μ and σ. However, to reflect the prior knowledge that errors in the model will be positively correlated, we take the prior for α to be uniform over the interval $[0, 1]$.

From (17.10), the log likelihood function will have the form

$$l(\theta, \phi) = -\tfrac{1}{2}T \log \sigma^2 + \tfrac{1}{2}\log(1 - \alpha^2) - \frac{\Psi(\mu, \alpha, \theta)}{2\sigma^2}. \quad (17.14)$$

The maximum likelihood estimates of the parameters of the hydrological model and the noise process can be found in the usual way by setting the differentials of (17.14) to zero and solving for the resulting optimal parameter values. Similarly standard likelihood techniques (Cox and Hinkley, 1974) can be used to develop confidence intervals for θ and ϕ. Our reason for not using such an approach is the resulting complex computation required to calculate profile likelihood based confidence intervals for future flows, which after all are the parameters of interest.

In a Bayesian analysis we wish to evaluate the whole distribution of likelihoods to allow the use of (17.1) in evaluating the predictive uncertainty. This leads to a critical difficulty with the use of the likelihood function (17.10) based on the error model assumed here. Equation (17.10) includes a function in which the error variance σ^2 is raised to the power $-\frac{1}{2}T$. In most hydrological examples T will be a large number, typically greater than 1000. Thus if σ is greater than $(2\pi)^{-1}$, the likelihood will approach zero rapidly, thereby eliminating that part of the parameter space from consideration. On the other hand, if σ is less than $(2\pi)^{-1}$, the likelihood will become very large very quickly, and it will be computationally difficult to evaluate the integral $L(z)$.

Here this difficulty has been circumvented by only evaluating the likelihoods over simulated flows for the order of 100 time steps, in which case $L(z)$ can still be calculated. Equation (17.2) is used to update the likelihood measures. The results should be equivalent to using a longer time series directly in that, for this likelihood function, applying Bayes' formula over successive time periods by recursive application of (17.12) means that the error summations $\Psi(\mu, \alpha, \theta)$ for the successive periods are effectively added within the multiplication of their exponentials. This is not true of the sum-of-squares likelihood function used by Beven and Binley (1992). However, this computational problem does not then arise.

One consequence of this extreme dependence of the calculated likelihood on σ, and the number of time steps, is that after only relatively few time steps (in hydrological terms), the posterior distribution will be dominated by the observations through the likelihood, and the prior distribution will have only a very minor effect. Thus the choice of the prior likelihood will not be very significant.

The parameters of the hydrological model, $\theta = [m, T_0]$, have some independent physical significance within the TOPMODEL structure, but in the full Bayesian analysis are given jointly with the error parameters ϕ through the posterior distribution. The marginal distribution of these θ may be evaluated by simply integrating the posterior distribution over the noise parameters ϕ, which is clearly much easier than obtaining profile likelihoods in this case.

17.6   COMPUTATIONAL ASPECTS

In the original GLUE procedure of Beven and Binley (1992) the likelihood function evaluation was carried out using Monte Carlo simulation to sample the parameter space. In their study the hydrological model used, based on a finite-element solution of the flow equation, was a major computational constraint. Even using a Meiko parallel processing computer system, only 500 simulations of the model were used in the calculation of the posterior associated with each parameter set.

In this study TOPMODEL has been used, because, as a semi-distributed model based on a simplified theory, the hydrological model is less of a constraint on the computations. However, it is still a computationally intensive procedure. The strategy followed was to discretise the (θ, ϕ) parameter space and evaluate the required integrals numerically as sums. However, even using a simplified approach, computational constraints and the dimensionality of the parameter space mean that the number of the discretisation intervals must be kept to a minimum. In this exploratory study we have used only two parameters in TOPMODEL (m and T_0) and the three error model parameters (μ, σ, α). The parameter space has been discretised into 48 increments within a specified range for each of the hydrological model parameters, and only 10 for each of the error model parameters. This therefore requires 2304 runs of the hydrological model for each set of the model parameters, each of which produces an error series, δ_t^*. Each error series is then used in 1000 likelihood function evaluations for each set of values in the ϕ space. The hydrological model predictions have again been carried out on a Meiko parallel processor. The final ranges of the hydrological and statistical model parameters were determined on the basis of using an initially coarse discretisation and refining the range on the basis of the first results.

In principle, discretisation of the parameter space could be avoided by the use of Gibbs sampling (Gelfand and Smith, 1990; Gelfand *et al.*, 1990). However, to form a sample from the true posterior involves obtaining convergence of the Gibbs sampler—thus many likelihood evaluations will be required before convergence is achieved. Since running the TOPMODEL would be required at each updating stage for θ, computations for this at present seem infeasible. On the other hand, Gibbs sampling is clearly the way to approach this estimation problem, avoiding arbitrary discretisation, the artificial ranges on parameters, and constrained model structure.

As will be shown in the following section, the main computational limitations of the technique used follow not from the evaluation of the posterior likelihood measures but from the evaluation of predictive uncertainty.

17.7 EVALUATION OF PREDICTIVE UNCERTAINTY

The posterior distribution $f(\theta, \phi | z)$ of the parameters can be used in estimating the predictive uncertainty of the hydrological model under the assumption that the distribution of the error series will be the same in prediction as in calibration. The distribution of predictions then depends upon the (nonlinear) hydrological model and the statistical error model. For the structure of the statistical model assumed here, the cumulative distribution of the error term at any time, given a particular set of hydrological and statistical model parameters θ and ϕ, will be given by

$$P(\delta_t^* < \delta | \theta, \phi) = \Phi\left(\frac{\delta - \mu}{\sigma/(1 - \alpha^2)^{\frac{1}{2}}}\right) \qquad (17.15)$$

where Φ is a standard normal distribution function. The predictive distribution of log discharges Z_t^* conditioned on the calibration data z will be then given by

$$P(Z_t^* < y | z) = \int_\theta \int_\phi \Phi\left(\frac{y - \mu - y_t^*}{\sigma/(1 - \alpha^2)^{\frac{1}{2}}}\right) f(\theta, \phi | z) \, d\phi \, d\theta \qquad (17.16)$$

where y_t^* is the hydrological model output at time t based on using hydrological model parameters θ. In discrete form, as used here,

$$P(Z_t^* < y | z) = \sum_\theta \sum_\phi \Phi\left(\frac{y - \mu - y_t^*}{\sigma/(1 - \alpha^2)^{\frac{1}{2}}}\right) f(\theta, \phi | z). \qquad (17.17)$$

We then seek, for chosen confidence limits, the lower and upper percentage points $[z_L^*, z_u^*]$ of the distribution of Z_t^*, for example, such that

$$P(Z_t^* < z_L^* | z) = 0.05,$$

$$P(Z_t^* < z_U^* | z) = 0.95$$

for a 90% confidence limit at time t. Clearly here (17.17) is the predictive distribution of log flows at time t and $[\exp(z_L^*), \exp(z_u^*)]$ a predictive 90% confidence interval for flows.

It can be seen from (17.17) that the evaluation of confidence limits at each time step involves intensive computations and large storage requirements. On top of the 2304×1000 discretisation of the TOPMODEL and noise parameter space, we have to add at least 50 discrete values of the white-noise distribution in solving for z_L^* and z_u^*. In this calculation it is possible to take advantage of the fact that the distribution of δ_t^* given θ is invariant to time as a result of the assumption that the ε_t are time-independent. This requires the evaluation of the distribution of the noise variable δ_t^* conditional on the TOPMODEL parameters θ. As the evaluation of that distribution need be done only once, it reduces both storage and the computing requirements considerably and allows us to separate the computations of confidence intervals from the posterior distribution function evaluation. The procedure followed here was to discretise the predicted log discharge at each time step and accumulate the incremental probabilities associated with the summation of hydrological and error model predictions for each discharge increment. This avoids the need to store all the hydrological model simulations — they can be dealt with one at a time.

An alternative approach, which is more consistent with the methods of obtaining confidence intervals for flows used in the GLUE procedure (Beven and Binley, 1992) is to estimate confidence intervals of the predictive distribution with the statistical errors removed. This amounts to

$$P(Y_t < y | z) = \int_\theta P(Y_t < y | \theta, z) f(\theta | z) \, d\theta, \qquad (17.18)$$

where $f(\theta|z)$ is the marginal posterior of θ. However, since Y_t is a deterministic function of the θ parameters, (17.18) reduces to

$$P_r(Y_t < y|z) = \int_B f(\theta|z)\,d\theta, \qquad (17.19)$$

where B is the set of θ such that Y_t is less than y, given both θ and z. Clearly in this case computational problems are simplified at the expense of obtaining confidence intervals for the hydrological model predictions rather than resultant flows.

17.8   RESULTS

As discussed earlier, there are two stages of modelling: calibration (model fitting) and validation (checking the adequacy of the fit). The calibration stage involves finding the parameters of a hydrological model which give the best fit of the simulated flows (model output) to the observed flow series, with rainfall and evaporation data as the model input. The criterion according to which the comparison of the simulated and observed flows is made is usually called the objective function. This criterion is also used to select a suitable model structure.

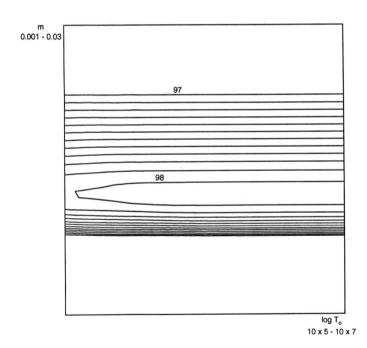

Figure 17.2   Response surface map for the maximum efficiency criterion (%) (November–December 1988).

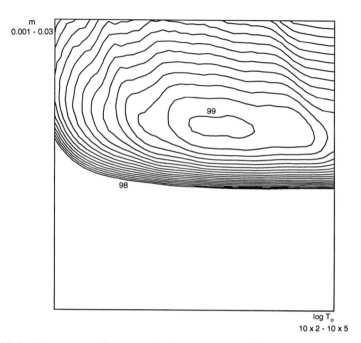

Figure 17.3 Response surface map for the maximum efficiency criterion (%) (end of September–October 1988).

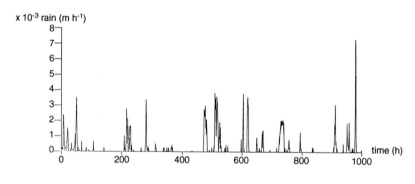

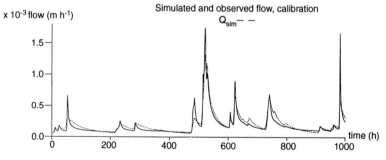

Figure 17.4 Rainfall series together with simulated and observed flows for the calibration period (November–December 1988).

Depending on the purpose of the modelling, and the available observations, different objective functions may be used. For example, (17.5) is the standard approach. However, for flood prediction purposes, functions that give more weight to fitting the maximum flows are used. The validation stage consists in running the model for a different time period and assessing the fit of aspects of interest.

For comparison of this work with standard methods, the parameters of TOPMODEL m and T_0 were chosen using the Powell non-gradient optimisation procedure (Powell, 1970) applied to (17.5). The response surface map for the entire 1000 h calibration period is given in Figure 17.2. It can be seen that the model becomes insensitive for the large T_0, but is sensitive to the recession parameter m throughout its parameter space. Only one of the examined 1000 h time periods gave a convex response surface (Figure 17.3). The hydrographs of the simulated and observed flows and rainfall series for the calibration and validation periods are given in Figures 17.4 and 17.5 respectively.

In this work the posterior distribution was evaluated for the same 1000 h calibration period, updating the posterior over ten 100 h intervals. As explained earlier, this procedure was dictated by the computational limitations in the evaluating the posterior distributions of the parameters for 1000 samples.

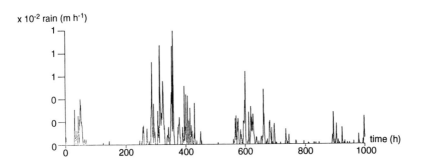

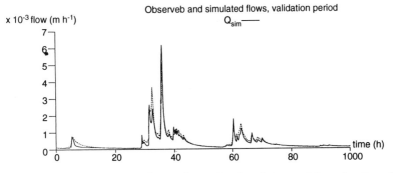

Figure 17.5 Rainfall series together with simulated and observed flows for the validation period (end of September–October 1988).

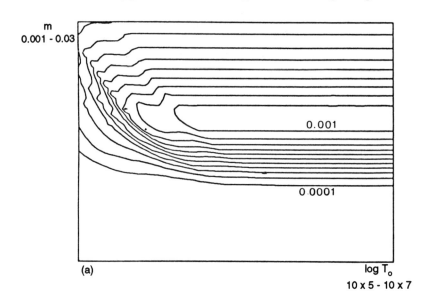

(a)

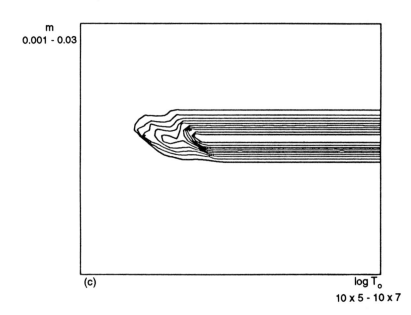

(c)

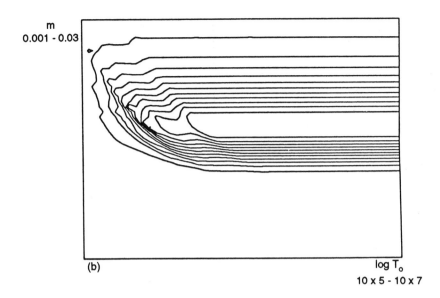

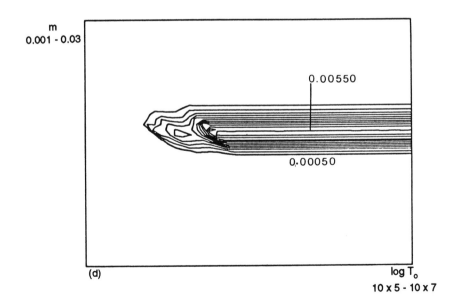

Figure 17.6 Updated marginal posterior probability contour maps for θ after (a) 100 h, (b) 200 h, (c) 400 h and (d) 1000 h of simulations.

TOPMODEL was run on the Meiko parallel transputer system for the continuous 1000 h period, and the results (efficiencies and the sums of simulated flows needed for the posterior evaluations) were saved after every 100 h period. The updating procedure is illustrated by Figure 17.6. Marginal posterior probability plots of the joint distribution of θ are obtained by summing over the noise parameters at each updating stage. The updating times shown are 100, 200, 400 and 1000 h. As the new data are taken into account, the region of the feasible parameters on the probability surface decreases. Note that the procedure of updating of posterior can also be used for 'event' data, which are often more easily available in hydrological practice than 'continuous' hourly data.

Now consider the confidence intervals for predictions of flows based on observed rainfalls. Following Section 17.7, we evaluate two forms for flows corresponding to TOPMODEL predictions with and without the inclusion of the statistical noise. Figure 17.7 shows confidence intervals for the predictive distribution for predictions, for a 100 h period, solely from TOPMODEL using (17.19). The tightness of these pointwise limits reflects the uncertainty measure being constrained to the TOPMODEL parameters θ, as with the GLUE procedure, and excluding errors in predictions. From a computational viewpoint, the evaluation of confidence intervals for the predictive distribution of actual flows proved more difficult owing to the dimensionality of the problem and the coarse discretisation

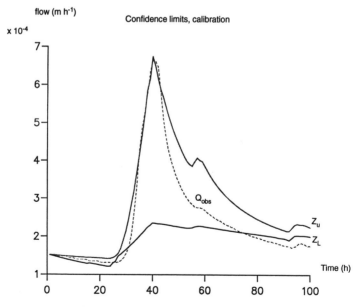

Figure 17.7   90% confidence limits for the simulated flow evaluated on the basis of marginal posterior probabilities for 100 h during the calibration period. Here Q_{obs} is the observed flow.

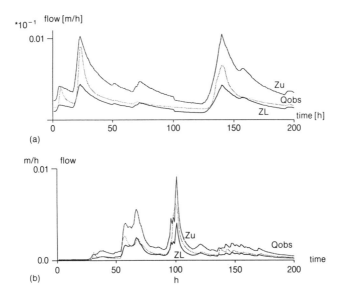

Figure 17.8 80% confidence limits for flow predictions evaluated with the noise model for 100 h during (a) the calibration and (b) the validation periods. Here Q_{obs} is the observed flow.

of the noise parameters. Thus the confidence limits shown in Figure 17.8, obtained from (17.17), involve crude computations. These intervals are notably wider than in Figure 17.7, since the additional uncertainty arising from the inclusion of the multiplicative noise is accounted for. In these calculations the noise distribution is time-independent, so errors in the time of the predicted peak flow lead to large variances in the noise distribution and hence wide confidence intervals. Such confidence intervals are of limited value. Clearly either more data are required, or the observed errors within the event need to be exploited in updating predictions, and hence confidence intervals, before they are of much practical use.

17.9   DISCUSSION

In the classical Bayesian approach the results of inference about the model parameters can give an indication as to any necessary changes of the error model structure. Hence an interactive inference procedure can be used (Box and Tiao, 1992). A change of model could be just a change of the error model itself, e.g. the distribution of the errors changed from lognormal to exponential, or it might be in the hydrological model that provides the flow predictions, which consequently may then also result in a change in the distribution of errors. The first change might follow from a statistical analysis of the data, whereas the second can be supported by the a priori knowledge of the modeller about the nature of the modelled hydrological event. Both these changes will lead to a change in the posterior distribution of the parameters of the model. Also, a change in the objective of the modelling, defined by the structure of the model observation equation, can impose the use of a different error model, thus changing the posterior distributions of the predicted variables as well as the feasible parameter region.

The Bayesian procedure outlined here is quite general, and does not depend on any particular likelihood measure. The likelihood function used here relies specifically on a model for the error structure. The confidence limits obtained consequently only quantify the uncertainty conditional on that statistical model being correct. Thus, to reduce the model effect in the calculations of predictive distribution confidence intervals, other models need to be examined. The alternative approach proposed by Beven and Binley (1992) amounts to being a Bayesian procedure with a likelihood measure, rather than a likelihood function, so need not be specific to a statistical model. They have listed a wide variety of likelihood measures typically corresponding to goodness-of-fit measures, including binary measures (behaviour/non-behaviour as used by Hornberger and Spear, 1980) and probabilistic measures from fuzzy set theory (as used e.g. by van Straten and Keesman, 1991). Their results, however, were based on a sum-of-squares likelihood measure defined as

$$L(z|\theta) = (\sigma_\varepsilon^2)^{-1}, \tag{17.20}$$

where σ_ε^2 is the variance of the error in predicted flows over a certain calibration period. In their study Bayes' formula was also used recursively, with the posterior distribution being updated after each calibration event. Note that with this likelihood measure, the recursive use of Bayes' formula means that the sum of squared errors enters multiplicatively into the posterior calculation. This effectively will give more weight to the most recent periods of the flow modelling in determining the posterior likelihood distribution, which seems an undesirable feature unless some change over time is expected.

There are a number of interesting questions that still require further work and analysis.

(i) Beven and Binley (1992) have shown how an analysis of uncertainty measures can be used to assess the value of additional data in the calibration process. It may be possible to define some criterion for stopping the updating procedure when additional data no longer lead to a reduction in predictive uncertainty. This is equivalent to the question of how many samples are needed for a required accuracy in the flow predictions.

(ii) It is also possible to consider the elimination of events that do not give distributions consistent with those already used, such as might be due to the occurrence of an extreme event or to different patterns of water storage in the catchment that are not properly predicted by the hydrological model. In many hydrological regimes, for example, the discharges resulting from the first storms after a long dry period are notoriously difficult to predict accurately.

(iii) A particularly interesting topic is the use of other, possibly spatially distributed, hydrological variables in the calibration process. Each data type could be associated with a certain likelihood measure (and error model) that can be used in Bayes' formula. Where the posterior distributions arising from different data types are markedly different, this may lead to a marked refinement in the parameter distributions. In other cases certain data can be shown to have little value in constraining uncertainty in the predictions (see e.g. Binley and Beven, 1991).

(iv) Binley *et al.* (1991) pointed out that in assessing the effects of man-induced land-use changes on the water balance of catchment areas, proper account should be taken of uncertainty in the predictions of any model used. The same is clearly true, of predictions of the effects of climate changes. Predictions into the future will also require that uncertainty in the input data be taken into account. The methodology proposed here allows such an assessment to be made and, using the recursive Bayesian approach outlined, also allows apparent changes in parameter distributions to be followed if data are available as such changes take place.

(v) Finally, at a technical level, our statistical analysis requires further work with the implementation of Gibbs sampling techniques, and the development of updated confidence intervals for flows.

One of the most important aspects of the work presented in this chapter is the way in which it focuses attention on the role of data in constraining model uncertainties. Estimation of the confidence limits of the model predictions should lead directly on to the next question of what would be the most valuable data to reduce the range of uncertainty. The answers may be unexpected.

18

Statistical Methods for the Evaluation of Hydrological Parameters in Landuse Planning

U. Busch
Dortmund University, Germany

H. M. Brechtel
Hessian Forest Research Centre, Germany

W. Urfer
Dortmund University, Germany

18.1 INTRODUCTION

In connection with environmental impact assessments for the purpose of landuse planning, investigations on the hydrological site conditions and their possible changes are of vital importance for environmental decision-making. In many cases long-term measurements must be carried out on various water balance components and parameters of hydrological processes, including substance deposition caused by air pollution. Such monitoring must take into account the large spatial and temporal variation and correlation of measured quantities (see Brechtel, 1989).

Data from a forest-ecologic investigation in the Rhine–Main Valley are used to demonstrate how patterns in field measurements over space and time can be analysed statistically. From a great number of methods used, two that proved

Statistics for the Environment 2: Water Related Issues. Edited by Vic Barnett and K. Feridun Turkman
© 1994 John Wiley & Sons Ltd.

useful for several major problems are chosen and explained: repeated measures analysis and time series models with interventions and input variables.

18.2 THE FOREST-ECOLOGIC PROOF DOCUMENTATION INVESTIGATION

During the years 1982–84, the Frankfurt International Airport was extended with a new runway. When the licence for building it was granted, the Flughafen AG was obliged to allow a forest-ecologic proof documentation to be made. This investigation was intended to determine possible negative changes of forest stands and site conditions caused by clearing 3.5 km² of forest, by the construction work, and by the later use of the runway. The Institute of Forest Hydrology of the Hessian Forest Research Centre (HFV) has been entrusted with leading the investigation and with carrying out its forest-hydrologic part (Hammes and Brechtel, 1986).

In the investigation area of 72 km² the HFV has fenced in 9 open area stations and 16 forest stand plots of 2500 m² each for observation, and built 36 wells where samples of groundwater can be taken. Twenty-five pine and spruce trees were chosen for taking needle samples.

The data were collected in the period from 1981 to 1991, divided into three investigational phases: a short phase to characterise the situation before the clearing was started, the phase of construction work (October 1981–April 1984) and the phase since May 1984, when the runway was put to use. The variables can be classified into four groups:

(i) Daily data from the airport weather station:
   air temperature,
   vapour pressure and atmospheric moisture,
   wind velocity,
   duration of sunshine,
   type and quantity of precipitation;
(ii) Monthly measurements of water balance components in the open and under forest stands:
   open area precipitation,
   throughfall and drip (canopy penetration),
   seepage water 10 cm below the soil surface,
   depth of the ground water level,
   soil moisture;
(iii) Bimonthly water analyses from the above components:
   pH, conductivity, hardness,
   anions (NO_3, SO_4, PO_4, BO_3, Cl, F),
   cations (NH_4, Ca, Mg, K, Na),
   metals (Cd, Pb, Mn, Al, Ni, Zn);

(iv) Needle analyses with regard to nutritive elements and toxic metals:
    half-yearly from 17 pine trees,
    once from 8 spruce trees, sorted by needle age.

For each group of data, there were several questions to be anwered by statistical methods. On the one hand, their evaluation was to contribute directly to the proof documentation. On the other hand, general forest-ecologic problems were to be examined.

(i) *Weather conditions*
    (a) Has the weather during the investigation period been representative for the normal conditions of the climate in this area? This is important mainly for the interpretation of possible changes in other variables.
    (b) Are there significant spatial differences in the quantity of open area precipitation? The amount of open area precipitation on the forest stand plots cannot be estimated directly, and therefore has to be estimated. The decision as to whether the mean from all open area stations should be used, or instead the measurements of the nearest stations only, depends on the answer to this question.
(ii) *Water balance*
    (a) Evapotranspiration is estimated using the Penman formula.
    (b) Which are the variables influencing canopy and litter interception–evaporation?
    (c) Vertical water movement in the soil is modelled.
    (d) Has there been a change in the reaction of the groundwater level depth to the amount of precipitation? Owing to the clearing of a large forest area, the amount of evapotranspiration in this area might be decreased. As a consequence, the groundwater level might rise.
(iii) *Water quality*
    (a) Are there any spatial or temporal changes during the three investigational phases? Especially those changes are of interest that might be caused by the construction or use of the new runway.
    (b) Do the concentrations of chemical substances in the groundwater exceed any of the tolerable highest concentrations in drinking water?
    (c) How is the quality of the water changing on the way from the open area precipitation to the groundwater?
    (d) Is there any interdependence between different chemical substances contained in the water?
(iv) *Needle analyses*
    (a) Are there any spatial or temporal changes in the element composition of the needles?
    (b) What is the development of element content with increasing age of the needles?
    (c) What is the interdependence between different chemical elements in the needles?

(d) How is the content of chemical elements in the needles influenced by the water quality?

In the evaluation of these data (Busch *et al.*, 1993) two groups of statistical models were of particular importance: *repeated measures analysis* and *time series models* (fixed or variable seasonal and trend components, autoregressive integrated moving average processes, and transfer functions).

A variety of descriptive methods have also been used, as well as regression and correlation analysis and kriging. Problems were caused mainly by simultaneous spatial and serial correlation of the data and by the strong interdependence of the variables.

In this chapter the use of repeated measures models and time series models will be explained and examples from our project will be provided. Both methods are very useful for evaluating ecological data sets.

18.3   REPEATED MEASURES ANALYSIS

The essential feature of repeated measures studies is that each subject or experimental unit is observed under two or more observational conditions. This means that the measurements are not independent. The units may or may not be divided into groups.

One important class of such studies are longitudinal studies, where successive time intervals correspond to the observational conditions. In this case, the repeated measures factor or within-units factor is time, and the data represent short time series. But totally different factors are possible as well, as will be shown in the following sections.

Koch *et al.* (1988) give several strategies for the analysis of repeated measures studies, including

  (i) univariate analysis of within-unit functions;
 (ii) multivariate analysis of within-unit functions;
(iii) repeated measures analysis of variance;
 (iv) nonparametric rank methods;
  (v) categorical data methods.

The specification of these statistical methods takes five considerations into account:

(a) the measurement scale;
(b) the number of units;
(c) the role of randomisation;
(d) the nature of the covariance structure of the repeated observations on each unit;
(e) the potential influence of carry-over effects.

We concentrate on models where the errors have Gaussian distributions. For this case Jones (1993) gives a state-space approach, which provides a convenient way to compute likelihoods using the Kalman filter. Diggle (1990) also used the method of maximum likelihood to estimate jointly the parameters which define the mean response and those which define the covariance structure. Bonney (1987) developed a methodology for serial observations with binary response, which was applied by Urfer (1993) for the analysis of data from permanent observation plots in forestry. Zeger and Liang (1992) review marginal, transition, and random effects models for the analysis of discrete and continuous longitudinal data.

18.3.1   The model

Table 18.1 shows the typical data structure for a repeated measures model.

Let us consider our first example: the manganese content in spruce needles. In spring 1987, $n = 15$ spruce branches were collected in the investigation area. They represent the experimental units. $n_1 = 9$ of them came from sites with slight soil acidification and $n_2 = 6$ from sites were the acidification is more severe. This is the group factor or between-units factor with $g = 2$ levels. For the chemical analyses, the needles were sorted by age. Needles grown in four different years were distinguished. So we have a repeated measures factor with $t = 4$ levels.

Concerning these data, there are three questions we want to answer by significance tests.

(i) Do the groups differ in the mean level of the values? (Here, does the soil acidification have an effect on the mean manganese content in the needles?)

Table 18.1   Data structure for repeated measures analysis

Group	Unit	1	2	$\cdots$	t
		\multicolumn{4}{c}{**Measurements under different conditions**}			
1	1	y_{111}	y_{112}	$\cdots$	y_{11t}
	2	y_{121}	y_{122}	$\cdots$	y_{12t}
	$\vdots$	$\vdots$	$\vdots$		$\vdots$
	n_1	y_{1n11}	y_{1n12}	$\cdots$	y_{1n1t}
2	1	y_{211}	y_{212}	$\cdots$	y_{21t}
	$\vdots$	$\vdots$	$\vdots$		$\vdots$
	n_2	y_{2n21}	y_{2n22}	$\cdots$	y_{2n2t}
$\vdots$	$\vdots$	$\vdots$	$\vdots$		$\vdots$
g	1	y_{g11}	y_{g12}	$\cdots$	y_{g1t}
	$\vdots$	$\vdots$	$\vdots$		$\vdots$
	n_g	y_{gng1}	y_{gng2}	$\cdots$	y_{gngt}

(ii) Is there an effect of the measurement conditions? (In our example, is the manganese content affected by the age of the needles?)

(iii) Do the groups differ in their reaction on the measurement conditions? (Here, does the soil acidification effect the development of the manganese content with increasing needle age?)

Disregarding the group factor for the moment, we immediately come to the second question. The simple analysis of variance seems to offer a first access to this problem (The corresponding model is Model 1 in Figure 18.1).

Since the measurements under the different conditions are made on the same experimental units, however, we cannot assume the error terms to be independent. This means that an important assumption for the F test is violated. Furthermore, often the measurement conditions not only affect the expected value, but also the variance. In this case we have no homogenity of variance either. And there is a third objection to this model. Its error terms become particularly large when the units differ in their mean levels, while only the differences within the units are of interest.

The repeated measures model takes these three aspects into account (see Model 2 in Figure 18.1). First, the measurements on each unit are allowed to be correlated, and their variance may depend on the measurement conditions. Only the units are assumed to be independent and the covariance structure of the measurements on all units to be equal. Secondly, the error term is split into the mean effect of the single units (a_j) and the remaining error (e_{jk}). The mean effects of the units are assumed to be normally distributed with mean 0 and variance σ_α^2. By splitting the error term, the error sum of squares becomes smaller, though at the same time its degrees of freedom are reduced by $n - 1$ or $n - g$ respectively.

Model 3 (in Figure 18.2) shows the repeated measures model with a group factor. The corresponding sums of squares (SS) and degrees of freedom (DF) are given together with the model equation. The restrictions are analogous to those in the simple analysis of variance model, the test assumptions are the same as in Model 2. The model is distinctly divided into two parts—one describing the differences in the mean between units and between groups, the other modelling the differences between the measurement conditions within the units.

In this model we can set up and test three hypotheses corresponding to the questions formulated above.

(i) $H_0^G: \beta_1 = \cdots = \beta_g = 0$ (the groups do not differ in the mean level of the values). The corresponding test statistic is

$$\frac{\text{MSG}}{\text{MSU}} \sim F_{n-g}^{g-1}. \tag{18.1}$$

This is the usual analysis of variance for the unit means.

(ii) $H_0^C: \tau_1 = \cdots = \tau_t = 0$ (there is no difference between the measurement conditions). If the covariance matrix Σ satisfies the sphericity condition (see below),

$y_{jk} = \mu + \tau_k + e_{jk}$,

    with $j = 1, \ldots, n$, $k = 1, \ldots, t$,

    the restriction $\sum\limits_{k=1}^{t} \tau_k = 0$,

    and the test assumptions

    $e_{jk} \sim N(0, \sigma^2)$

    and $\operatorname{cov}(e_{j_1 k_1}, e_{j_2 k_2}) = 0$ for $(j_1, k_1) \neq (j_2, k_2)$;

    that is, $\Sigma_e = \begin{pmatrix} \sigma^2 & & 0 \\ & \ddots & \\ 0 & & \sigma^2 \end{pmatrix}$

Model 1: ANOVA-model (repeated measures factor only)

$y_{jk} = \mu + a_j + \tau_k + e_{jk}$,

    with $j = 1, \ldots, n$, $k = 1, \ldots, t$,

    $a_j \sim N(0, \sigma_\alpha^2)$, $\sum\limits_{k=1}^{t} \tau_k = 0$,

    and the test assumptions

    $e_{jk} \sim N(0, \sigma_k^2)$,

    $\operatorname{cov}(e_{jk_1}, e_{jk_2}) = \sigma_{k_1 k_2}$

    and $\operatorname{cov}(e_{j_1 k_1}, e_{j_2 k_2}) = 0$ for $j_1 \neq j_2$;

    that is, $\Sigma_e = \begin{pmatrix} \Sigma & & 0 \\ & \ddots & \\ 0 & & \Sigma \end{pmatrix}$, $\Sigma = \begin{pmatrix} \sigma_1^2 & \sigma_{12} & \cdots & \sigma_{1t} \\ \sigma_{21} & \sigma_2^2 & & \vdots \\ \vdots & & \ddots & \vdots \\ \sigma_{t1} & \cdots & \cdots & \sigma_t^2 \end{pmatrix}$

Model 2: repeated measures model (repeated measures factor only)

Figure 18.1   Models for ANOVA and repeated measures

this hypothesis can be tested by an F test as well:

$$\frac{\text{MSC}}{\text{MSE}} \sim F^{t-1}_{(n-g)(t-1)}. \tag{18.2}$$

If the sphericity condition is not satisfied, a T^2 test can be used instead. The distribution of its test statistic is equivalent to an F distribution with less

$$
\begin{array}{llll}
& y_{ijk} - \mu = & \beta_i & + & a_{ij} \\
\text{SS:} & \text{SST} = & \text{SSG} & + & \text{SSU} \\
& \text{total} & \text{groups} & & \text{units} \\
\text{DF:} & nt - 1 = & (g - 1) & + & (n - g)
\end{array}
$$

$$\underbrace{}$$
between units
$(n - 1)$

$$
\begin{array}{lllll}
+ & \tau_k & + & (\beta\tau)_{ik} & + & e_{ijk} \\
+ & \text{SSC} & + & \text{SSGC} & + & \text{SSE} \\
& \text{conditions} & & \text{groups} \times \text{conditions} & & \text{error} \\
+ & (t - 1) & + & (g - 1)(t - 1) & + & (n - g)(t - 1)
\end{array}
$$

$$\underbrace{}$$
SSWU
within units
$n(t - 1)$

Figure 18.2 Model 3: repeated measures model with groups.

degrees of freedom in the denominator. This means the test is less powerful:

$$T^2_{t-1}(n - g) \cong F^{t-1}_{n-g-(t-1)+1}. \tag{18.3}$$

The sphericity condition is satisfied if and only if the contrasts of each complete system of standardised orthogonal contrasts in the measurement conditions are independent and have the same variance. Applied to the covariance matrix Σ, this means that $\sigma^2_{k_1} + \sigma^2_{k_2} - 2\sigma_{k_1k_2} = 2\sigma^2_c$ has to be constant for each pair (k_1, k_2) of measurement conditions.

A test for this sphericity condition is given by Anderson (1958). Alternatively, a coefficient ε measuring the deviation from sphericity can be estimated. This coefficient can be used to correct the degrees of freedom in the F test (cf. e.g. Greenhouse and Geisser, 1959).

(iii) H_0^{GC}: $(\beta\tau)_{11} = \cdots = (\beta\tau)_{gt} = 0$ (the groups do not differ in their reaction to the measurement conditions). As for the second hypothesis, the F test is possible if the sphericity condition is satisfied:

$$\frac{\text{MSGC}}{\text{MSE}} \sim F^{(g-1)(t-1)}_{(n-g)(t-1)}. \tag{18.4}$$

Otherwise, a multivariate test can be used (e.g. Wilks' lambda). If this third hypothesis is rejected, the results of the first two tests have to be interpreted with great care, because the existence of an interaction implies that there is an effect of both the single factors. But the direction of the repeated measures effect and/or how strong it is depends on the level of the group factor.

This method is described in Busch (1990, 1994), Elashoff (1986), Koch *et al.* (1988), Urfer and Pohlmann (1992) and others. The multivariate approach is given in Davidson (1980).

In Section 18.3.2 we apply repeated measures analysis to two of the problems from our project. The necessary calculations have been done by means of the SAS-procedure GLM (SAS, 1987).

18.3.2 Manganese in spruce needles

As described above, the manganese content in the dried needles of 15 spruce branches in two groups (group 1, with slight soil acidification; groups 2, with severe soil acidification) was analysed. The needles grown in four successive years were examined separately. So we have $n = 15$ units in $g = 2$ groups ($n_1 = 9$, $n_2 = 6$), and the repeated measures factor consists of $t = 4$ measurement conditions.

In a preceding test the hypothesis of sphericity was rejected (χ^2-approx. = 19.9, 5 degrees of freedom, $p = 0.0013$). For this reason, we use the multivariate tests for the hypotheses concerning the repeated measures factor needle age. Since there are only two groups, the T^2 test can be used in both cases.

Table 18.2 shows the results of the three tests. All the hypotheses are clearly rejected.

If we estimate the parameters in the corresponding model and plot the results together with the original values, we get the picture in Figure 18.3. The effects are so obvious that the tests are hardly required to prove them.

Where the soil acidification is severe, the manganese content of the needles is always higher than in places with slight acidification. This means that the groups differ in the mean level of the values. The mean manganese content increases with the age of the needles. Here we see the effect of the repeated measures factor. This increase is not equal in the two groups. While in the group with slight soil acidification the manganese content is nearly constant and increases only slightly in the fourth year, in the group with severe soil acidification there is a pronounced accumulation of manganese, especially in the first years. So the reaction to the measurement conditions is different in the two groups.

These conclusions add to the results of a study by Gärtner *et al.* (1990), who found a positive relation between the manganese content in the needles of spruce trees and their loss of needles—an indicator for their state of health.

Table 18.2   Results for the manganese content of the spruce needles

	F statistic		d.f.	p-value	Reject H_0
Group	132.49		1, 13	0.0001	Yes
Needle age	99.21	$(\hateq T^2)$	3, 11	0.0001	Yes
Group × needle age	52.95	$(\hateq T^2)$	3, 11	0.0001	Yes

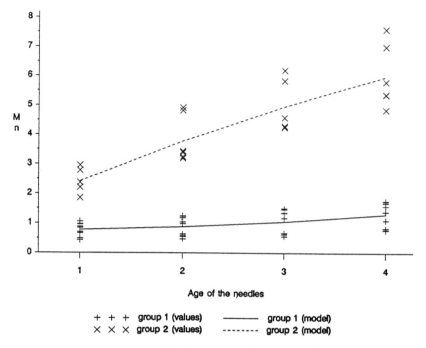

Figure 18.3 Manganese content of the spruce needles (in g per kg dry matter weight): group 1 with slight soil acidification; group 2 with severe soil acidification.

18.3.3 Spatial distribution of precipitation

Our data set on the distribution of the precipitation in the investigation area consists of monthly measurements of the precipitation amount on nine different open area stations. With these data, we want to find out whether there are systematic spatial or seasonal differences in the quantity of precipitation and whether the seasonal component differs from one station to the other.

The variance of the data is large. At first glance, a plot shows no recurrent seasonal component, and the order of the stations is changing from month to month. In contrast to the first example, the results of the tests cannot be predicted easily in this case.

A closer look at the data shows that the deviations between the various stations and their variance are approximately proportional to the mean of all stations. This made us choose a multiplicative model, or, to be precise, to use the logarithm of the data in a linear model. This means that the differences between the stations are seen relative to the mean amount of precipitation from that particular month.

When formulating the model, it seems natural to interpret the stations as units and the months as repeated measurements. But the covariance structure of the data makes another interpretation more plausible. There is hardly any

Table 18.3 Results for the spatial distribution of the precipitation

	F statistic		d.f.	p-value	Reject H_0
Month	1.15		11, 108	0.3300	No
Station	12.97	$(\hat{=} T^2)$	8, 101	0.0001	Yes
Month × station	1.06	$(\approx \Lambda)$	88, 671.8	0.3320	No

autocorrelation in the time series; the correlation between one month and the next is only -0.08 to -0.15 (standard error 0.09). So we can assume independence here. In contrast, the correlation between the stations is very high, with values from 0.97 to 0.99, and must not be ignored. Furthermore, there are large differences in the level of the values between months with low and high precipitation. These are of interest only as far as they represent a systematic seasonal component.

For these reasons, we obtain a suitable model if we regard the single months as units ($n = 120$) and divide them into $g = 12$ groups according to the calendar month, with $n_i = 10$ measurements each from the 10 years. The repeated measures factor is the station with $t = 9$ levels. The model equation is

$$y_{ijk} = \mu \beta_i a_{ij} \tau_k (\beta\tau)_{ik} e_{ijk} \qquad (18.5)$$

$$\Leftrightarrow y_{ijk}^* = \log y_{ijk} = \mu^* + \beta_i^* + a_{ij}^* + \tau_k^* + (\beta\tau)_{ik}^* + e_{ijk}^*,$$

where y_{ijk} is the amount of precipitation in station k in the month j of the year i. Again the preceding test shows a significant deviation from the sphericity condition (X^2-approx. $= 133.5$, 35 degrees of freedom, $p < 0.0001$); therefore the multivariate test is used. The results are given in Table 18.3.

Table 18.4 Estimates of the parameters in the precipitation model

Station	$\overline{y_{\cdot\cdot k}^*}$	$\hat{\tau}_k^* = \overline{y_{\cdot\cdot k}^*} - \overline{y_{\cdot\cdot\cdot}^*}$	$\hat{\tau}_k = e^{\hat{\tau}_k^*}$	Deviation from the mean
F1	3.800	0.058	1.060	+ 6.0%
F2	3.748	0.007	1.007	+ 0.7%
F3	3.694	− 0.048	0.953	− 4.7%
F4	3.712	− 0.029	0.971	− 2.9%
F5	3.708	− 0.033	0.967	− 3.3%
F6	3.770	0.029	1.029	+ 2.9%
F7	3.713	− 0.029	0.972	− 2.8%
F8	3.781	0.040	1.041	+ 4.1%
F9	3.746	0.004	1.004	+ 0.4%

$$\hat{\mu}^* = \overline{y_{\cdots}^*} = 3.741 \qquad \hat{\mu} = e^{\hat{\mu}^*} = 42.1 \text{ mm month}^{-1}$$

The differences between the open area stations are significant. Seasonal differences can be proved neither in the mean amount of precipitation ('month') nor in its spatial distribution ('month × station'). For this reason, the group factor month has been eliminated from the model, and we obtain as final model

$$y_{ijk} = \mu a_{ij} \tau_k e_{ijk} \tag{18.6}$$
$$\Leftrightarrow y_{ijk}^* = \log y_{ijk} = \mu^* + a_{ij}^* + \tau_k^* + e_{ijk}^*.$$

In this model we have estimated the parameters μ and τ_k (Table 18.4). The estimated percentages of the single station's deviation from the mean of all

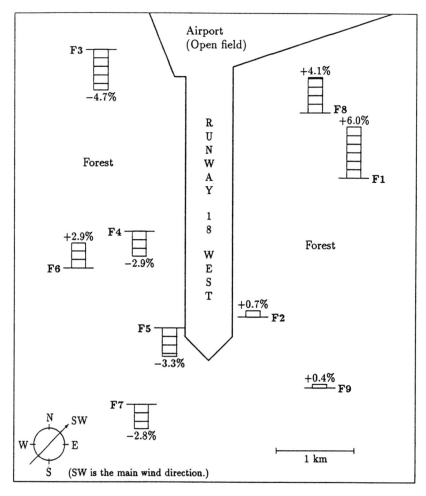

Figure 18.4 Deviation of the amount of open area precipitation from the mean of all stations in comparison with the location of the different stations F1–F9 in open areas within forest.

stations are shown graphically in Figure 18.4. According to these estimates, there can be differences between the stations of more than 10% of the mean precipitation in the area (stations F1 and F3), in spite of the small distances of not more than 4 km. We find the highest precipitation in the north-east of the runway and the lowest along its western side. We conclude that it is not sufficient to use the mean amount of precipitation in the area in water balance models for all the forest stand plots. For each plot, the values of an open area station nearby or the mean of the surrounding stations should be used instead.

18.4 TIME SERIES ANALYSIS

This section deals with the question of whether the chemical quality in water samples of open area precipitation, canopy penetration, seepage water and groundwater has been affected by the construction and the later use of the new runway.

Unfortunately, there are several reasons why it is impossible to prove that changes in the water quality are caused by the runway. Firstly, there are no data from the period before the construction work began. So we can only compare the construction phase and the phase when the runway was put to use. Secondly, the investigation area lies near to the big city of Frankfurt, and is surrounded by diverse industries, refuse incinerators, motorways and other sources of pollution of all kinds. Of course, their emissions could have changed over time in quantity and quality, and we do not have sufficient data to model these changes. Thirdly, the quality of the precipitation is influenced by the weather conditions. However, the data is not suitable to take this into account, because the water has always been collected in aggregated form over the whole month, while the weather changes much more rapidly.

What we can do is to look for a development of the chemical quality of water sampled in the above-mentioned four compartments parallel to the changes in the use of the area, particularly for a change of the level when the runway was put to use and for a trend after that. These changes can hint at a connection between the water quality and the construction and later use of the runway.

At least the amount of precipitation can be used as a covariate. In most cases its correlation with the concentrations is negative, while it is positively correlated with deposition rates. For the groundwater, we do not know the absolute quantity but only its level. This means that we cannot calculate anything analogous to deposition, but we can examine the concentrations, and the change of the groundwater level can be used as a covariate.

The relations of adjacent water balance components ('compartments') can be analysed by transfer functions.

Another problem is the number of the time series to be analysed. There are data from 20 variables in 4 compartments in 16 forest stand plots, i.e. 1280 time series. We solved this problem for the present study by choosing only four forest stand plots and omitting the least important variables. Transfer functions were analysed for one plot only. Of course this is not a satisfactory solution.

We shall now explain the time series and transfer function models we have used for these analyses, and then show the results for nitrate.

18.4.1   A time series model for the water quality data

On the open area station F4 and the forest stand plot No. 4 the water samples of open area precipitation, canopy penetration, seepage water, and groundwater have been analysed most throughly. There are data from the even months from October 1981 to August 1991, and, beginning in December 1983, the water has been analysed monthly. As the regularly missing values in the construction phase (phase 1, up to April 1984) might cause a bias in the estimators, we used only data from even months for the time series model. For the transfer functions, we used the complete data from December 1983 to the end, but omitted the data from the first two years. We can do this because we do not expect any great changes in the relation between the compartments from one investigational phase to the other.

On the other forest stand plots, the water samples have been analysed in the even months in phase 1 and in the odd months in phase 2. Whichever model we use, this can bias the estimates. The safest way to analyse these data seemed to be to use half-year concentration means, weighted by the water quantity, combining May–September and October–April (growing season and dormant season).

For our time series analyses we choose a traditional model consisting of deterministic trend P_t and seasonal S_t components and an irregular or stochastic component X_t as described in Wei (1990) and others. Another component is added to model the influence of the covariate C_t:

$$Y_t = P_t + S_t + C_t + X_t. \tag{18.7}$$

The components are assumed to be additive and independent. In the following we shall discuss the models for plot No. 4 only, but those for the other plots are quite similar except for the seasonal component, which consists only of a single indicator variable for growing or dormant season. For the full model (Model 4), see Figure 18.5.

The modelling of the *trend component* is complicated slightly by an intervention: the opening of the new runway and at the same time the end of the construction work. The consequence could be a change in the level of the values. Furthermore, there might be a gradual increase or decrease after this intervention, caused by the continuously increasing number and size of aircraft starting from the runway or by more effective measures controlling the pollution. These two possible effects of the intervention are modelled by two variables with their corresponding parameters. They are chosen in such a way that they can be

$$Y_t^* = \underset{P_t}{\underbrace{\mu + \alpha N_t + \beta T_t^{(2)}}} + \underset{S_t}{\underbrace{\gamma_1 \sin ht + \gamma_2 \cos ht + \gamma_3 \sin 2ht + \gamma_4 \cos 2ht + \gamma_5 \cos 3ht}} + \underset{C_t}{\underbrace{\delta Q_t^m}} + X_t$$

The parameters and variables in this model are

t	the number of the measurement (every second month, counted from February 1981),
Y_t^*	the **measured value** at time t, transformed if necessary,
P_t	the **trend component**,
μ	the overall mean,
α	the change in the mean level of the data from phase 1 (construction) to phase 2 (use of the runway),
N_t	$= \begin{cases} -n_2/n & \text{for } t \leqslant 20 \text{ (April 1984)} \\ +n_1/n & \text{for } t \geqslant 21 \text{ (June 1984),} \end{cases}$
n	the overall number of measurements,
n_1	the number of measurements in phase 1,
n_2	the number of measurements in phase 2,
β	the trend parameter (increase per year in phase 2),
$T_t^{(2)}$	$= \begin{cases} 0 & \text{for } t \leqslant 20 \text{ (April 1984)} \\ \frac{1}{6}(t - \bar{t}^{(2)}) & \text{for } t \geqslant 21 \text{ (June 1984),} \end{cases}$
$\bar{t}^{(2)}$	the mean of all values t in phase 2,
S_t	the **seasonal component**,
h	$= \frac{1}{3}\pi$
$\gamma_1 \sin ht + \gamma_2 \cos ht$	a sine-cosine function with one cycle per year,
$\gamma_3 \sin 2ht + \gamma_4 \cos 2ht$	a sine–cosine function with two cycles per year,
$\gamma_5 \cos 3ht$	a cosine function with three cycles per year,
C_t	the **effect of the covariate**,
δ	the parameter for the linear influence of the water quantity,
Q_t^m	the mean-adjusted water quantity at time t,
X_t	the **irregular component**,
X_t	$= \left(\sum_{j=1}^{p} \phi_p(B) \right)^{-1} a_t = \left(1 - \sum_{j=1}^{p} \phi_j B^j \right)^{-1} a_t = \sum_{j=1}^{p} \phi_j X_{t-j} + a_t,$
	a pth-order autoregressive process AR (p),
ϕ_j	the parameters of the AR (p) process,
a_t	white noise.

Figure 18.5 Model 4: time series model for the bimonthly data.

estimated and tested independently from each other and from the overall mean, which formes the third part of the trend component.

The *seasonal component* is modelled as a linear combination of sine and cosine functions. Since the measurements were made bimonthly, there are six measurements a year. Therefore we need five parameters to describe an arbitrary but constant seasonal component.

For the *covariate component*, we assume a linear dependence on the (mean-adjusted) water quantity. The assumption of independence of the other components is plausible, because the amount of water in the compartments shows neither trend or systematic seasonal fluctuation nor autocorrelation (see Section 18.3.3). The change in groundwater level has to be seasonally adjusted before it can be used as an independent covariate, but there is no autocorrelation either.

The *irregular component* is an AR (*p*) process or merely white noise. Moving-average components were not ruled out from the beginning, but the estimated autocorrelation functions gave no hints of their presence. Especially in the open area precipitation and canopy penetration data, there often was hardly any autocorrelation at all. For the seepage water and groundwater, mostly auto-regressive processes of order 1 or 2 were sufficient.

Prior to all estimation and testing, we still had to think about the homogenity of variance and the distribution of the residuals. Usually a Box–Cox transformation is a good method for stabilising the variance and the normal distribution of the residuals at the same time. The problem in our case is that the optimal transfor-mation parameter differs from one compartment to the other. As we wanted to compare the results of all the compartments, we had to choose one transform-ation that is reasonably good for all of them. In order to save time, we did this by trying only three common transformations: identity, square-root and logarithm of $Y + 1$.

We started by estimating the full model (Model 4) for every water balance component. After that, we eliminated one by one the variables with the highest *p* values until either all remaining variables had a significant effect ($p < 0.05$) or Akaike's information criterion (AIC) reached its minimum. We have restricted this backward elimination by the following rules: $\sin ht$ and $\cos ht$ as well as $\sin 2ht$ and $\cos 2ht$ are treated as unseparable pairs, and if $T_t^{(2)}$ stays in the model then N_t stays as well.

The estimation is done by maximum-likelihood estimation, because for some of the variables there are many missing values. In this case it is more robust than least-squares estimation (see Wei, 1990).

In those cases where there is no autocorrelation, the maximum-likelihood estimates are the same as the well-known least-squares estimates. In the other cases the noise models were AR (1) and AR (2) processes, and their parameters can be estimated by using the exact log-likelihood functions as described in Box and Jenkins (1976). For the calculations, we have used the SAS procedure AUTOREG (SAS, 1988).

Another approach would be to fit an ARIMA model with intervention variables. Then the seasonal fluctuation and the trend (if any) would be thought of as stochastic and would be removed by differencing. Possibly the influence of the intervention could be modelled more precisely in such a model.

There are two reasons why we preferred the more deterministic model. Firstly, we were reluctant to remove the data from the first year by differencing, because the construction phase is rather short. Secondly, the hydrologists were interested in the shape of the seasonal component as well, so we did not want to remove it from the data.

The disadvantage of a large number of parameters in the model did not seem too severe, because, after eliminating the least important ones, most models contained no more than four to six parameters and none more than eight out of eleven.

18.4.2   Transfer functions for adjacent compartments

For the compartments regarded in this study, we can assume that there is a causal relationship between each pair of adjacent compartments: the amount of a chemical substance in one compartment will, sooner or later, influence the amount in the compartment below, and there is no relevant feedback. With transfer function models, we can try to find out for which compartments and chemicals this relationship is significant, whether there is a delay, how long it is, and how long the influence lasts.

Let $Y_t^{(I)}$ be the input series and $Y_t^{(O)}$ the output series. What we look for is a so-called ARMAX (b, r, s) model

$$Y_t^{(O)} = \left(\sum_{j=0}^{\infty} v_j B^j \right) Y_t^{(I)} + X_t = \frac{(w_0 - w_1 B - \cdots - w_s B^s) B^b}{(1 - \psi_1 B - \cdots - \psi_r B^r)} Y_t^{(I)} + n_t, \quad (18.8)$$

where B is the backshift operator, n_t is an arbitrary stationary noise series, and

$$v(B) = \sum_{j=0}^{\infty} v_j B^j = \frac{(w_0 - w_1 B - \cdots - w_s B^s) B^b}{(1 - \psi_1 B - \cdots - \psi_r B^r)} \quad (18.9)$$

is called the transfer function for the input time series $Y_t^{(I)}$. In this model b is the delay and r and s are analoguous to the parameters p and q in the ARMA (p, q) model (see Model 5 in Figure 18.6). If $r = 0$, s is the length of the time interval that influences the value at one time. This model can be identified and estimated as described by Wei (1990).

Prior to identifying and estimating the transfer function, the input and output series have to be analysed, because

(i) the time series have to be stationary, otherwise parallel seasonal fluctuations or trends might be mistaken for a causal relationship;
(ii) the autocorrelation structure of the input time series has to be known in order to calculate the cross-correlation function that is needed for identifying the transfer function.

In order to get stationary time series without seasonal fluctuations, we differenced the original series by a period of 12 months. A possible linear trend becomes an intercept in the differenced series. Then autoregressive and moving average components were added to these series. This means that we fitted so-called SARIMA $(p, d, q) \times (P, D, Q)_{12}$ models (Model 5, Figure 18.6).

The orders of the processes can be identified with the help of the empirical autocorrelation and partial autocorrelation functions. We always had $d = 0$ and

$$(1 - B^{12})^D(1 - B)^d Y_t = \mu + \frac{\theta_q(B)}{\phi_p(B)} \frac{\Theta_Q(B^{12})}{\Phi_P(B^{12})} a_t,$$

The variables and components in this model are

B	the backshift operator,
$(1 - B)^d$	differencing with period 1 (d times),
$(1 - B^{12})^D$	differencing with period 12 (D times),
μ	the intercept of the differenced series,
$\theta_q(B)$	$= 1 - \sum\limits_{j=1}^{q} \theta_j B^j$, the moving-average factor,
$\phi_p(B)$	$= 1 - \sum\limits_{j=1}^{p} \phi_j B^j$, the autoregressive factor,
$\Theta_Q(B^{12})$	$= 1 - \sum\limits_{j=1}^{Q} \Theta_j B^{12j}$, the seasonal moving-average factor,
$\Phi_P(B^{12})$	$= 1 - \sum\limits_{j=1}^{P} \Phi_j B^{12j}$, the seasonal autoregressive factor,
a_t	white noise.

Figure 18.6 Model 5: the SARIMA time series model.

$D = 1$. Usually either P or Q was 1, the other one being 0, and p and q were 0, 1 or 2. The parameters can be estimated by maximum-likelihood estimators as before.

If the input series is white noise then there is a close relationship between the transfer function and the cross-correlation function:

$$v_j = \frac{\sigma_O}{\sigma_I} \rho_{IO}(j), \tag{18.10}$$

where $\rho_{IO}(j)$ is the correlation between $Y_t^{(O)}$ and $Y_{t-j}^{(I)}$. If the input series is not white noise, its autocorrelation structure makes this relationship much more complicated.

Therefore we start by prewhitening the input series and transforming the output series by the same filter. Then the cross-correlation function for these two transformed time series is estimated. From its pattern, b, r, and s can be chosen, and its values can be used for a preliminary estimate of the transfer function.

This enables us to estimate the noise series and to identify a noise model. In our case all the transfer functions were of the ARMAX $(b, 0, s)$ type with b being 0 to 7 and s being 0 to 3, and the noise models were SARIMA processes as before. Finally, the parameters of the transfer function and of the noise model are estimated simultaneously by maximum-likelihood estimation, using the original differenced input and output series.

The necessary calculations were done with the SAS procedures AUTOREG and ARIMA (SAS, 1988).

18.4.3   Nitrate in precipitation and groundwater

Nitrate can emerge from nitrogen oxides as they are emitted by aircraft. Another possible source on the airport is the urea that is used as a frost protection and defrosting agent on the runway. Thus it is one of the most important variables for the proof documentation.

In order to give an overview of the quantities and concentrations of nitrate in the four compartments on the examined forest stand plots, Table 18.5 shows the mean concentration (unweighted and weighted by water quantity) as well as the mean deposition and the mean water quantity per month.

It seems reasonable to use the weighted mean of the concentration rather than the unweighted one for the upper three compartments, because of the strong correlation with the water quantity (see below), and because otherwise the influence of measurements in small samples with often high concentrations would be too large.

We see that as the water quantity is decreased by evaporation and transpiration, the nitrate concentrations and deposition rates increase on the way from the open area precipitation to the canopy penetration and especially to the seepage water. Here the concentration is up to eight times higher than in the compartment above. In the groundwater the concentration is smaller again.

In the open area precipitation there are hardly any differences between the different stations. They are to some extent larger in the canopy penetration. Here

Table 18.5   Mean nitrate concentration and deposition

Plot No.	Compartment[a]	Concentration (mg l^{-1})		Deposition rate (mg m^{-2}month^{-1})	Water quantity (mm month^{-1})
		Unweighted	Weighted		
3	OP (F5)	3.89	3.63	186.7	51.41
(oak)	CP	7.75	6.11	254.0	41.55
	SW	39.98	32.94	1186.9	36.03
	GW	1.37	—	—	—
4	OP (F4)	3.79	3.38	171.5	50.72
(pine)	CP	8.48	6.84	268.0	39.21
	SW	64.70	56.93	1923.6	33.79
	GW	46.75	—	—	—
9	OP (F7)	3.51	3.37	165.0	49.02
(beech/	CP	10.38	7.87	282.0	35.85
ash)	SW	42.74	36.83	1273.4	34.58
	GW	5.00	—	—	—
13	OP (F3)	3.69	3.68	179.7	48.07
(pine)	CP	10.03	6.05	247.6	40.29
	SW	46.49	35.03	1214.1	34.15
	GW	7.67	—	—	—

[a]OP, open area precipitation; CP, canopy penetration (throughfall and drip); SW, seepage water; GW, groundwater.

Table 18.6   Spearman rank correlations for nitrate after seasonal adjustment of all variables

| | Correlation between water quantity and | |
Compartment	concentration	deposition rate
Open area precipitation (OP)	− 0.433	+ 0.581
Canopy penetration (CP)	− 0.535	+ 0.518
Seepage water (SW)	− 0.360	+ 0.601
Groundwater (GW)	− 0.167[a]	—

[a]Correlation with the change groundwater level.

we find the highest concentrations and deposition rates on the oak plot, where the water quantity is smallest. The mean quantity of the seepage water is roughly equal on all plots, but on plot No. 4 the concentrations and deposition rates are about 60% higher than on the other plots. The greatest differences can be found in the groundwater. The concentration is smallest on plot No. 3, where it is even smaller than in the open area precipitation. On plots Nos. 9 and 13, it is about as high as in the canopy penetration water, and on plot No. 4 it stays at about 80% of that in the seepage water.

In the upper compartments the Spearman rank correlation between the water quantity and nitrate concentration is negative, while the correlation for the deposition rate is positive (see Table 18.6). Only in the groundwater is this correlation not significant. This means that the concentration decreases and the deposition rate increases, if there is more precipitation input or if the groundwater level rises.

To stabilise the variance, we used the square root of the original data. This transformation was best for the data from seepage water and groundwater, and satisfactory for open area precipitation and canopy penetration. It also improves the approximation to the normal distribution.

The fitted time series for the nitrate concentrations on plot No. 4 in the whole period (even months only) are given in Table 18.7.

The time series have been calculated for the deposition rates as well. The results were very similar to those for the concentrations, and therefore we do not give them here. This similarity corresponds to the fact that the *water quantity* shows neither seasonal fluctuation nor trend. As a covariate, the water quantity significantly reduces the variance in the upper three compartments, but, according to Akaike's information criterion (AIC), it does not improve the model for the nitrate concentration in the groundwater.

In both cases we find that in the open area precipitation there is a significant *seasonal fluctuation*, with one sine–cosine wave a year, with maximum concentration and deposition in April and the minimum in October. In the canopy penetration the seasonal component is not significant, though the variable $\cos 3ht$ improves the model for the deposition according to AIC. In the seepage water both

Table 18.7   Time series models for the nitrate concentrations on forest stand plot No. 4 (even months only)

Transformation: $Y_t^* = \sqrt{Y_t}$

Open area precipitation (F4):

$$C_t^{(OP)*} = \begin{array}{llll} 1.91 & +0.18\sin ht & -0.21\cos ht & -0.0087 \quad Q_t^{(OP)m} + a_t \\ (\pm 0.06) & (\pm 0.09) & (\pm 0.09) & (\pm 0.0023) \end{array}$$

$\hat{\sigma}_a^2 = 0.248 \qquad \hat{\sigma}_a = 0.498 \qquad R^2 = 0.32$

Canopy penetration:

$$C_t^{(CP)*} = \begin{array}{llll} 2.64 & +0.68Nt & +0.16T_t & -0.0232 \quad Q_t^{(CP)m} + a_t \\ (\pm 0.12) & (\pm 0.28) & (\pm 0.07) & (\pm 0.0053) \end{array}$$

$\hat{\sigma}_a^2 = 0.906 \qquad \hat{\sigma}_a = 0.952 \qquad R^2 = 0.36$

Seepage water (10 cm):

$$C_t^{(SW)*} = \begin{array}{lllll} 7.26 & +0.27N_t & +0.54T_t & -2.29\sin ht & -0.46\cos ht \\ (\pm 0.21) & (\pm 0.48) & (\pm 0.12) & (\pm 0.31) & (\pm 0.30) \end{array}$$
$$\begin{array}{l} -0.0501 \quad Q_t^{(SW)m} + a_t \\ (\pm 0.0101) \end{array}$$

$\hat{\sigma}_a^2 = 2.685 \qquad \hat{\sigma}_a = 1.639 \qquad R^2 = 0.63$

Groundwater:

$$C_t^{(GW)*} = \begin{array}{ll} 6.58 & -2.37N_t \\ (\pm 0.25) & (\pm 0.55) \end{array}$$
$$\begin{array}{llll} +0.40\sin ht & -0.54\cos ht & +0.55\sin 2ht & -0.54\cos 2ht + a_t \\ (\pm 0.34) & (\pm 0.36) & (\pm 0.34) & (\pm 0.35) \end{array}$$

$\hat{\sigma}_a^2 = 3.446 \qquad \hat{\sigma}_a = 1.856 \qquad R^2 = 0.33$

---

models show a relatively strong one-wave season with minimum in February and maximum in August. In the groundwater a sine–cosine function with two waves a year superimposes an annual one. The concentration is maximum in February, decreases until June, rises again in August and reaches a minimum in December.

In the open area precipitation we could not prove any systematic *changes over time*. But in the canopy penetration the mean concentration and deposition rate in phase 2, when the runway was put to use, is significantly higher than in the construction phase. There also is a significant rise within the second phase. In the seepage water this trend is even stronger, but the mean level rises only slightly, so that most values at the beginning of phase 2 are lower than those of phase 1. In contrast to the upper compartments, the mean nitrate concentration in the groundwater is significantly lower in phase 2 than in phase 1. There is no trend within the second phase.

None of the bimonthly time series show any relevant *autocorrelation*. The *fit* of the models lies between $R^2 = 0.32$ and 0.63. It is best for the seepage water. In the open area precipitation it is slightly better for the deposition model, in the canopy penetration it is about equal for both models, and in the seepage water the fit of the concentration model is a little better.

Table 18.8 shows the main results for all four examined forest stand plots. Since the experience from plot No. 4 shows that considering the deposition rates instead of the concentrations does not change the results, we have refrained from calculating separate time series for the deposition rates on the other plots.

On all the forest stand plots we find that the mean nitrate concentrations in the canopy penetration and the seepage water are higher in the growing season than in the dormant season. Though this difference is not always significant, taking it into account improves the models according to AIC. In the open area precipitation and the groundwater no seasonal fluctuation can be found in the half-year data.

The results for plot No. 3 (oaks) are very similar to those for plot No. 4. For canopy penetration and seepage water, we find that the mean level of the data is

Table 18.8 Significance of the time series parameters for the nitrate concentration on forest stand plots Nos 3, 4, 9 and 13

Plot No.	Compartment	α(level)	β(trend)	γ(season)
3 (oak)	OP (F5)			
	CP	+ +	+ +	+
	SW	+ +	+ +	+
	GW	−	+ +	
4 (pine)	OP (F4)			**
	CP	+ +	+ +	*
	SW	+	+ +	**
	GW	− −		**
9 (beech/ash)	OP (F7)			
	CP		+ +	+ +
	SW			+ +
	GW	− −		
13 (pine)	OP (F3)			
	CP			+
	SW	+ +	+	+ +
	GW			

+ +, significant at 5% level, positive;
+, improves the model according to AIC, positive;
− −, significant at 5% level, negative;
−, improves the model according to AIC, negative;
**, significant;
*, improves the model according to AIC.
A positive value for γ signifies that the concentrations in the growing season are higher than those in the dormant season.

higher in phase 2 than in phase 1 and that there is a positive trend within the second phase. In the groundwater the concentration level is slightly diminished in the second phase. In contrast to plot No. 4, we find a significant upward trend in this compartment.

On forest stand plot No. 9 (beeches and ashes) only two effects are significant: a rising concentration within the second phase for the canopy penetration, and a decrease in the groundwater from phase 1 to phase 2.

On the second forest stand plot with pines (plot No. 13) the only compartment in which we could find changes in time was the seepage water. The mean level of the concentration in this compartment is significantly higher in phase 2 than in phase 1, and according to AIC it increases also within the second phase.

Summing up, we can say that we cannot prove any changes in the open area precipitation concerning nitrate concentrations and deposition rates. Apart from stochastic fluctuations, the nitrate input seems to have remained approximately constant.

In contrast to these findings, on all the examined plots there is some degree of significant change in the canopy penetration and/or in the seepage water, and in all these cases the nitrate concentration is rising. In the groundwater the situation is different again: here the concentration level has decreased in the second phase; only on plot No. 3 we find a positive trend within this phase.

These differences between the compartments might be a hint to changes in the circulation of nitrate in the ecological system. This might be caused, for instance, by changes in the concentrations of other chemical substances or in the weather conditions.

The transfer functions discussed in the remainder of this chapter are only a first step to modelling this circulation. Important matters like chemical reactions and the upward movement by capillary ascent and absorption by plants are ignored. The time series are too short yet to examine changes in the transfer rates. But the forest stand plot No. 13 is still being observed, and it is hoped that after a few more years we can tackle this question as well.

For the transfer functions, the time series had to be calculated anew, because a different data set and a different model were used: monthly data from December 1983 to July 1991 and SARIMA model. Table 18.9 contains both time series and transfer functions for the concentrations.

Though a simple AR (1) process seemed to fit better for the groundwater series, we given a model with differencing as for the other compartments. It was needed for the transfer function, because the seepage water series shows large seasonal fluctuations and therefore had to be differenced before it could be used as an input time series.

Since the time interval between the data is now smaller, we find serial correlation in all the series. The *autocorrelation* of lag 1 is always positive ($+0.16$ to $+0.53$). In most cases, the best model is an AR component; for the canopy penetration an MA component fits better. In all the series there is a strong negative seasonal autocorrelation as well (-0.49 to -0.59). The best fit could be

Table 18.9 Time series models and transfer functions for the nitrate concentration on forest stand plot No. 4 (monthly values from December 1983 to July 1991)

Transformation: $Y_i^* = \sqrt{Yt}$

Open area precipitation:

$$(1 - B^{12})C_t^{(OP)*} = \underset{(\pm 0.03)}{0.01} + \frac{\overset{(\pm 0.32)}{(1 - 0.89B^{12})}}{\underset{(\pm 0.10)}{(1 - 0.19B)}} a_t \qquad \begin{array}{l} \hat{\sigma}_a^2 = 0.28 \\ \hat{\sigma}_a = 0.53 \end{array}$$

Canopy penetration:

$$(1 - B^{12})C_t^{(CP)*} = \underset{(\pm 0.10)}{0.17} + \frac{\overset{(\pm 0.11)}{(1 + 0.22B)}}{\underset{(\pm 0.09)}{(1 + 0.58B^{12})}} a_t \qquad \begin{array}{l} \hat{\sigma}_a^2 = 1.15 \\ \hat{\sigma}_a = 1.07 \end{array}$$

$$(1 - B^{12})C_t^{(CP)*} = \underset{(\pm 0.08)}{0.18} + \underset{(\pm 0.15)}{(1.01} \underset{(\pm 0.16)}{- 0.43B} \underset{(\pm 0.15)}{- 0.46B^3)}(1 - B^{12})C_t^{(OP)*}$$

$$+ \frac{\overset{(\pm 0.11)}{(1 + 0.33B)}}{\underset{(\pm 0.11)}{(1 + 0.49B^{12})}} a_t \qquad \begin{array}{l} \hat{\sigma}_a^2 = 0.66 \\ \hat{\sigma}_a = 0.81 \end{array}$$

Seepage water (10 cm):

$$(1 - B^{12})C_t^{(SW)*} = \underset{(\pm 0.19)}{0.30} + \frac{\overset{(\pm 0.18)}{(1 - 0.69B^{12})}}{\underset{(\pm 0.10)}{(1 - 0.50B)}} a_t \qquad \begin{array}{l} \hat{\sigma}_a^2 = 3.61 \\ \hat{\sigma}_a = 1.90 \end{array}$$

$$(1 - B^{12})C_t^{(SW)*} = \underset{(\pm 0.20)}{0.15} + \underset{(\pm 0.17)}{(0.77} \underset{(\pm 0.16)}{+ 0.41B)}(1 - B^{12})C_t^{(CP)*}$$

$$+ \frac{\overset{(\pm 0.12)}{(1 + 0.38B^2)}\overset{(\pm 0.16)}{(1 - 0.68B^{12})}}{\underset{(\pm 0.11)}{(1 - 0.41B)}} a_t \qquad \begin{array}{l} \hat{\sigma}_a^2 = 2.71 \\ \hat{\sigma}_a = 1.65 \end{array}$$

Table 18.9 *(Continued)*

Groundwater:

$$(1 - B^{12})C_t^{(GW)*} = \quad \underset{(\pm 0.30)}{0.03} \quad + \frac{1}{(1 - 0.44B)(1 + 0.58B^{12})}a_t \qquad \hat{\sigma}_a^2 = 5.26$$
$$\underset{(\pm 0.10)}{} \quad \underset{(\pm 0.09)}{} \qquad \hat{\sigma}_a = 2.29$$

$$(1 - B^{12})C_t^{(GW)*} = \quad \underset{(\pm 0.32)}{0.11} \quad + \underset{(\pm 0.13)}{0.29}B^7(1 - B^{12})C_t^{(SW)*}$$

$$+ \frac{1}{(1 - 0.42B)(1 + 0.45B^{12})}a_t \qquad \hat{\sigma}_a^2 = 3.47$$
$$\underset{(\pm 0.11)}{} \quad \underset{(\pm 0.11)}{} \qquad \hat{\sigma}_a = 1.86$$

achieved with an MA factor for open area precipitation and seepage water and an AR factor for canopy penetration and groundwater. No more parameters were needed in the time series models.

The *noise components* of the transfer functions are very similar to the corresponding time series. Only for the seepage water did another factor have to be added to the model: an MA component with lag 2.

The intercept estimates in the time series models are very small in comparison with their standard errors. Only in the transfer function for the canopy penetration is it larger than twice its standard error (0.18 ± 0.08), indicating a slight upward *trend* of the data.

Now we come to the *influence of the upper compartments on those below*. In all cases we can find such a dependence between adjacent compartments, and taking it into account in the transfer functions noticeably reduces the variance of the white-noise component in all the models.

As the cross-correlation functions showed, the nitrate concentrations in the canopy penetration and in the seepage water depend mainly on the corresponding values for the same month in the compartment above. In both cases the correlation is positive.

Apart from this, we find a significant negative influence of the concentrations in the open area precipitation one and three months before on the concentration in the canopy penetration. So far, we can not find any explanation for this.

The results for canopy penetration and seepage water seem to be more plausible: the nitrate content in the seepage water is influenced positively by the concentrations in the canopy penetration of both, for the same month and the month before. This can be easily explained by the seepage process through the upper layer of the soil.

The only relevant cross-correlation between the seepage water and the ground-water is that for a shift of seven months. This effect is positive and significant. Considering that the groundwater on this plot is only 0.6–2.1 m below the surface and that nitrate is not usually absorbed by the soil in large quantities, the delay of more than half a year seems rather long, and certainly cannot be explained by the seepage process alone. But the explanation might be much more complex, because nitrate often reacts with other chemical substances in the water and these reactions are influenced by many other variables, such as moisture, temperature and acidity.

19

Statistics and Probability: Wrong Remedies for a Confused Hydrologic Modeller

V. Klemeš
Victoria, BC, Canada

19.1 INTRODUCTION

Confusion seems to be the main asset that one acquires from the run-of-the-mill graduate education in hydrology. One embarks on it equipped, in most cases, with an undergraduate degree in engineering (with a very few exceptions, one cannot get an undergraduate degree in hydrology) and, after being taught (often by amateurs) rudiments of linear algebra, mathematical statistics, probability theory, systems theory and computer programming, is academically certified as an expert in hydrology—a geophysical science. The common result is a perma- nent or, at any rate, deeply planted inability to see the difference between technology, science and mathematics, and the mixing of them in ways ranging from amusing to dangerous. I have previously (Klemeš 1988a) remarked that

> Suspended between a technology he does not practice and a science for which he has not been trained, his 'research' is naturally guided to performing elaborate pirouettes on the high wire of [mathematical] techniques connecting the two distant poles and holding him in place.

Statistics for the Environment 2: Water Related Issues. Edited by Vic Barnett and K. Feridun Turkman
© 1994 John Wiley & Sons Ltd.

Having myself suffered such education and remembering the lessons it inadvertently taught me about its inherent dangers to both the practice and research in hydrology, I have tried, alongside a few other colleagues (e.g. Philip, 1975, 1991; Nash *et al.*, 1990; Bras and Eagleson, 1987), to help correct the situation by repeatedly exposing these dangers over the past 20 years (Klemeš, 1971, 1974, 1978, 1982, 1986, 1987a, b, 1988a, b, 1989, 1991, 1993). Although some recent developments indicate that not all the efforts were in vain, the old attitudes to hydrology are still holding fast in the academe and elsewhere and keep polluting hydrological waters.

It is no coincidence that relatively few of the hydrologists of note were university professors but were practicing professionals working for hydrological and water-resource services, agencies and research facilities (a few of these the past generation were Hazen, Horton, Theis and Langbein in the USA; Hurst in the UK; and Sokolovskiy, Kritskiy and Menkel in Russia). In the academic world hydrology used to be (and, as a rule, still is) only an appendix to some other discipline, and professors taught it as a sideline. Of late, this has been changing— though not always in desirable ways. While hydrology now may nominally be the main line of a professor, it need not be hydrology but merely 'blackboard hydrology' (J. E. Nash, personal communication) that is being taught. In view of the remarks in the first paragraph, this of course should be no surprise.

The confusion is particularly evident in hydrologic modelling. In essence, most hydrologic models are developed as tools for aiding decisions in various water-related technologies. However, since their creators consider themselves scientists, they often confuse these models with scientific models and expect from them new hydrological insights, forgetting that they had already prescribed what they hope to learn. The other confusion reflects the aforementioned way in which hydrologic scientists have been metamorphosed out of technologists. Since this has been achieved by teaching them mostly mathematical techniques, they tend to equate scientific substance with mathematical skill; hence a thicker layer of mathematical rigour and polish is thought to translate into a superior hydrologic validity of a model.

This is true in particular in the area of statistical hydrologic modelling, where the fallacy is less easy to expose than it is in its deterministic counterpart. It is, for instance, relatively simple to find out how much better a flood forecasts is in a given river section when, based on streamflows 50 km upstream, it is calculated by a complex or a simple hydraulic model of the river channel. On the other hand, it is virtually impossible to tell whether replacing biased parameter estimates with unbiased ones in a distribution model based on a 20-year record improves the 100-year flood estimate or makes it worse. The point is that a 20-year long flood record does not contain the information necessary to make the judgement. Practically everything about the model is a guess, and the whole modelling exercise is not science but merely a hydrologically motivated and statistically executed rationale masquerading as science on the pretext of the (often spurious) mathematics of its formulation.

It is poor consolation that hydrology is not the only science where attempts are frequently made to create knowledge about nature by misguided manipulation of mathematical formulae. Professor George E. P. Box called such use of mathematics *mathematistry* in his R. A. Fisher Memorial Lecture of 1974 (for a written version: see Box, 1976), where he lamented as follows:

> In such areas as sociology, psychology, education, and even, I sadly say, engineering, investigators who are not themselves statisticians sometimes take mathematistry seriously. Overawed by what they do not understand, they mistakenly distrust their own common sense and adopt inappropriate procedures devised by mathematicians with no scientific experience.

I would only add—and hope to be able to demonstrate in the following sections—that many of these procedures may have been quite appropriate for the purposes for which they originally had been developed, but, to paraphrase Professor Box, overawed by what they do not understand, many investigators, and I sadly say, hydrologists among them, freely abuse them while being convinced that they are advancing science.

It is surprising how widespread has been the following paradox: when an investigator is at the end of his wits in understanding the particular segment of reality on which he is an expert, he turns for help to formalisms of mathematics—an area most removed from reality and explicitly disclaiming any relation to it and, moreover, one in which he is not an expert!

It seems to me that statisticians, probability theorists and mathematicians have been much more aware of the futility of such approach than have been practitioners in the natural, engineering and social sciences. Let me quote some examples:

> Derivations based upon postulates which have no concrete physical interpretation are of little use to the scientist. (Fry, 1928)

> In economic and social sciences ... any statistical analysis must be closely knit with as full a theoretical specification as possible. (Bartlett in 1953 as quoted in Bartlett, 1962)

> ... unless the statistician has a well defined and realistic model of the actual process he is studying, his analysis is likely to be abortive. (Bartlett in 1954 as quoted in Bartlett, 1962)

> The ultimate object of analysis of a time series—as of statistical analysis as a whole—is to arrive at a deeper understanding of the causal mechanisms which generated it. (Kendall and Stuart, 1966)

> ... the modern apparatus of the theory of small samples [as] a method for positive statistical inference ... does not inspire one with confidence, unless it is applied by a statistician by whom the main elements of the dynamics of the situation are either explicitly known or implicitly felt. (Norbert Wiener as quoted in Bartlett, 1962).

> ... the form of the [flood peak] distribution is not known and any distribution used must be guessed ... since the part of the distribution we are interested in is well away

from the part where observations provide some information ... [This presents
a difficulty that] cannot be overcome by mathematical sleight of hand. (Moran,
1957)

I am totally scornful of the idea that one can understand by pure thought, whatever
that is, and to get some understanding of processes, you should open your mind to
real situations, not the stupid hypothetical situations one finds in many papers.
(Kempthorne, 1971)

I appreciate the opportunity to summarize my views on this and related issues
as they pertain to hydrology here at the Rothamsted Statistics Department, whose
founder, the illustrious statistician Sir Ronald Fisher, felt very strongly the need
for a solid scientific foundation of every statistical analysis and held in contempt
sterile mathematical sleight of hand.

19.2   ANALYTICAL VERSUS SYNTHETIC MODELS

In my opinion (Klemeš, 1986), it is because of the fact that the basic training of
hydrologists is in other disciplines for which hydrology is just one of many tools
needed for their own purposes that **the vast majority of hydrologic models are,
consciously or subconsciously, user-oriented. They are not meant to help hydrology but to make hydrology help someone else.** Implicitly, and often explicitly,
their makers take the attitude that the hydrology underlying the model is already
known and, if it is not, it should have been—apparently, somebody has not done
his job and it is not their business to do it for him: their business is to build a model!
All they can afford to do in such circumstances is to plug the hydrologic hole,
come up with some quick fix and go ahead with the modelling—because the user
is waiting!

This user attitude tends to persist in their thinking even long after the modellers
have considered themselves genuine hydrologists–scientists. After all, such a trait
is by no means unique to hydrologists.

Until about 40 years ago, the situation was clear since the species of 'hydrologic
modeller' did not yet exist and, when some representation of hydrologic information was needed, convenient ways of doing it—models—were developed by those
in need: engineers, water resource managers, foresters, planners, agronomists etc.

The problem started in the 1950s, when, well before hydrology had a chance to
establish itself as a scientific discipline in its own right (it still is far from it today),
the type of training of hydrologic specialists mentioned in the introductory
paragraph was introduced. The muddle was made worse by the fact that this
happened simultaneously with the 'discovery' of mathematical statistics and
probability theory by the applied sciences on the one hand and with the advent of
the computer and the 'random number generator' on the other.

This historical collusion created, almost overnight, a new discipline—'stochastic
hydrology'—by infusing old ingenious techniques of Allen Hazen and Charles

Sudler (Klemeš, 1981) with 'mathematical rigour' and computer efficiency. Only a few individuals clearly saw the danger and sounded early warnings. Among them was one of the originators of stochastic hydrology, the late Myron Fiering, who introduced the term *operational hydrology* in 'an effort to remind the user that hydrologic sequences generated by recursive models, of whatever sort, are meaningless unless transformed into some metric and then ranked to aid and abet in the exercise of a decision' (Fiering, 1966).

Similarly devoid of hydrologic content has become 'statistical hydrology', especially its all-important branch of distribution modelling and the jewel in its crown known as flood frequency analysis (FFA). I have discussed this in detail elsewhere (Klemeš, 1971, 1986, 1987b, 1988a, 1989, 1993), and it suffices here to say that their only connection with hydrology is that the numbers to which various distributional models are fitted have hydrological names. Their true nature is of no consequence, because absolutely nothing of what is done with these numbers requires any hydrologic information, knowledge or experience. As I once put it (Klemeš, 1971), the data '... are treated as a collection of abstract numbers that could pertain to anything or to nothing at all'.

Ten years after I had made this assertion, an opportunity presented itself to test its validity, when a leading American FFA expert asked me for a set of flood peak discharge data to test his new regional FF model, and explicitly requested that I give him no other information except the numbers. So I sent him a set of fabricated numbers from which his model duly produced the required regional flood distribution. Since I still have the computer printout with the results in my files, I can tell you with confidence that, for example, the 10 000-year flood for the region is 2.5181. While the units are not known (as requested, I supplied numbers only), the magnitude of the flood must be quite accurate, because the parameters of the distributional model are given to eight decimal places. To the credit of the investigator, I must say that he detected some peculiarities in my data, namely that some of the computed model parameters pointed to a small homogeneous region, others to a large heterogenous region. I assured my friend he was right on both counts: the small region was my desk, the large region my imagination.

There is of course nothing wrong with employing various computational schemes for processing data for specific purposes, even if the schemes are in some sense obviously wrong, provided that the purpose is well served and the result is not unduly affected by the wrong aspect of the scheme. A typical case involves the use of the normal distribution, on which Box (1976) commented as follows:

> ... the statistician knows, for example, that in nature there never was a normal distribution, there never was a straight line, yet with normal and linear assumptions, known to be false, he can often derive results which match, to a useful approximation, those found in the real world.

This is often true in water management decisions, where distributions of hydro-logic variables like monthly flows, annual flows or extreme flows are routinely used as weighting functions for some associated benefits or costs. Given the fact that, on the one hand, economic criteria usually employ expected values and quadratic loss functions and, on the other, the basic rule of distribution fitting is to preserve the mean and variance of the data, a satisfactory performance of the normal distribution (or other simple two-parameter models) is not surprising (Klemeš, 1977). An illustration of this was provided by Slack *et al.* (1975), who analysed the effect on optimum design of the distributional model used for the representation of the distribution of floods. They concluded that

> . . . the use of the normal distribution . . . is generally better than either the Gumbel, lognormal, or Weibull distributions. Nothing is gained in terms of reducing expected opportunity design losses if the underlying distribution . . . is identified over and above simply using the normal as the assumed distribution.

The important point is that, in such and similar cases, the purpose of the investigation was not to find a scientifically (i.e. hydrologically) correct type of flood distribution itself but to use a simple approximate representation of its shape that would adequately serve some other purpose. The reason why I decided to play the aforementioned trick on my American friend was not to question the reasonableness and usefulness of his model for some decision-related application, but because he presented it to me as a *hydrological* model capable of providing information about probabilities of real floods even without any relevant hydrologic information. In other words, I just wanted to demonstrate that, as Myron Fiering might have put it, flood parameters generated by a distributional model unrelated to a specific hydrological situation are, by themselves, meaningless. But more on this later.

In short, the difference is one between an investigation where the objective is an insight into hydrology itself and an investigation where a description of hydrology is used only as a stepping stone for insight into some other problem. In the older hydrologic jargon, the former activity used to be labelled *pure* hydrology, while the hydrological component of the latter was, and still often is, referred to as *applied* hydrology (the connotation of the latter term being that, before hydrologic knowledge can be applied, it must be available). Transposing the difference into the sphere of modelling, one might perhaps refer to investigative and descriptive hydrologic models, or analytic and synthetic models, respectively.

It seems obvious that the two types of models require different approaches, which cannot be freely interchanged. It is a prerequisite of effective modelling to be clear about which of the two objectives is being pursued. When I started on the crusade aimed at a clarification of this difference, I posed the problem as follows (Klemeš, 1971):

> Pure hydrology is concerned with hydrological processes as such, should strive for explanations of how things happen and why they behave as they do, and its methods should be independent of any eventual practical use of the acquired knowledge. In applied hydrology, on the other hand, the major concern should be to

know to what extent our findings about hydrological processes are relevant to the practical decision making process in water resource management, to what extent a more precise knowledge can make the decisions more rational, the results more predictable, and the means of achieving them more economical.

(At the time, I had no idea that the same point was simultaneously being raised in regard to the practice of statistics by Oscar Kempthorne: 'It is critical in my opinion to make a differentiation between the acquisition of knowledge on the one hand and the making of decisions on the other hand' (Kempthorne, 1971).)

I further pointed out that, 'Logical as this concept seems to be, it is far from being implemented in hydrology in general and in statistical hydrology in particular'. This, to a large extent, is still true today, in part owing to the fact that students of hydrology have been led to believe that applied hydrology becomes pure hydrology and, implicitly, a synthetic model becomes analytical, simply by infusion of more mathematics. This is as wrong as it can be; the difference lies elsewhere.

An analytical (investigative, scientific) model asks questions about nature, while a synthetic model describes the obtained answers (or, in their absence, modeller's guesses) in a way useful to human endeavours. The analytical model is a component of the so-called scientific method, i.e. of the iterated loop between observation (data), hypothesis (theory, model), new observation, etc. Its intent is to have its result subjected to possible falsification by new data. Falsification is an aim, and it is regarded as a success when it occurs because only then something new can be learned, a new, deeper question can be asked. With an analytical model, we are in the 'knowledge business', to use Kempthorne's (1971) phrase.

On the contrary, the synthetic model is not meant as an element of a learning loop but as a 'final report' when all the testing has been done and the modelling result verified to agree with reality. At this stage, its falsification would not be a success but a disaster! One does not want to test the adequacy of a 10 000-year flood estimate by the collapse of a dam built to withstand such a flood! We are not in the knowledge business any more, we are out of school, in the real world, in the 'survival business', and we want to be as sure as possible that our model can be relied upon.

With this background, it is easy to find out whether a particular hydrologic modeller is a pure hydrologist (scientist) or an applied hydrologist (technologist)— just ask him what would please him more: if his model were falsified or verified?

It is easy to see why stochastic and statistical models are so popular with applied hydrologists—they are, as a rule, practically unfalsifiable. For when and how can anybody prove by observations additional to, say, a 30-year historic record employed that an estimate of a 1000-year flood, or a 3% probability of a reservoir running dry, was significantly wrong even if the reservoir ran dry in each of the very first five years after its completion and collapsed in a flood in the sixth year?

Does this mean that the traditional statistical and stochastic hydrologic models are not scientific in the true (Popperian) sense? I am convinced of it, and have tried to convince fellow hydrologists for almost quarter of a century that such models are merely expedient rationales (Yevjevich, 1968) and do not have the carrying capacity for all the mathematical rigour with which they have been invested. The success of my efforts can be judged by the exponential growth in both the number and the rigour of unfalsifiable stochastic and statistical hydrologic models over the past two decades.

19.3   THE FALLACY OF 'THEORETICAL MODELS' IN STATISTICAL AND STOCHASTIC HYDROLOGY

A good analogy for much reserach in statistical and stochastic hydrology is the anecdote about the drunk searching, at night, for his lost keys under a lamp post—not because he believes he had lost them there but because it is the only place where he can see anything. But there is a twist. The stochastic hydrologist finds the exploration of this well-lighted foreign place so fascinating that he soon forgets about his keys and his house and makes his camp right there, under the lamp post. The story about lost keys becomes just a habitual excuse to avoid looking silly, a façade of false dignity for the irrelevance of his research.

It is this phenomenon that gives rise to mathematistry, which, as Box (1976) pointed out, redefines a problem rather than solving it. Such redefinition may be 'justified' by the most tenuous similarities and in the face of glaring dissimilarities between the two problems, as long as the new problem can be treated by a rigorous mathematical theory—because, where there is a mathematical theory behind a solution, the solution is deemed scientific and therefore correct.

A strong incentive for backing a solution by a rigorous mathematical theory that admits no challenge and falsification comes from the explicitly or implicitly 'applied' nature of most statistical and stochastic hydrologic models and from the tacit assumption that they provide a backbone for crucial decisions that society cannot afford to let depend on fallible human judgement. We thus see that the tendency towards the replacement of scientific rigour with sterile mathematical rigour goes hand in hand with the tendency to replace (or confuse) analytical models with synthetic models.

An amusing consequence of this confusion is the following. While everybody knows that realistic estimation of probabilities of extreme hydrologic events, namely precipitation, flood flows, lake levels, snowpacks etc., **is not possible even in principle** (see the quotation from Professor Moran in Section 1), scientific papers, textbooks, handbooks and computer algorithms on **how to do it rigorously** by using 'theoretical models' comprise the bulk of the literature on statistical hydrology.

The following examples illustrate this paradox.

A comprehensive review of statistical models for flood frequency estimation (Cunnane, 1986) covers everything from Bayes' theorem to the effect of log transformations to a warning by one hydrologic modeller not to use historic information on rare floods because it 'may cause a degradation rather than improvement in the estimates' (for which, read they may render the estimates based on "theoretical models" ridiculous'). Among its 140 references, the report does not list the only one which treated the problem on the basis of its dynamics (Eagleson, 1972), nor does it list Moran (1957), who pointed out the fundamental weakness in the whole approach.

When discussing, in his comprehensive book on *Frequency and Risk Analysis in Hydrology* (Kite, 1977), the problem of how to assign a probability of exceedance (or an average return period) to, say, the annual maximum discharge in a 10-year observation record, the author gives an honest answer: 'We do not know'. But he adds: 'And yet, for practical reasons, some probability of occurrence must be assigned to this flood'. In his introduction he indicates how this impossible task is to be accomplished: 'An assumption must be made of a theoretical frequency distribution for the population of events and the statistical parameters of the distribution must be computed from the sample data'. He says that the objective of his book is 'to provide sufficient background information to enable an hydrologist to intelligently select a distribution to use in frequency analysis'. Characteristically and inevitably, the book contains little hydrologically relevant information, but it is a very good guide to the fitting of some simple distributions to small samples of numbers assumed to have been generated from them.

The rationale behind the theory of flood probability estimation can be summarised as follows. First an foremost, we (hydrologic modellers, applied hydrologists) *must* get the probabilities of extreme floods because the user (often our employer) needs them. Secondly, we know it cannot be done because the available information is insufficient. Thirdly, we know that a rigorous probabilistic theory exists for an entity called a random variable, which is defined in the way necessary to make the theory valid. Fourthly, we know that real floods do not satisfy the assumptions underlying this theory. Fifthly, we adopt this theory regardless, because we *must* get the probabilities at any price (if we do not, the user will get them from somebody else). Sixthly, to compensate for the small inconsistency involved, we *must* construct the theoretical models with the highest mathematical rigour and polish their every detail to perfection.

When talking about statistical models, Box (1976) observed that 'It is inappropriate to be concerned about mice when there are tigers abroad'. However, many statistical hydrologists (meteorologists, climatologists, ...) seem to subscribe to the maxim that 'It is easier to catch mice than tigers'.

Two of the many tigers roaming in the dark and easily avoiding the light from the lamp post of flood frequency theory are exhibited in Figure 19.1 (for more see Klemeš, 1986, 1987b). While the numerical and geographical specifics of the example are fictitious, it illustrates two real-life situations. In case *A*, a substantial part of the basin of river R_1 is controlled by a lake that seldom overflows, and the shaded basin area starts contributing to floods recorded in station S only when the lake is full. A similar situation occurs in the Santa Anna River basin in California, with about one third of it controlled by Lake Elsinor and known to have

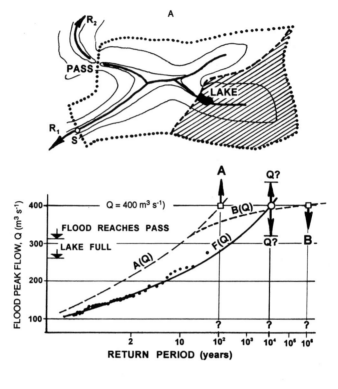

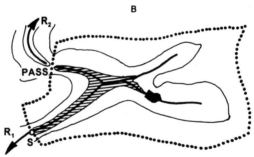

Figure 19.1 Example of a possible influence on the flood-distribution tail of different physical conditions not effective during smaller floods on which records in station S are available: A, small floods generated by the shaded part of the basin are detained in the lake; large floods fill it up and render it ineffective; B, after the flood level exceeds the elevation of the pass (shaded area), part of the flood is diverted into river R_2.

contributed to downstream floods only a couple of times during the 80-or-so years of record. In case B, when the flood level along river R_1 rises above the elevation of the pass (shaded area), part of the flood flow is diverted into river R_2. This is similar to the situation in the western border area between Canada and the USA, where floods from the US Nooksack River occasionally overflow into the Canadian Sumas River (Klemeš, 1987b).

Suppose that an existing 30–40-year record of annual flood peaks in station S is fitted by a 'theoretical' distribution model $F(Q)$ as shown and that the modeller has used all the mathematical sophistication available to fit the model. Also assume, as it is often the case, that he has used only the 'numbers' and has no knowledge about (or interest in) the hydrologic or other conditions affecting the extreme floods in the basin. Now suppose that a large flood occurs with an estimated peak flow $Q = 400\,\mathrm{m}^3\,\mathrm{s}^{-1}$. The model will assess it as a 10 000-year flood. However, it may well be only a 100-year flood if the actual conditions resemble case A, since the lake may have been rendered ineffective and the proper distribution to use for that flood would be the 'no-lake model' $A(Q)$. Or it may well have been a million-year flood if, in addition to the lake, there was also a potential of flood overflow into river R_2 so that the proper model was $B(Q)$. Moreover, given the notorious inaccuracy in the estimated flood peak discharges (they are practically never measured and, even if they were, their accuracy would not be better than about $\pm 25\%$), the actual uncertainty in the return period of the flood would be even higher, despite all the rigour invested in the 'theoretical' model $F(Q)$.

A most interesting development in the attitude to theoretical models for flood probabilities is indicated in a recent authoritative report on *Estimating Probabilities of Extreme Floods* produced by a high-level US committee (Committee on Techniques for Estimating Probabilities of Extreme Floods, 1988). After explaining the bureaucracy behind the committee's existence, the report continues with two theoretical chapters on *Improving the Theoretical Basis* and *Flood-Based Statistical Techniques*. The first starts with these words:

> Extreme or rare floods, with probabilities in the range of 10^{-3} to 10^{-7} (more or less) chance of occurrence per year, are of continuing interest to the hydrologic and engineering communities for purposes of design and planning. When compared to the very long return periods of interest, the historical record of such events is small; thus opportunities to test or compare estimated flood quantiles with experienced events almost never occur. Nevertheless, the need to design or plan for the occurrence of extreme floods is real. The committee believes that advances in the probabilistic modelling and statistical analysis of extreme events have been made and that these advances can be applied to improve extreme-event hydrologic analyses so that estimates of the probability of extreme hydrologic events will become possible.

Since the authors recognise that 'estimating the probabilities of extreme floods will always require extrapolation well beyond the data set', they have '...identified three principles for improving extreme flood estimation, (1) "substitution of space for time"; (2) introduction of more "structure" into the models; and (3) focus on extremes or "tails"...' In the next chapter they define modelling thus: 'Modelling here refers to distributional assumptions regarding the underlying random variables'. With that background, the authors get down to business: 'Let Y denote the random variable representing the annual peak discharge...', etc., etc. The reader learns many things about the small-sample theory of **random variables** possessing a few simple distributional forms, about asymptotic unbiasedness, consistency, maximum-likelihood estimation whose 'asymptotic theory exists and enables one (under regularity conditions) to obtain an approximate standard error of an estimator or . . . confidence interval', etc., etc., but learns absolutely nothing about the original problem, namely how to estimate the probabilities of **extreme floods**.

However, the following three chapters on runoff modelling, data characteristics and research needs are written in a completely different spirit. In fact, they undermine the credibility of the expounded theories. For example, one reads that 'It is particularly important to determine whether there exist physical processes which are critical to very large floods but which do not operate at lower discharges—as shown in Figure 19.1, a most important point, but nowhere reflected in the 'theoretical' chapters. Or, 'Streamflow and precipitation are complex multidimensional processes. They exhibit strong and complicated patterns of variability in both space and time'—true, but disregarded in the underlying 'theoretical' assumptions. Or 'Proper statistical treatment of historical and paleo-flood data requires the use of auxiliary information in addition to the flood magnitudes'. Or even 'Research is needed on the mechanisms by which paleoflood and historical flood records are produced and preserved with an emphasis on constructing statistical models that reflect these mechanisms, as well as consideration of climatological changes that may have occurred between the historical event and the contemporary data record'. All this is essentially equivalent to a warning that the theories presented in the first two chapters may well be useless. And, indeed, in the final short chapter on developmental issues and research needs we read: 'There is little in this report that can be taken by the practitioner and applied without further development and or research.'

In its very last paragraph, the report proclaims that 'Many valuable lessons might be learned from study of this work'. Indeed! Perhaps the most valuable lesson can be learned from the inconsistency described above. The reasons behind it are hinted at in the committee chairman's introduction: 'The history of research and practice in flood probability estimation . . . has been marked by sharp disagreements and well-defined schools of thought. All camps were represented on the committee . . . ' Having the privilege to know most of its members, I can readily distinguish two camps, one making itself comfortable under the aforementioned lamp post, the other eager to get a good searchlight and set out after the tigers hiding in the dark.

19.4   TWO COROLLARIES

19.4.1   What's in a name!

The suggestive power of names is well known. Had, for instance, Hamburg rather than Berlin been a divided city, President Kennedy would, no doubt, have thought twice before proudly proclaiming 'Ich bin ein Hamburger!'

Ten years ago, in an extremely stimulating televised interview (NOVA, 1983), the world-renowned American physicist, the late Richard Feynman, told a story how, as a boy, he learned about inertia from his father. When he once asked him why it was that the ball in his toy wagon rolled to the back of the wagon when he pulled the wagon forward and to its front when he stopped it,

> ... he says nobody knows. He said, 'The general principle is that things that are moving try to keep moving and things that are standing still tend to stand still unless you push on them hard.' And he says this tendency is called *inertia*, but nobody knows why it's true. Now that's a deep understanding. He doesn't give me a name. He knew the difference between **knowing the name of something** and **knowing something**, which I learned very early.

(emphasis added). I believe that this early knowledge was one of the most important keys to Feynman's later accomplishments as a scientist and a teacher. I also believe that the lack of this knowledge is one of the main roots of confusions in statistical and stochastic hydrologic modelling. This is why I am taking the liberty of elaborating on the following point, which may seem trivial.

For example, if x is defined as a random variable with a prescribed form of probability distribution then a mathematical function $p = f(x)$ describing this form is routinely called a 'probability distribution $f(x)$'—it is just a convenient terminological shortcut for 'mathematical expression describing the form of the probability distribution of x'. The necessary condition for this to be true is that the entity p whose numerical value the expression $f(x)$ specifies in terms of the value of x is **a priori** known to be a probability. The mathematical expression $f(x)$ has nothing intrinsically 'probabilistic' about it, and its probabilistic name is not derived from its mathematical form. The same expression may describe the form of a perfectly deterministic attribute of some entirely non-random variable denoted by x.

A few examples will illustrate how the lack of appreciation of such and similar differences, and the sheer power of names, affect statistical hydrology.

It is well known that the gamma distribution fits reasonably well the histograms of historic frequencies of annual flow totals of many rivers. For now, let us not dwell on the logical jump required to equate these frequencies with probabilities, and let us accept the common wisdom that annual flows of many rivers have gamma probability distributions. It is also well known in hydrology that if we route a unit inflow through a cascade of linear reservoirs (release from each is a linear function of its instantaneous water storage) then the shape of the downstream hydrograph (the so-called *Unit Hydrograph*, i.e. the unit response function of the cascade) also happens to have the form of gamma distribution, in which case the latter has no probabilistic connotation. This mathematical similarity once ruined a good part of an evening for me when a distinguished Australian hydrologist tried to persuade me that the Unit Hydrograph is a probability distribution because the gamma distribution is obviously a probability distribution and the Unit Hydrograph is obviously a gamma distribution.

In probability theory one often reads that something has one or other theoretical distribution, for example that the theoretical model of the sum of gamma-distributed random variables is again a gamma-distributed variable. The connotation of the term 'theoretical' is that the form of the resultant distribution has been arrived at (from the given original propositions and descriptions) by applying the theory governing the operation in question, in this case the summation of random variables. However, the notion of 'theoretical probability distribution model' has been transposed in hydrology to mean any mathematical function that has ever been used in probability theory or statistics to describe a probability distribution. Thus any such function, when fitted to a frequency distribution of floods (or of any hydrological variable), automatically assumes the status of its 'theoretical probability distribution', although its selection may be completely arbitrary and have nothing whatsoever to do with any theory related to the process that generated the floods. Such a fitted distribution model is then used for the estimation of the 'theoretical' values of flood probabilities.

An example of this practice was cited in (Klemeš, 1988a). In that case 'theoretical' values of seasonal level fluctuations in an African equatorial lake, obtained by a fitted Pearson III distribution (which '... has proved to be the distribution of best fit') indicated that, in every year, seasonal flooding of a better part of Uganda by that lake has a small but real probability, while, in terms of the climatic, hydrologic and topographic realities, these 'theoretical' values are pure fiction.

Because of their name, extreme value distributions are still among the favourite 'theoretical' models for extreme hydrologic events. It was to no avail that the illustrious late Australian statistician and probabilist P.A.P. Moran exposed the fallacy already in the late 1950s when he wrote

> A great deal has been written on the subject, most of which is wrong in principle. . . . Distributions which may be particularly useful in this connection are the type III and the Logarithmic Normal . . . These particular distributions are here chosen solely for practical reasons and there is no theoretical reason why they should fit observed series . . . It has been pretended that better results can be obtained by the use of Gumbel's 'extreme value' distribution. This is the asymptotic form of the distribution of the largest member of a sample from a given distribution. If we knew the latter exactly we could estimate the parameters of Gumbel's distribution. However, Gumbel's distribution depends solely on the form of the tail of $f(x)$ which . . . is usually outside the range of observations and can only be guessed at. No amount of mathematical presdigitation can remove this uncertainty. Another defect in this approach is the fact that Gumbel's extreme value distribution is attained very slowly—in the case of the normal distribution only for samples of size at least 10^{12}. (Moran, 1959)

Even the humble arithmetic mean is a frequent victim of the connotation of its name. In a process that does have a central tendency the mean is its useful measure, and can be used as a basis of other process characteristics: its 'expected value', central moments etc. The problem is that the mean can be easily computed regardless of the existence of any central tendency, and it is often implied that a central tendency follows from the fact that the mean has been computed. If an analysis proceeds on this basis and a central tendency does in fact not exist, various unexpected things can happen, as I once had an opportunity to demonstrate in connection with the 'Hurst phenomenon' (Klemeš, 1974).

The most blatant abusers of the arithmetic mean are of course climatologists and meteorologists, who dare to call the often unrealistic meteorological conditions described by long-term means the 'normals' and the normal actual conditions 'anomalies'. Economists and politicians are not much better—but potentially far more dangerous by implying that below-average conditions are intrinsically undesirable, a sort of social evil to be eradicated. Below-average harvests have already been accepted as legitimate reasons for subsidies, and, given the unmitigated power of the media to create illusions of reality out of words repeated *ad nauseam*, 'normal' weather conditions and above-average income may soon be added to the list of basic human rights.

19.4.2 What's not in a name!

Another problem that mathematical modellers often do not appreciate is the following. A model, and a mathematical operation in general, even when it is properly named and valid in the sense that it models what it is supposed to, does not tell the modeller the physical limits of its validity, nor does it tell him whether a physical mechanism it implies does in fact exist.

Suppose that one has a theoretically reasonable model—i.e. one based on a physical theory of flood formation—for flood probabilities on the Manitou River on the Manitoulin Island in Lake Huron. Suppose, further, that the model was fitted to a 50-year historic flood record and, after an additional 50 years of observations, it is found to agree perfectly with the whole century-long evidence. I once asked an esteemed American stochastic hydrologist whether the model could be used to make a realistic estimate of a million-year flood peak flow on the river. 'Of course it can' was his reply. I disagreed on the following grounds. It may well be that under the conditions necessary to produce such a flood, the whole Manitou River, and even the Manitoulin Island itself, would disappear so that no such peak flow could ever arise there (Klemeš, 1989).

Earlier this year, I was confronted with a similar problem in a much more concrete setting. In order to reach high safety standards for its dams, a hydropower corporation in Canada computed the so-called probable maximum flood (PMF) that they are supposed to withstand, as a flood produced by the combination of a 1000-year snowpack and a 1000-year air temperature sequence; both these values were obtained by extrapolation of simple distributional models (e.g. cube-root normal for the snowpack) fitted to short historic records (20–50 years). Among my reservations regarding this approach were the following. First, in northern Canada, when the snowpack reaches a certain thickness, it may not melt during one summer season and may start accumulating, perhaps even initiating glaciation of the area. As a result, the tail of the distribution may be very different from that of the distribution fitting the few independent historical annual accumulations, and it may be necessary to invoke the storage theory to guess its likely form. Secondly, should such a situation arise, it may conceivably lead to one of two diametrally different outcomes:

(i) an almost continuous availability of a snow (+ ice) layer identified now as having only a 10^{-3} probability and thus perhaps a 100-fold increase of the probability of the computed PMF and a proportional decrease of dam safety; or

(ii) more likely, the situation might be a consequence of climatic cooling, in which case not only the probability of the temperature sequence now used may be practically zero but the river may cease to exist because of glaciation, thus making any dam on it 100% safe from flood damage.

However, even such simple mathematical constructs like the arithmetic mean may imply unrealistic physics and invalidate modelling results. The physical equivalent of replacing in a model the actual daily precipitation with, say, its monthly mean would be first to store the total monthly precipitation volume somewhere (no computer will ever be able to do that) and, at the end of the month, go back in time and start continuous spraying at the mean rate from day one. This is actually 'going on' in most general circulation models (GCMs), which produce no flash floods but abound in drizzle (especially in the dry summers on Canadian prairies, I suppose). Needless to say, many hydrologic modellers eagerly use outputs of these models for predicting effects of climate changes on probabilities of floods and other hydrologic phenomena.

Running means, customarily plotted in the middle of the averaging windows, clairvoyantly anticipate future trends of physical processes long before Mother Nature has any idea of them. For example, the five-year running means of precipitation in Figure 19.2 (together with similar plots for temperature and discharge not shown here) have been found to provide an 'indication of underlying sustained variations in climatic conditions'. They also show that 'precipitation appears to have increased marginally before suffering a decline' (Collins, 1984). As can be seen, this underlying increase coincides with the actual (presumably anomalous?) drop in 1968. However, this could be readily remedied by taking, for instance, 10-year running means which have been plotted as a dotted line. Their other advantage would be that the phase of the 'sustained underlying variations' could be made better to coincide with the actual ones and that the precipitation increase at the end of the period (caused by a sustained underlying variation?) could be anticipated almost three years before the trend reveals itself in the five-year means and, remarkably, while the actual precipitation is still drastically (but anomalously?) declining.

The last example that I will present here concerns the running integral of deviations from the sample mean ('residual mass curve')—a favourite tool for analysis of hydrological and climatic records and time series in general. Indeed,

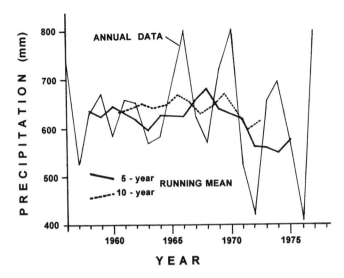

Figure 19.2 Annual precipitation totals at Sion, Switzerland, and their 5- and 10-year running means (annual data and 5-year means after Collins, 1984).

many 'significant' periodicities and climate changes have been discovered with its help. Thus Williams (1961) reached the following conclusion by analysing records between 60 and 70 years long: 'If cumulative deviations from the mean are computed for hydrologic data, continuous periods of 10 years to 35 years or more will be revealed in which hydrologic records are consistently below or above their means'. One problem is that, as Feller (1966) showed, such cumulative series exhibit long swings even when the underlying process is completely random and the length of the swings grows with the length of the series. A related problem is that residual mass curves of higher orders of virtually any series converge to a single sine wave extending over the whole sample, whatever its size may be (Klemeš and Klemeš, 1988). This convergence, caused by amplification of the first Fourier frequency by successive integrations, is very rapid and robust, and is virtually completed in the fourth order (Figure 19.3).

However, the point that I want to emphasise in the present context is that there is a considerable difference in who is doing the integrating, the analyst on his computer or Mother Nature in the field. Suppose that precipitation forms a stationary random series x with a mean $\bar{x}$. The cumulative series $y_t = \int_0^t (x - \bar{x})\, dt$ may well exhibit suggestive quasi-cyclic patterns, but they have no hydrologic or other physical significance, since the storing of the differences from the mean has been done only symbolically, with numbers in a computer or on paper as shown in (Figure 19.3).

Now suppose that the same precipitation falls onto a large lake into which it is the only water input and that the only output is evaporation, which is relatively constant over time and approximately equal to $\bar{x}$. Then the series of water-level

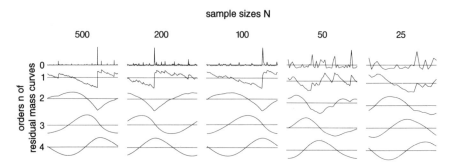

Figure 19.3   Residual mass curves of higher orders for random series of different lengths generated from a lognormal distribution (Klemeš and Klemeš, 1988).

fluctuations in the lake will have essentially the same shape as the above series y_t; but now the series is a record of a real physical process, and its longer and shorter epochs of higher and lower levels will represent epochs with relatively uniform, but from epoch to epoch genuinely different, environment (or 'climate') for the fauna and flora within different elevation bands of the lake shore. But it is important to appreciate that such a real 'microclimate' cyclicity (inferred, say, from lake varves) may have no macroclimatic significance. Hence, to use such a varve record as proxy data for precipitation may be misleading.

One can, of course, easily compute a second-order residual mass curve of the x series as $z_t = \int_0^t (y - \bar{y})\, dt$, which will have an even more pronounced sinusoidal shape but will have no physical significance. On the other hand, nature may produce a series with practically the same shape by, say, allowing for a small fraction of the lake output to take place as seepage proportional to the water stage, accumulate it in another lake downstream (or in an aquifer), which again may have a constant output equal to the mean input. Conceivably, processes of still higher cumulative orders may well arise that will exhibit real sinusoidal periods over some time interval N during which the output of a particular system is approximately equal to the mean (over N) of the input—and all this may be embedded in a perfectly random controlling environment as in Figure 19.3.

As a result, it is not possible to judge the significance of cyclic patterns in a computed residual mass curve of a natural process without knowing the physical structure of the process—to what extent it itself is a cumulative process and, if so, of what order. For example, a hydrologist may refrain from using a residual mass curve of lake levels because of their cumulative nature. On the other hand, he seldom hesitates to use it for streamflow records, which are regarded as simple stochastic series. However, streamflow may well be a cumulative process owing to the aforementioned mechanisms (the St Lawrence River, draining the Great Lakes, is a good example), and then the significance of the mass curve cyclicity may be grossly over-rated. Even more important in the present era, obsessed with the detection of climatic changes, is the possibility that some paleohydrologic and paleoclimatic reconstructions and modelling based on proxy data may be flawed.

19.5 HYDROLOGIC MODELLING AS 'KNOWLEDGE BUSINESS'

Hydrology and statistics were not among the sciences that motivated me when I was a student. I wanted to be an engineer and design dams. There was no contradiction, because no statistics course was required in my engineering curriculum, and hydrology was a marginal affair from which I remember three things: the Thiessen Polygon, the Rational Formula and condoms; the latter, stretched over a hollow metal cylinder and weighted down in the middle, were used by a graduate student to model the shape of the drawdown cone around a well (he soon had to switch to an electrical analogy because his professor found it difficult to get research money for condoms).

On the job, I soon realised that the most important thing in the design of a dam is to correctly specify the storage capacity of the reservoir: that determines the size of the dam, which controls its cost. With great enthusiasm, I plunged into the theory of storage only to find out that, unfortunately, its methods rested on probability theory and its inputs on statistical hydrology. The latter was always considered known ('Given a gamma-distributed random input...', etc., etc.). However, since it was not known to me, I set out to find out.

Allas, there was little to be found out! There was virtually no 'accumulated knowledge' about statistical and stochastic behaviour of hydrological phenomena, no theory. There was only accumulated knowledge about techniques: how to calculate the moments, the correlation coefficient, fit a 'probability' distribution, etc.

Now I know that this was not unique to statistical hydrology. I was struck when, about 10 years after it was published, I came across Oscar Kempthorne's paper on 'Probability, statistics and the knowledge business' (Kempthorne, 1971), where he claimed that such attitudes were in fact imported into the sciences by statisticians, that this was how many statisticians saw their own work! To quote:

> How does a consulting statistician work? Does he say to the scientist: You must do this. You must collect data in this way. You must analyze the data in this way. And so on. You must make a *t*-test... The answers are: Of course not! How idiotic can you be? And yet, the great bulk of workers on foundations of statistical inference seem to think in terms of such answers... I have met scientists who have been brainwashed by statisticians to the view that their problems amount to the calculation of a linear discriminant and the scientists want to know how to do this.

Wherever the responsibility may have lain, this certainly was the situation in statistical hydrology in the early 1960s when I turned to it to find answers to questions like why should streamflows follow one or other distribution, be random or serially correlated, etc. There were no answers. Worse, there were not even such questions! The Big Names, whose books I was reading and some of whom I later had a chance to consult in person, often did not even seem to understand the problem I was raising. For example, when I once asked the then dean of American hydrology what could be the reason that the distributions of some annual streamflow series were negatively skewed, his answer was 'When it is so, you may be able to fit it well with some positively skewed model if you flip it over'. Later, when I proposed an answer to my question at a symposium (Klemeš, 1970), a leading Australian hydrologist commented 'If you took the square of the skew coefficient you would get rid of your problem'. They did not seem to understand that my problem was not how to fit something or how to get rid of something that does not fit, but to get a scientific insight into something.

In short, statistical hydrology as a knowledge business was not yet defined. From those days, I remember only two instances fitting into this category. One was Kalinin's (1962) attempt to trace the tendency to gamma distribution in annual runoff to the binomial-like alternation of wet and dry periods within the year; the other was Yevjevich's (1963) model explaining the origin of serial correlation in time series of annual runoff. I tried to formulate the concept of statistical hydrology as a knowledge business in the late 1970s, when I called it *physically based stochastic hydrologic analysis* (Klemeš, 1978); by that time, I could already cite about two dozen references dealing with its different specific aspects.

Nowadays, the problem is no longer the recognition that hydrology, including its statistical and stochastic aspects, is a physical science rather than a mere input to water management decisions (see e.g. Committee on Opportunities in the Hydrologic Sciences, 1991). The problem is rather the continuing obfuscation of the difference and the evasion of doing hydrologic science by citing difficulties in securing the funding (no, condoms no longer appear in hydrology research proposals—on reflection, nowadays this could be helpful for the same reasons that it was not 40 years ago). The tendencies to 'push the hydrology ball to the end of the forward-moving wagon of science' reflect the inertia of the historical professional and educational structures within which hydrology as a whole has been developing and within which it still operates. It is still typical that the academic home of hydrology is in Engineering rather than Science departments of universities, and the bulk of hydrologic work is carried out by organisations whose chief purpose (or 'mandate') is some kind of Resource Management.

For example, under the Canadian Federal Government, the National Hydrology Research Institute, which now operates within the Conservation and Protection (formerly Environmental Management) Service of the Environment Department, gradually evolved from a water resource engineering unit attached, in sequential order over the past 40 years or so, to the following Departments (R. H. Clark, personal communication): Resources and Development; Northern Affairs and Natural Resources; Mines and Technical Surveys; Energy, Mines and Resources; Environment and Fisheries; Environment. The word 'hydrology' first appeared in its designation in 1962 when a Hydrology Branch was formed by my friend and former colleague, R. H. Clark. Ironically, after his departure, and several years after the word 'research' had been proudly added to its name, the role of the National Hydrology Research Institute was officially defined in 1986 as *developing, applying and advising on the technology required by clients managing Canadian waters*. And, as late as two years ago, an official report of the Canadian Associate Committee on Hydrology, entitled *Canadian Hydrological Science* (Associate Committee on Hydrology, 1991), contains these three laconic sentences buried in its 13 pages of heroic prose on scientific challenges, deepening of knowledge about basic processes, favourable political and social climate, etc.: 'Hydrological research expertise is by and large focused on applied, not fundamental science'; 'The research priorities of most Canadian hydrological establishments address technology'; and 'Should funding become available, scientists currently conducting research in applied hydrology, water management and allied sciences would likely be enticed to shift into basic research'.

So it goes! (as the narrator in Kurt Vonnegut's (1969) novel *Slaughterhouse-Five* comments when something sad but apparently inevitable happens).

19.6 CONCLUSIONS

In a nutshell, my thesis has been that if a hydrologic modeller is confused about the difference between mathematics, physical science and decisions, statistics and probability theory will not make his models better, and, quite possibly, can make them worse, even if the techniques he may be employing are formally correct and rigorous. The obvious remedy is first to clear up the confusions, some of which were discussed in the preceding sections, and only then turn to the two disciplines. The essentials of the clean-up process may be summarised as follows.

To learn the difference

(i) between something and the name of something—in particular between a physical process and its mathematical model;
(ii) between science and technology—in particular between hydrology and water management decisions;
(iii) between the aims of analytical (investigative) and synthetic (descriptive) models;
(iv) between a scientific statement (falsifiable theory) and a rationale (unfalsifiable theory).

To appreciate that

(i) what is a scientific statement in one context may be only a rationale in another;
(ii) mathematical formulation does not transform a rationale into a scientific statement;
(iii) rationale is useless as a tool for advancing knowledge;

(iv) rationale may be useful, even inevitable, in a decision context;

 (v) a mathematical formula is not responsible for the validity of its application.

It would also be helpful if prospective hydrologic modellers were taught hydrology, as well as statistics and probability theory, by professors who had already mastered what was said in the two preceding paragraphs.

ACKNOWLEDGEMENT

This essay is dedicated to the memory of the late Professor P. A. P. Moran, a wise statistician, kind man and a sadly missed friend.

References

Addiscott, T. M. (1977) A simple computer model for leaching in structured soils. *J. Soil Sci.*, **28**, 554–563. (262, 263, 266, 267)

Addiscott, T. M. (1993) Simulation modelling and soil behaviour. *Geoderma*, **60**, 15–40. (264)

Addiscott, T. M. and Bailey, N. J. (1990) Relating the parameters of a leaching model to the percentages of clay and other components. In Roth, K., Flühler, H., Jury, W. A. and Parker, J. C. (eds) *Field-Scale Solute and Water Transport Through Soil*, pp. 209–221. Birkhäuser, Basel. (265)

Addiscott, T. M. and Bland, G. J. (1988) Nitrate leaching models and soil heterogeneity. In Jenkinson, D. S. and Smith, K. A. *Nitrogen Efficiency in Agricultural Soils*, pp. 394–408. Elsevier Applied Science, Barking. (263, 265)

Addiscott, T. M., Rose, D. A. and Bolton, J. (1978). Chloride leaching in the Rothamsted Drain Gauges: influence of rainfall pattern and soil structure. *J. Soil Sci.*, **29**, 305–314. (263)

Addiscott, T. M. and Wagenet, R. J. (1985) Concepts of solute leaching in soils: a review of modelling approaches. *J. Soil Sci.*, **36**, 411–424. (261)

Addiscott, T. M. and Whitmore, A. P. (1991) Simulation of solute leaching in soils of differing permeabilities. *Soil Use Manage.*, **7**, 94–102. (262, 263, 265, 266)

Addiscott, T. M., Whitmare, A. P. and Powlson, D. S. (1991) *Farming, Fertilizers and the Nitrate Problem*. CAB International, Wallingford. (252)

Aitchison, J. and Brown, J. A. C. (1957) *The Lognormal Distribution*. Combridge University Press, Cambridge. (254, 255)

Anderson, B. G. (1980) Aquatic invertebrates in tolerance investigations from Aristotle to Naumann. In Buikema, A. L. Jr and Cairns, J. Jr (eds) *Aquatic Invertebrate Bioassays*, pp. 3–35. American Society for Testing and Materials, Philadelphia, ASTM STP 715. (286)

Anderson, C. W. and Nadarajah, S. (1993) Environmental factors affecting reservoir safety: In Barnett, V. and Turkman, K. F. (eds) *Statistics for the Environment*, pp. 163–182. Wiley, Chichester. (79, 80)

Anderson, C. W. and Turkman, K. F. (1991) The joint limiting distribution of sums and maxima of stationary sequences. *J. Appl. Prob.*, **28**, 33–44. (180)

Anderson, T. W. (1958) *An Introduction to Multivariate Statistical Analysis*. Wiley, New York. (326)

Arkin, P. A. and Ardanuy, P. E. (1989) Estimating climate-scale precipitation from space: a review. *J. Climate*, 1229–1238. (47)

Associate Committee on Hydrology (1991) *Canadian Hydrological Science*. National Research Council Canada, Ottawa. (365)

Athanassoulis, G. A. (1992) *Description of wave parameters variability for different time scales*. Report, National Technical University, Athens. (211)

Baier, W., Dyer, J. A. and Sharp, W. R. (1979) *The versatile soil moisture budget model.* Technical Bulletin No. 87, Agrometeorology Section, Land Resource Research Institute, Agriculture Canada, Ottawa, Ontario. (265)

Barbee, G. C. and Brown, K. W. (1986) Comparison between suction and free-drainage solution samplers. *Soil Sci.,* **141**, 149–154. (253)

Barkham, J. P., MacGuire, F. A. S. and Jones, S. J. (1992) *Sea-Level Rise and the UK, a Report for Friends of the Earth.* Friends of the Earth Trust Ltd. (147, 177, 178, 179)

Barnett, T. P. (1984) The estimation of 'global' sea-level change: a problem of uniqueness. *J. Geophys. Res.,* **89**, 7980–7988. (166, 173)

Barnett, V. (1973) *Comparative Statistical Inference,* p. 122. Wiley, New York. (255)

Barnett, V. and Turkman, F.K. (eds) (1993) *Statistics for the Environment.* Wiley, Chichester.

Barraclough, D. (1989) A usable mechanistic model of nitrate leaching. I. The model. *J. Soil Sci.,* **40**, 543–554. (262)

Barry, R. G. and Perry, A. H. (1973) *Synoptic climatology.* Methuen, London. (58)

Bartlett, M. S. (1962) *Essays on Probability and Statistics.* Methuen, London. (347)

Bates, B. C. and Townley, L. R. (1988) Nonlinear discrete flood event models. 1. Bayesian estimation of parameters, *J. Hydrol,* **99**, 61–76. (298)

Belford, R. K. (1979) Collection and evaluation of large-scale monoliths for soil and crop studies. *J. Soil Sci.,* **30**, 363–373. (258)

Beran, R. (1987) Prepivoting to reduce level error of confidence sets. *Biometrika,* **74**, 457–468. (111)

Berndtsson, R. (1989) Topographical and coastal influence on spatial precipitation patterns in Tunisia, *Int. J. Climatol.,* **9**, 357–369. (54)

Beven, K. (1987) Towards the use of catchment geomorphology in flood frequency predictions. *Earth Surf. Processes and Landforms,* **12**, 69–82. (143)

Beven, K. J. (1989) Changing ideas in hydrology—the case of physically-based models. *J. Hydrol.* **105**, 157–172. (266, 267, 298)

Beven, K. J. (1993) Prophecy, reality and uncertainty in distributed hydrological modelling. *Adv. Water Resources,* **16**, 41–51. (298, 301)

Beven, K. J. and Binley, A. (1992) The future of distributed models: model calibration and uncertainty prediction. *Hydrol. Process.,* **6**, 279–298. (250, 268, 298, 299, 300, 302, 306, 308, 316)

Beven, K. J., Kirkby, M. J. (1979) A physically based variable contributing area model of basin hydrology. *Hydrol. Soc. Bull.,* **24**, 43–69. (300)

Beven, K. J., Kirkby, M. J., Schoffield, N. and Tagg, A. (1984) Test of a physically-based flood forecasting model (TOPMODEL) for three UK catchments. *J. Hydrol.,* **69**, 119–143. (300)

Binley, A. M. and Beven, K. J. (1991) Physically based modelling of catchment hydrology: a likelihood approach to reducing predictive uncertainty. In Farmer, D. J. and Rycroft, M. J. (eds) *Computer modelling in the Environmental Sciences,* pp. 75–88, Clarendon press, Oxford. (317)

Binley, A. M., Beven, K. J., Calver, A. and Watts, L. (1991) Changing responses in hydrology: assessing the uncertainty in physically-based predictions. *Water Resources Research,* **27**, 1253–1262. (317)

Bird, E. C. F. (1993) *Submerging Coasts the Effects of a Rising Sea-Level on Coastal Environments.* Wiley, Chichester. (177)

Biggar, J. W. and Nielsen, D. R. (1976) Spatial variability of the leaching characteristics of a field soil. *Water Resources Research,* **12**, 78–94. (263)

Blackman, D. L. and Graff, J. (1978) Analysis of maximum sea-levels in southern England. In *Proceedings of 16th Coastal Engineering Conference, Hamburg, 1978*, Vol. 1, pp. 931–947. American Society of Civil Engineers. (180)

Bonney, G. E. (1987) Logistic regression for dependent binary observations. *Biometrics*, **43**, 951–973. (323)

Boorman, D. B., Hollis, J. and Lilly, A. (1991) The production of the Hydrology of Soil Types (HOST) data set. In *Proceedings of 3rd National Hydrology Symposium, Southampton, September 1991*, pp. 6.7–6.13. British Hydrological Society. (142)

Bouma, J. (1993) Measuring soil behaviour. *Geoderma*, **60**, 1–14. (250, 263)

Box, G. E. P. (1976) Science and statistics. *J. Amer. Statist. Assoc.*, **71**, 356, 791–799. (347, 349, 352, 353)

Box, G. E. P. and Cox, D. R. (1964) An analysis of transformations (with discussion). *J. R. Statist. Soc.*, **B26**, 211–252. (304)

Box, G. E. P. and Jenkins, G. M. (1976) *Time Series Analysis, Forecasting and Control*, revised edn. Holden-Day, San Francisco. (334)

Box, G. E. P. and Tiao, G. C. (1992) *Bayesian Inference in Statistical Analysis*. Wiley, Chichester. (299, 315)

Bradu, D. and Mundlak, Y. (1970) Estimation in lognormal linear models. *J. Amer. Statist. Assoc.*, **65**, 198–211. (114)

Bras, R. and Eagleson, P. S. (1987) Hydrology, the forgotten earth science. *Eos*, **68**, 227. (346)

Brechtel, H. M. (1989) *Monitoring wet deposition in forest; quantitative and qualitative aspects*. Air Pollution Report Series of the Environmental Research Programme of the Commission of the European Communities, Brussels, Report No. 21, pp. 39–63. (319)

Bresler, E., Bielorai, H. and Laufer, A. (1979) Field test of solution flow models in a heterogeneous irrigated cropped soil. *Water Resources Research*, **15**, 645–652. (262)

Bruneau, P. (1993) Influence of space and time resolution on optimal parameter estimation for TOPMODEL. Paper presented, at SPRUCE II International Conference, Rothampstead Experimental Station, September 1993. (46)

Buishand, T. A. (1984) Bivariate extreme-value data and the station-year method. *J. Hydrol.*, **69**, 77–95. (132)

Buishand, T. A. (1993) Rainfall depth–duration–frequency curves; a problem of dependent extremes. In Barnett, V. and Turkman, K. F. (eds) *Statistics for the Environment*, pp. 183–197, Wiley, Chichester. (44)

Bulich, A. A. (1979) Use of luminescent marine bacteria for determining toxicity. In *Aquatic Toxicology: Second Conference*, pp. 171–180. American Society for Testing and Materials, Philadelphia, ASTM STP 667. (286)

Burns, D. H. (1990) Evaluation of regional flood frequency analysis with a Region of Influence Approach. *Water Resources Research*, **26**, 2257–2265. (128)

Burns, I. G. (1974) A model for predicting the redistribution of salts applied to fallow soils after excess rainfall or evaporation. *J. Soil Sci.*, **25**, 165–178. (263, 266)

Burns, I. G. (1975) An equation to predict the leaching of surface-applied nitrate. *J. Agric. Sci. Camb.*, **85**, 443–454. (263)

Busch, U., Berke, O., Bonke, R. and Kunert, J. (1993) *Mathematisch-statistische Endauswertung zum forstlich-ö kologischen Beweissicherungsverfahren Startbahn 18 West des Frankfurter Flughafens*. Unpublished report, Dortmund University. (322)

Busch, U. (1990) *Statistische Auswertung eines Versuchs aus der Waldschadensforschung mit Expositionskammern*. Degree dissertation. University of Munich. (326)

Busch, U. (1994) Statistische Aspekte eines forstlich-ökologischen Beweissicherungsverfahrens. In Kreienbrock, L. and Kublin, E., (eds); *Tagungsberichte der Arbeitsgruppe Ökologie*, Vol. 5, pp. 10–22. Deutsche Region der Internationalen Biometrischen Gesellschaft. (362)

Campbell, N. A. (1980) Robust procedures in multivariate analysis I: Robust covariance estimation. *Appl. Statist.*, **29**, 231–237. (289)

Cannell, R. Q., Goss, M. J., Harris, G. L., Jarvis, M. G., Douglas, J. T., Howse, K. R. and Le Grice, S. (1984) A study of mole drainage with simplified cultivation for autumn-sown crops on a clay soil. 1. Background, experiment and site details, drainage systems, measurement of drainflow and summary of results. *J. Agric. Sci. Camb.*, **102**, 539–559. (261)

Cochran, W. G. (1977) *Sampling Techniques*, 3rd edn. Wiley, New York.

Coleman, E. A. (1946) A laboratory study of lysimeter drainage under controlled soil moisture tension. *Soil Sci.*, **62**, 365–382. (259)

Coles, S. G. (1994) A temporal study of extreme rainfall. In Barnett, V. and Turkman, K. F. (eds) *Statistics for the Environment 2: Water Related issues*. Wiley, Chichester, pp. 61–78. (24)

Coles, S. G. and Tawn, J. A. (1990) Statistics of coastal flood prevention. *Phil. Trans. R. Soc. Lond.* **A332**, 457–476. (221, 225)

Coles, S. G. and Tawn, J. A. (1990) Statistics of coastal flood prevention. *Phil. Trans. R. Soc. Lond.* **A332**, 457–476. (178, 180)

Coles, S. G. and Tawn, J. A. (1991) Modelling extreme multivariate events. *J. R. Statist. Soc.*, **B53**, 377–392. (68, 80, 225)

Coles, S. G. and Tawn, J. A. (1992) *Modelling extremes of the areal rainfall process*. University of Nottingham Technical Report, 92–12. (75)

Coles, S. G. and Tawn, J. A. (1994) Statistical methods for multivariate extremes: an application to structural design (with discussion). *Appl. Statist.*, **43**, 1–48. (80)

Coles, S. G., Tawn, J. A. and Smith, R. L. (1994) A seasonal Markov model for extremely low temperatures. Submitted. (61, 64, 68)

Collins, D. N. (1984) Climatic variation and runoff from Alpine glaciers. *Z. Gletscherkunde und Glaziologie*, **20**, 127–145. (360, 361)

Committee on Techniques for Estimating Probabilities of Extreme Floods (1988) *Estimating Probabilities of Extreme Floods*. National Academy Press, Washington, DC. (355)

Committee on Opportunities in the Hydrologic Sciences (1991) *Opportunities in the Hydrologic Sciences*. National Academy Press, Washington, DC. (364)

Cooke, G. W. (1976) A review of the effects of agriculture on the chemical composition and quality of surface and underground waters. *Tech. Bull. Minist. Agric. Fish. Fd*, **32**, 5–57. (283)

Corwin, D. L., Waggoner, B. L. and Rhoades, J. D. (1991) A functional model of solute transport that accounts for bypass. *J. Environ. Qual.*, **20**, 647–658. (267)

Cox, D. R. and Hinkley, D. V. (1974) *Theoretical Statistics*. Chapman and Hall, London. (305)

Cox, D. R. and Isham, V. (1980) *Point Processes*. Chapman & Hall, London. (101, 102)

Cox, D. R. and Isham, V. (1988) A simple spatial-temporal model of rainfall. *Proc. R. Soc. Lond.*, **A415**, 317–328. (8, 10, 11, 44)

Cox, D. R. and Isham, V. (1994) Stochastic models of precipitation. In Barnett, V. and Turkman, K. F. (eds) *Statistics for the Environment 2: Water Related Issues*. Wiley, Chichester, pp. 3–18. (19, 41, 44, 58)

Crabb, J. (1980) Synthesis of a directional wave climate. In Count, B. (ed.) *Power from Sea Waves*, pp. 41–74. Academic Press, London. (209, 216, 217)

Cressie, N. A. C. (1991) *Statistics for Spatial Data*. Wiley, New York. (36, 40, 181)

Croft, P. J. and Shulman, M. D. (1989) A five-year radar climatology of convective precipitation for New Jersey. *Int. J. Climatol.*, **9**, 581–600. (47)

Cunnane, C. (1986) Review of statistical models for flood frequency estimation. Presented at International Symposium on Flood Frequency and Risk Analyses, Louisiana State University, Baton Rouge. (352)

Cunnane, C. (1989) *Statistical distributions for flood frequency analysis*. World Meteorological Organization, Operational Hydrology Report No. 33, WMO Publication No. 718, Geneva. (129)

Cuttle, S. P., Hallard, M., Daniel, G. and Scurlock, R. V. (1992) Nitrate leaching from sheep-grazed grass/clover and fertilized grass pastures. *J. Agric. Sci. Camb.*, **119**, 335–343. (252, 254, 256, 257)

Dales, M. Y. and Reed, D. W. (1989) *Regional flood and storm hazard assessment*. Report No. 102, Institute of Hydrology, Wallingford. (131, 132, 133, 140, 141)

Daley, D. J. and Vere-Jones, D. (1988) *An Introduction to the Theory of Point Processes*. Springer, New York. (101)

Davidson, M. L. (1980) *The multivariate approach to repeated measures*. BMDP Technical Report 75, BMDP Statistical Software Ltd, Los Angeles. (326)

Davis, A. W. (1979) On certain ratio statistics in weather modification experiments. *Technometrics*, **21**, 283–289. (113)

Davison, A. C., Hinkley, D. V. and Schechtman, E. (1986) Efficient bootstrap simulation. *Biometrika*, **73**, 555–563. (116)

Davison, A. C. and Smith, R. L. (1990) Models for exceedances over high thresholds (with discussion). *J. R. Statist. Soc.*, **B52**, 393–442. (65, 127)

Dawdy, D. R. and Lichty, R. W. (1968) Methodology of hydrologic model building. In *Digital Computers in Hydrology*, International Association of Scientific Hydrology Publi-cation No. 81. pp. 123–137. (304)

De Jong, R., Dukanski, J. and Bootsma, A. (1992) Implications of spatial averaging weather and soil moisture data for broad-scale modelling activities. *Soil Use Manage.*, **8**, 74–79. (265)

Deltacommissie (1960) *Delta rapport*. Den Haag. (195)

DiCiccio, T. and Tibshirani, R. (1987) Bootstrap confidence intervals and bootstrap approximations. *J. Amer. Statist. Assoc.*, **82**, 163–170. (111)

DiCiccio, T. J. and Romano, J. P. (1988) A review of bootstrap confidence intervals (with discussion). *J. R. Statist. Soc.*, **B50**, 338–354. (110)

Diggle, P. J. (1990) *Time Series. A Biostatistical Introduction*. Clarendon Press, Oxford. (323)

Dilks, D. W., Canale, R. P. and Meier, P. G. (1992) Development of Beyesian Monte Carlo techniques for water quality model uncertainty. *Ecol. Modell.*, **62**, 149–162. (299)

Dixon, M. J. and Tawn, J. A. (1992) Trends in UK extreme sea-levels: a spatial approach. *Geophys. J. Int.*, **111**, 607–616. (166, 180)

Dixon, W. J. (ed.) (1990) *BMDP Statistical Software Manual*. University of California Press, Berkeley. (289)

Do, K. and Hall, P. (1991) On importance resampling for the bootstrap. *Biometrika*, **78**, 161–167. (116)

Douglas, B. C. (1992) Global sea-level acceleration. *J. Geophys. Res.*, **97**, 12 699–12 706. (167)

Doviak, R. J. and Zrnic, D. S. (1984) *Doppler Radar and Weather Observations*. Academic Press, London. (125)

Draper, L. (1988) Has the NE Atlantic become rougher? *Nature*, **322**, 494. (212, 216)

Duband, D., Michel, C., Garros, H. and Astier, J. (1988) Estimating extreme value floods and the design flood by the Gradex method. In *ICOLD-16, San Francisco, Q.63 R.67*, pp. 1009–1047. International Commission on Large Dams, Paris. (142)

Duan, Q., Sorooshian, S. and Gupta, V. K. (1992) Effective and efficient global optimisation for conceptual rainfall-runoff models. *Water Resources Research*, **28**, 1015–1031. (298)

Eagleson, P. S. (1972) Dynamics of flood frequency. *Water Resources Research*, **8**, 878–897. (353)

Efron, B. (1979) Bootstrap methods: another look at the jackknife. *Ann. Statist.*, **7**, 1–26. (110)

Efron, B. (1982) *The jackknife, the bootstrap and other resampling plans*. CBMS 38, SIAM-NSF. (110, 111)

Efron, B. (1987) Better bootstrap confidence intervals (with discussion). *J. Amer. Statist. Assoc.*, **82**, 171–185. (111)

Efron, B. (1990) More efficient bootstrap computations. *J. Amer. Statist. Assoc.*, **85**, 79–89. (116)

Efron, B. and Gong, G. (1983) A leisurely look at the bootstrap, the jackknife and cross validation. *Amer. Statistician*, **37**, 36–48. (114)

Elashoff, J. D. (1986) *Analysis of repeated measures designs*. BMDP Technical Report 83, BMDP Statistical Software Ltd, Los Angeles. (326)

Emery, K. O. and Aubrey, D. G. (1991) *Sea-levels, land levels and tide gauges*. Springer, New York. (163)

England, C. B. (1974) Comments on 'A technique using porous cups for water sampling at any depth in the unsaturated zone' by Warren W. Wood. *Water Resources Research*, **10**, 1049. (253)

Evans, D. V. (1988) The maximum efficiency of wave-energy devices near coastlines. *Appl. Ocean Res.*, **10**, 162–164. (218)

Fedra, K. (1980) Mathematical modelling—a management tool for aquatic ecosystems? *Helgol. Meersunters.*, **34**, 15–23. (299)

Feller, W. (1966) *An Introduction to Probability Theory and Its Applications*, 2nd edn, Vol. 1. Wiley, New York. (361)

Fiering, M. B. (1966) Synthetic hydrology: an assessment. In Kneese, A. V. and Smith, S. C. (eds) *Water Research*, pp. 331–341. John Hopkins University Press, Baltimore. (349)

Finney, D. J. (1941) On the distribution of a variate whose logarithm is normally distributed. *R. Statist. Soc. Lond. J. Suppl.* **7**, 155–161. (249, 255)

Flather, R. A. (1987) Estimates of extreme conditions of tide and surge using a numerical model of the north-west European continental shelf. *Estuarine Coastal Shelf Sci.*, **24**, 69–93. (180, 222)

Flueck, J. A. (1986) Principles and prescriptions for improved experimentation in precipitation augmentation research. In *Precipitation enhancement—a scientific challenge*. Meteor. Monogr., No. 21, Amer. Meteor. Soc., pp. 155–171. (113)

Flueck, J. A. and Holland, B. S. (1976) Ratio estimators and some inherent problems in their utilization. *J. Appl. Meteor.*, **15**, 535–543. (110)

Foy, R. H., Smith, R. V., Jordon, C. and Lennox, S. D. (1993) Upward trend in soluble phosphorus loadings to Lough Neagh. *Water Research*, submitted. (272, 283)

Foy, R. H., Smith, R. V., Stevens, R. J. and Stewart, D. A. (1982) Identification of factors affecting nitrogen and phosphorus loadings to Lough Neagh. *J. Environ. Mgt*, **15**, 109–129. (272, 273)

Franke, J. and Roeder, A. (eds) (1992) *Mathematical Modelling of Forest Ecosystems*. J. D. Sauerländer's Verlag, Frankfurt/Main.

Fry, T. C. (1928) *Probability and Its Engineering Uses*. Van Nostrand, New York. (347)

Gabriel, K. R. (1967) The Israeli artificial rainfall stimulation experiment. Statistical evaluation for the period 1961–1965. In Lecam, L. M. and Neyman, J. (eds) *Proceedings of 5th Berkeley Symposium on Mathematical Statistics and Probability*, Vol. 5; *Weather Modification*, pp. 91–114. University of California Press. Berkeley. (109)

Gabriel, K. R. (1970) *The Israeli rainmaking experiment 1961–1967, final statistical tables and evaluation* (tables prepared by M. Baras). Tech. Rep., Jerusalem, Hebrew University. (109)

Gabriel, K. R. (1979) Some statistical issues in weather experimentation. *Commun. Statist.: Theor. Meth.*, **A8**, 975–1015. (110)

Gabriel, K. R. and Feder, P. (1969) On the distribution of statistics suitable for evaluating rainfall stimulation experiments. *Technometrics*, **11**, 149–160. (110)

Gabriel, K. R. and Rosenfeld, D. (1990) The second Israeli rainfall stimulation experiment: analysis of precipitation on both targets. *J. Appl. Meteor.*, 1055–1067. (114)

Gagin, A. and Neumann, J. (1981) The second Israeli randomized cloud seeding experiment: evaluation of the results. *J. Appl. Meteor.*, **20**, 1301–1311. (109)

Garen, D. G. and Burges, S. J. (1981) Approximate error bounds for simulated hydrographs. *J. Hydraul. Div. Amer. Soc. Civ. Engrs*, **107**, 1519–1534. (298)

Gärtner, E. J., Urfer, W., Eichhorn, J., Grabowski, H. and Huss H. (1990) Mangan—ein Bioindikator für den derzeitigen Schadzustand mittelalter Fichten in Hessen. *Forstarchiv*, **61**, 229–233. (327)

Gelfand A. E., Hills, S. E., Racine-Poon, A. and Smith, A. F. M. (1990) Illustration of Bayesian inference in normal data models using Gibbs sampling. *J. Amer. Statist. Assoc.*, **85**, 972–985. (307)

Gelfand, A. and Smith, A. F. M. (1990) Sampling-based approaches to calculating marginal densities. *J. Amer. Statist. Assoc.*, **85**, 398–409. (177, 307)

Gibson, C. E. (1986) Preliminary results on phosphorus reduction in Lough Neagh—assessing the effect against a background of change. *Hyrobiol. Bull.*, **20**, 173–182. (273)

Gibson, C. E. and Fitzsimons, A. G. (1982) Periodicity and morphology of planktonic blue-green algae in an unstratified lake (Lough Neagh, N. Ireland). *Int. Rev. Gen. Hydrobiol.*, **67**, 459–476. (271)

Golbey, L. A. and O'Hagan, A. (1993) *ABLE Manual, Version 1*. Nottingham University Consultants Ltd. (238, 244, 245)

Golding, B. (1980) Computer calculations of waves from wind fields. In Count, B. (ed.) *Power from Sea Waves*, pp. 115–134. Academic Press, London. (213)

Golding, B. (1983) A wave prediction system for real-time sea state forecasting. *Q. J. R. Met. Soc.*, **109**, 393–416. (213)

Goldstein, M. and O'Hagan, A. (1993) *Bayes linear methods for systems of expert posterior assessments*. Nottingham Statistics Group Research Report 93–2. (245)

Goldstein, M. and Wooff, D. A. (1993) Robustness measures for Bayes linear analyses. *J. Statist. Planning and Inference*, in press. (247)

Gornitz, V., Lebedeff, S. and Hansen, J. (1982) Global sea-level trend in the past century. *Science*, **215**, 1611–1614. (162)

Graff, J. (1981) An investigation of the frequency distributions of annual sea-level maxima at ports around Great Britain. *Estuarine. Coastal Shelf Sci.*, **12**, 289–449 (178)

Gray, A. V. (1984) The Lough Neagh rivers—monitoring and water quality. Part II Sewage treatment works. In *Lough Neagh and its Rivers, Department of the Environment (NI) Conference, Antrim, N. Ireland*, pp. 1–6. (274)

Grayson R. B., Moore, I. D. and McMahon, T. A. (1992) Physically-based hydrologic modelling. 2. Is the concept realistic? *Water Resources Research*, **28**, 2659–2666. (298)

Greenhouse, S. W. and Geisser, S. (1959) On methods in the analysis of profile data. *Psychometrika*, **24**(2, 6/1959), 95–112. (326)

Greenhow, M. (1989) A probability distribution of breaking wave crest height based on a crest-acceleration threshold method. *Ocean Engng.*, **16**, 537–544. (212)

Haggett, C. M. (1988) Thunderstorms over north-west London—8 May 1988. *Weather*, **43**, 266–267. (47)

Haines, B. L., Waide, J. B. and Todd, R. L. (1982) Soil solution nutrient concentrations sampled with tension and non-tension lysimeters: report of discrepancies. *Soil Sci. Soc. Amer. J.*, **46**, 658–661. (260)

Hall, D. G. M. (1993) An amended functional leaching model applicable to structured soils. I. Model description. *J. Soil Sci.*, **44**, 579–588. (267)

Hall, P. (1988) Theoretical comparison of bootstrap confidence intervals (with discussion). *Ann. Statist.*, **16**, 927–953. (110, 111, 122)

Hall, P. (1989) On efficient bootstrap simulation. *Biometrika*, **76**, 613–617. (116)

Haltiner, G. J. and Martin, F. L. (1957) *Dynamical and Physical Meteorology*. McGraw-Hill, New York. (83)

Hammes, W. and Brechtel, H.-M. (1986) Verfahrensgang und erste Ergebnisse der forstlich-ökologischen Beweissicherung im Raum der Startbahn 18 West des Frankfurter Flughafens. *Verhandlungen der Gesellschaft für Ökologie*, **16**, 497–508. (320)

Hansen, E. A. and Harris, A. R. (1975) Validity of soil-water samples collected with porous ceramic cups. *Soil Sci. Soc. Amer. Proc.*, **39**, 528–536. (252, 253)

Haslett, J. and Raftery, A. E. (1989) Space–time forecasting with long-memory dependence: assessing Ireland's wind power resource (with Discussion). *J. R. Statist. Soc.*, **A38**, 1–50. (220)

Hastie, T. J. and Tibshirani, R. J. (1990) *Generalised Additive Models*. Chapman & Hall, London. (172)

Hatch, W. L. (1983) *Selective Guide to Climatic Data Sources*. US Department of Commerce, National Climatic Data Center, Asheville, NC. (48)

Henderson, K. G. and Robinson, P. J. (1994) Relationships between the Pacific/North American teleconnection patterns and precipitation events in the southeastern United States. *Int. J. Climatol.*, **14**, 307–323. (46, 52, 53)

Hendriks, J. *et al.* (1989) *Sea level rise*. PIANC-bulletin No. 66. (199)

Hevesi, J. A., Istok, J. D. and Flint, A. L. (1992) Precipitation estimation in mountainous terrain using multivariate geostatistics. Part I: Structural analysis. *J. Appl. Meteorol.*, **31**, 661–676. (54)

Hillel, D. (1991) Research in soil physics: a re-view. *Soil Sci.*, **151**, 30–34. (250, 264)

Hornberger, G. M. and Spear, R. C. (1981) An approach to the preliminary analysis of environmental systems. *J. Environ. Manag.*, **12**, 7–18. (299, 316)

Hosking, J. R. M. (1990) L-moments: analysis and estimation of distributions using linear combinations of order statistics. *J. R. Statist. Soc.*, **B52**, 105–124. (129)

Hosking, J. R. M. and Wallis, J. R. (1988) The effect of intersite dependence on regional flood frequency analysis. *Water Resources Research*, **24**, 588–600. (128)

Hosking, J. R. M., Wallis, J. R. and Wood, E. F. (1985) Estimation of the generalised extreme value distribution by the method of probability weighted moments. *Technometrics*, **27**, 251–261. (129)

Houghton, J. T., Callander, B. A. and Varney, S. K. (eds) (1992) *Climate Change 1992: The supplementary report to the IPCC Scientific Assessment.* Cambridge University Press, Cambridge. (46)

Houghton, J. T., Jenkins, G. J. and Ephraums, J. J. (eds) (1990) *The IPCC Scientific Assessment.* Cambridge University Press, Cambridge. (147, 162, 174, 178)

House of Lords Select Committee on the European Communities (1989) *Nitrate in Water.* HMSO, London. (250)

Hromadka, T. V. and McCuen, R. H. (1988) Uncertainty estimates for surface runoff models. *Adv. Water Resources*, **11**, 2–14. (298)

Hughes, C., Strangeways, I. C. and Roberts, A. M. (1993) Field evaluation of two aerodynamic raingauges. *Weather*, **48**, 66–71. (47)

Ibbitt, R. P. and O'Donnell, T. (1971) Fitting methods for conceptual catchment models. *J. Hydraul. Div. Amer. Soc. Civ. Engrs*, **97**, 1331–1342. (298)

International Norm ISO 6341 (1989) *Qualité de l'eau—détermination de l'inhibition de la mobilité de Daphina magna Straus (Clacodera, Crustacea)*. (287)

IOC (Intergovernmental Oceanographic Commission) (1985) *Manual on sea-level measurement and interpretation.* Intergovernmental Oceanographic Commission, Manuals and Guides No. 14. (161)

IPPC (1990) *Strategies for adaptation to sea level rise*. (201)

Jeffrey, D. C., Keller, G. J., Mollison, D., Richmond, D. J. E., Salter, S. H., Taylor, J. R. M. and Young, I. A. (1978) *Edinburgh Wave Power Project Fourth Year Report*, **3**. (208)

Joe, H., Smith, R. L. and Weissman, I. (1992) Bivariate threshold methods for extremes. *J. R. Statist. Soc.*, **B54**, 171–183. (80)

Johns, V. (1988) Importance sampling for bootstrap confidence intervals. *J. Amer. Statist. Assoc.*, **83**, 709–714. (116)

Johnson, N. L., Kotz, S. and Kemp, A. (1993) *Discrete Distributions*, 2nd eds. Wiley, New York. (99, 100)

Johnston, J. (1984) *Econometric Methods*, 3rd edn. McGraw-Hill, New York. (169)

Jones, R. H. (1993) *Longitudinal Data with Serial Correlation: A State-Space Approach.* Chapman and Hall, London. (323)

Jordan, C. and Dinsmore, P. (1985) Determination of biologically available phosphorus using a radiobioassay technique. *Freshwat. Biol.*, **15**, 597–603. (272)

Jury, W. A. (1982) Simulation of solute transport using a transfer function model. *Water Resources Research*, **18**, 363–368. (263)

Kalinin, G. P. (1962) On the basis of runoff distributions (in Russian). *Meteorologiya i Gidrologiya*, **6**, 20–27. (364)

Karr, A. (1986) *Statistical Inference for Point Processes.* Marcel Dekker, New York. (102)

Kemp, C. D. and Kemp, A. W. (1988) Rapid estimation for discrete distributions. *The Statistician*, **37**, 243–255. (96)

Kendall, M. G. and Stuart, A. (1966) *The Advanced Theory of Statistics.* Vol. 3. Griffin, London. (347)

Kempthorne, O. (1971) Probability, statistics and the knowledge business. In Godambe, V. P. and Sprott, D. A. (eds) *Foundations of Statistical Inference*, pp. 470–492. Holt, Rinehart and Winston, Toronto. (348, 351, 363)

Kite, G. W. (1977) *Frequency and Risk Analyses in Hydrology.* Water Resources Publications, Fort Collins. (353)

Klein, W. H. and Bloom, H. J. (1987) Specification of monthly precipitation over the United States from the surrounding 700 mb height field. *Mon. Weather Rev.,* **115**, 2118–2132. (46)

Klemeš, V. (1970) Negatively skewed distribution of runoff. In *Proceedings of Symposium of Wellington (NZ),* pp. 219–236. Publication No. 96, IAHS. (364)

Klemeš, V. (1971) Some problems in pure and applied stochastic hydrology. In *Proceedings of Symposium on Statistical Hydrology, University of Arizona,* pp. 2–15. US Department of Agriculture, Miscellaneous Publication No. 1275, 1974, Washington, DC. (346, 349, 350)

Klemeš, V. (1974) The Hurst phenomenon–a puzzle?. *Water Resources Research,* **10**, 675–688. (346, 359)

Klemeš, V. (1977) Value of information in reservoir optimization. *Water Resources Research,* **13**, 837–850. (350)

Klemeš, V. (1978) Physically based stochastic hydrologic analysis. In Chow, V. T. (ed.) *Advances in Hydroscience,* Vol. 11, pp. 285–355. Academic Press, New York. (346, 364)

Klemeš, V. (1981) Applied storage reservoir theory in evolution. In Chow, V. T. (ed.) *Advances in Hydroscience,* Vol. 12, pp. 79–141. Academic Press, New York. (349)

Klemeš, V. (1982) Empirical and causal models in hydrology. In *Scientific Basis of Water Resource Management,* pp. 95–104. National Academy of Sciences, Washington, DC. (346)

Klemeš, V. (1986) Dilettantism in hydrology: Transition or density? *Water Resources Research,* **22**, 177S–188S. (346, 348, 349, 353)

Klemeš, V. (1987a) Empirical and causal models in hydrologic reliability analysis. In Duckstein, L. and Plate, E. J. (eds) *Engineering Reliability and Risk in Water Resources,* pp. 391–403. Nijhoff, Dordrecht. (346)

Klemeš, V. (1987b) Hydrological and engineering relevance of flood frequency analysis. In: Singh, V. P. (ed.) *Hydrologic Frequency Modelling,* pp. 1–18. Reidel, Dordrecht. (346, 349, 353, 355)

Klemeš, V. (1988a) A hydrological perspective. *J. Hydrol.,* **100**, 3–28. (345, 346, 349, 358)

Klemeš, V. (1988b) Hydrology and water resources management: The burden of common roots. In *Proceedings of 6th IWRA Congress on Water Resources,* Vol. 1, pp. 368–376. IWRA, Urbana. (346)

Klemeš, V. (1989) The improbable probabilities of extreme floods and droughts. In Starosolszky, O. and Melder, O. M. (eds) *Hydrology and Disasters,* pp. 43–51. James and James, London. (346, 349)

Klemeš, V. (1991) The Science of Hydrology: Where Have We Been? Where Should We Be Going? What Do Hydrologists Need To Know?. In *Proceedings of International Symposium To Commemorate the 25 Years of IHD/IHP,* pp. 41–50. UNESCO, Paris. (346)

Klemeš, V. (1993) Probability of extreme hydrometeorological events—a different approach. In Kundzewicz, Z. W., Rosbjerg, D., Simonovic, S. P. and Takeuchi, K. (eds) *Extreme Hydrological Events: Precipitation, Floods and Droughts,* pp. 167–176. IAHS Publication No. 213. (346, 349)

Klemeš, V. and Klemeš, I. (1988) Cycles in finite samples and cumulative processes of higher orders. *Water Resources Research,* **24**, 93–104. (361, 362)

Koch, G. G., Elashoff, J. D. and Amara, I. A. (1988) Repeated measurements—design and analysis. In Johnson, N. L. and Kotz, S. (eds) *Encyclopedia of Statistical Sciences,* Vol. 8, pp. 46–73. Wiley, New York. (322, 326)

Kocherlakota, S. and Kocherlakota, K. (1986) Goodness of fit tests for discrete distributions. *Commun. Statist.: Theory Meth.*, **A15**, 815–829. (96)

Konikow, L. F. and Bredehoft, J. D. (1992) Groundwater models cannot be validated. *Adv. Water Resources*, **15**, 75–83. (298)

Krug, A. (1993) Drainage history and land use pattern of a Swedish river system—their importance for understanding nitrogen and phosphorus load. *Hydrobiologia*, **251**, 285–296. (283)

Kuczera, G. (1988) On the validity of first order prediction limits for conceptual hydrologic models. *J. Hydrol.*, **103**, 229–247. (298)

Lamb, H. H. (1982) *Climate History and the Modern World*. Methuen, London. (183)

Lawes, J. B., Gilbert, J. H. and Warington, R. (1882) On the amount and composition of drainage water collected at Rothamsted. III. The quantity of nitrogen lost by drainage. *J. R. Agric. Soc. Engl., 2nd Ser.*, **18**, 43–71. (252)

Leadbetter, M. R., Lindgren, G. and Rootzén, H. (1983) *Extremes and Related Properties of Random Sequences and Series*. Springer, New York. (66, 72)

Ledford, A. W. and Tawn, J. A. (1994) Statistics for near independence in multivariate extreme values. Submitted. (68)

Longuet-Higgins, M. S. (1957) The statistical analysis of a random, moving surface. *Phil. Trans. R. Soc. Land*, **A249**, 321–387. (207)

Longuet-Higgins, M. S. (1984) Statistical properties of wave groups in a random sea state. *Phil. Trans. R. Soc. Lond.*, **A312**, 219–250. (211)

Malmo, O. and Reitan, A. (1986) Development of the Kvaerner multiresonant OWC. In Evans, D. V. and Falcão, A. F. de O. (eds) *Hydrodynamics of Ocean Wave-Energy Utilization*, pp. 57–67. Springer, Heidelberg. (218)

Manabe, S. and Broccoli, A. J. (1990) Mountains and arid climates of middle latitudes. *Science*, **247**, 192–195. (46)

Margoum, M. and Oberlin, G. (1991) *AGREGEE: un modèle opérationnel d'estimation des crues rares et extrêmes*. Informations Techniques No. 84-4, CEMAGREF, Antony, Paris. (142)

Mason, J. (1986) Numerical weather prediction. *Proc. R. Soc. Lond.*, **A407**, 51–60. (3)

May, B. R. (1988) Progress in the development of PARAGON. *Meteorol. Mag.*, **117**, 79–86. (127)

May, B. R. and Hitch, T. J. (1989) Improved values of 1-hour M5 rainfalls in the United Kingdom. *Meteorol. Mag.*, **118**, 45–50. (127)

Mehlum, E. (1986) Tapered channel wave power plants. In Evans, D. V. and Falcão, A. F. de O. (eds) *Hydrodynamics of Ocean Wave-Energy Utilization*, pp. 51–55. Springer, Heidelberg. (218)

Melching, C. S. (1992) An improved first-order reliability approach for assessing uncertainties in hydrologic modelling. *J. Hydrol.*, **132**, 157–177. (298)

Microbics Inc. (1988) *How to run toxicity tests using the Microtox Model 500*. Microbics Instructions 55 H 501. Carlsbad, USA. (287)

Mitsuyasu, H. (1975) Observations of the directional spectrum of ocean waves using a cloverleaf buoy. *J. Phys. Oceanog.*, **5**, 750–760. (209)

Mollison, D. (1980) The prediction of device performance. In Count, B. (ed.) *Power from Sea Waves*, pp. 135–172. Academic Press, London. (208, 217, 220)

Mollison, D. (1983) Wave energy losses in intermediate depths. *Appl. Ocean Res.*, **5**, 234–237. (207, 219)

Mollison, D. (1984) The assessment of wave power by the Department of Energy. In *Second Report from the Energy Committee, 28 November 1984*, pp. xvii–xviii. House of Commons **75**, HMSO, London. (220)

Mollison, D. (1986) Wave climate and the wave power resource. In Evans, D. V. and Falcão, A. F. de O. (eds) *Hydrodynamics of Ocean Wave-Energy Utilization*, pp. 133–156. Springer, Heidelberg. (205, 218)

Mollison, D. (1991) The UK wave power resource. In *Wave Energy*, pp. 1–6. Institute of Mechanical Engineers. (213, 214, 215, 216, 217)

Mollison, D., Buneman, O. P. and Salter, S. H. (1976) Wave power availability in the NE Atlantic. *Nature*, **263**, 223–226. (207)

Mollison, D and Pontes, M. T. (1992) Assessing the Portuguese wave power resource. *Energy*, **17**, 255–268. (217)

Moore, R. J. (1993) Integrated systems for the hydrometeorological forecasting of floods. In *Proceedings of EUROPROTECH Conference on Science and Technology for the Reduction of Natural Risks*. CSIM, Udine, Italy. (125)

Moran, P. M. P. (1957) The statistical treatment of flood flows. *Trans. AGU*, **38**, 519–523. (348, 353)

Moran, P. M. P. (1959) *The Theory of Storage*. Methuen, London. (358)

Morton, T. G., Gold, A. J. and Sullivan, W. M. (1988) Influence of over-watering and fertilization on nitrogen losses from home lawns. *J. Environ. Qual.*, **17**, 124–130. (260)

Mulvaney T. J. (1850) On the use of self registering rain and flood gauges. *Trans. Inst. Civ. Engrs. Ire. Proc.* **4**(2) p, 1–8. (297)

Murphy, J. and Riley, J. R. (1962) A modified single solution method for the determination of phosphate in natural waters. *Analytic. Chim. Acta*, **27**, 31–36. (274)

Nadarajah, S. (1994a) Discussion of paper by S. G. Coles and J. A. Tawn. *Appl. Statist.*, **43**, 34–35. (80)

Nadarajah, S. (1994b) *Multivariate extreme value methods with applications to reservoir flood safety*. University of Sheffield Ph.D. thesis. (84, 86)

Nakamura, M. and Pérez-Abreu, V. (1993a) Exploratory data analysis for counts using the empirical probability generating function. *Commun. Statist.: Theory Meth.*, **A22**, 827–842. (96, 100, 102)

Nakamura, M. and Pérez-Abreu, V. (1993b) Use of an empirical probability generating function for testing a Poisson model. *Canadian J. Statist.*, **21**, 149–156. (96)

Nakamura, M. and Pérez-Abreu, V. (1993c) Empirical probability generating function. An overview. *Insurance: Math. Econ.*, **12**, 287–295. (96)

Nash, J. E. and Sutcliffe, J. V. (1970) River flow forecasting through conceptual models. 1. A discussion of principles. *J. Hydrol.*, **10**, 282–290. (301)

Nash, J. E., Eagleson, P. S., Philip, J. R. and Van Der Molen, W. H. (1990) The education of hydrologists. *Hydrol. Sci. J.*, **35**, 597–607. (346)

NERC (1975) *Flood Studies Report* (in five volumes). Natural Environment Research Council, London. (127, 141)

Nielsen, D. R., Biggar, J. W. and Erh. K. T. (1973) Spatial variability of field-measured soil-water properties. *Hilgardia*, **42**, 215–260. (262)

Nirel, R. (1993) *Estimation of the effect of operational seeding on rain amounts in Israel*. Ph.D. dissertation, The Hebrew University of Jerusalem, Department of Statistics. (109, 114)

NOVA (1983) *The Pleasure of Finding Things Out*. WGBH Transcripts, Boston. (357)

O'Connell, P. E. (1991) A historical perspective. In Bowles, D. S. and O'Connell, P. E. (eds) *Advances in the Modelling of Hydrologic Systems*, pp. 3–30. Kluwer, Dordrecht. (298)

O'Hagan, A. (1993) Robust modelling for asset management. *J. Statist. Planning and Inference*, in press. (247)

O'Hagan, A., Glennie, E. B. and Beardsall, R. E. (1992) subjective modelling and Bayes linear estimation in the UK water industry. *Appl. Statist.* **41**, 563–577. (236)

O'Hagan, A. and Wells, F. S. (1993) Use of prior information to estimate costs in a sewerage operation. In Gatsonis, C., Hodges, J. S., Kass, R. E., and Singpurwalla, N. D. (eds) *Case Studies in Bayesian Statistics*, pp. 118–152. Springer, New York, 118–152. (236, 237, 238, 246)

Parkin, T. B., Meisinger, J. J., Chester, S. T., Starr, J. L. and Robinson, J. A. (1988) Evaluation of statistical estimation methods for lognormally distributed variables. *Soil Sci. Soc. Amer. J.*, **52**, 323–329. (254, 255, 256)

Parr, M. P. and Smith, R. V. (1976) The identification of phosphorus as a growth limiting nutrient in Lough Neagh using bioassays. *Water Research*, **10**, 1151–1154. (271)

Peltier, W. R. and Tushingham, A. M. (1989) Global sea-level rise and the greenhouse effect: might they be connected? *Science*, **244**, 806–810. (162, 163, 173)

Phelan, M. and Goodall, C. (1990) An assessment of a generalised Waymire–Gupta–Rodriguez–Iturbe model for GARP Atlantic Tropical Experiment rainfall. *J. Geophys. Res.*, **95**, 7603–7615. (9)

Philip, J. R. (1975) Some remarks on science and catchment prediction. In *Prediction in Catchment Hydrology*, pp. 23–30, Australian Academy of Sciences, Canberra. (346)

Philip, J. R. (1991) Soil, natural science and models. *Soil Sci.*, **151**, 91–98. (346)

Pickands, J. (1975) Statistical inference using extreme order statistics, *Ann. Statist*, **3**, 119–131. (65)

Pierson, W. and Moskowitz, L. (1964) A proposed spectral form for fully developed wind seas based on the similarity theory of S. A. Kitaigorodskii. *J. Geophys. Res.*, **69**, 5181–5190. (217)

Pirazzoli, P. A. (1987) Recent sea-level changes and related engineering problems in the Lagoon of Venice (Italy). *Prog. Oceanog.*, **18**, 323–346. (163)

Pontes, M. T., Mollison, D., Cavalieri, L., Carlos, J. and Athanassoulis, G. A. (1993) *Wave studies and resource evaluation methodology*. Report to EC, DG XII. (213)

Pontes, M. T. and Pires, H. O. (1992) Assessment of shoreline wave energy resource. In *Proceedings of 2nd International Offshore and Polar Engineering Conference (ISOPE 92)*, San Francisco, Vol. III, pp. 114–118. (219)

Popper, K. R. (1959) *The Logic of Scientific Discovery*. Hutchinson, London. (250)

Powell, M. (1970) A survey of numerical methods for unconstrained optimisation. *SIAM Rev.*, **12**, 79–97. (301)

Prandle, D. and Wolf, J. (1978) The interaction of surge and tide in the North Sea and River Thames. *Geophys. J. R. Astron. Soc.*, **55**, 203–216. (222)

Pugh, D. T. (1987) *Tides, Surges and Mean Sea-Level: A Handbook for Engineers and Scientists*. Wiley, Chichester. (147, 160, 163, 222)

Pugh, D. T. (1990) Is there a sea-level problem? *Proc. Instn Civ. Engrs, Part 1*, **88**, 347–366. (147, 162)

Pugh, D. T., Spencer, N. E. and Woodworth, P. L. (1987) *Data Holdings of the Permanent Service for Mean Sea-Level*. Permanent Service for Mean Sea-Level, Bidston. (160)

Pugh, D. T. and Vassie, J. M. (1980) Applications of the joint probability method for extreme sea level computations. *Proc. Instn. Civ. Engrs, Part 2*, **69**, 959–975. (221)

Quinn, P. F. and Beven, K. J. (1993) Spatial and temporal predictions of soil moisture dynamics, runoff variable source areas and evapotranspiration for Plynlimon, Mid-Wales. *Hydrol. Process.*, **7**, 425–448. (300)

Rao, C. R. (1973) *Linear Statistical Inference and its Applications*, 2nd edn. Wiley, Chichester. (86)

Rawlings, J. O. (1989) *Applied Regression Analysis: A Research Tool*. Wadsworth and Brooks/Cole. (85, 89)

Reed, D. W. and Stewart, E. J. (1989) Focus on rainfall growth estimation. In *Proceedings of 2nd National Hydrology Symposium, Sheffield, September 1989*, pp. 3.57–3.65. British Hydrological Society. (128)

Reed, D. W. and Stewart, E. J. (1991) Discussion on 'Dam safety: an evaluation of some procedures for design flood estimation'. *Hydrol. Sci. J.*, **36**, 499–502. (132)

Resnick, S. I. (1987) *Extreme Values, Point Processes and Regular Variation*. Springer, New York. (67)

Richards, J. D., Windle, S. A., Blackman, D. L., Flather, R. A. and Woodworth, P. L. (1993) *Analysis of high waters and tides at Lawyers Sluice in the Wash, eastern England*. Proudman Oceanographic Laboraroty Report, No. 31. (165)

Richards, L. A., Neal, O. R. and Russell, M. B. (1939) Observations on moisture conditions in lysimeters. *Soil Sci. Soc. Amer. Proc.*, **4**, 55–59. (259)

Robinson, P. J. (1994) Precipitation regime changes over small watersheds. In Barnett, V. and Turkman, K. F. (eds) *Statistics for the Environment 2: Water Related Issues*. Wiley Chichester, pp. 43–59. (19, 20, 22, 30, 32, 37, 38, 40, 41)

Robinson, P. J. and Finkelstein, P. L. (1991) The development of impact-oriented climate scenarios. *Bull. Amer. Meteorol. Soc.*, **72**, 481–490. (46)

Robinson, P. J. and Henderson, K. G. (1992) Precipitation events in the South-east United States of America. *Int. J. Climatol.*, **12**, 701–720. (44, 46, 48)

Robinson, P. J., Samel, A. N. and Madden, G. (1993) Comparisons of modelled and observed climate for impact assessments. *Theor. Appl. Climatol.*, **48**, 75–87. (46)

Rodriguez-Iturbe, I., Cox, D. R. and Eagleson, P. (1986) Spatial modelling of total storm rainfall. *Proc. R. Soc. Lond.*, **A403**, 27–50. (7, 9)

Rodriguez-Iturbe, I., Cox, D. R. and Isham, V. (1987) Some models for rainfall based on stochastic point processes. *Proc. R. Soc. Lond.*, **A410**, 269–288. (5, 10, 13, 15)

Rodriguez-Iturbe, I., Cox, D. R. and Isham, V. (1988) A point process model for rainfall: further developments. *Proc. R. Soc. Lond.*, **A417**, 283–298. (5, 6, 7, 10, 13, 17)

Rodriguez-Iturbe, I. and Eagleson, P. (1987) Mathematical models of rainstorm events in space and time. *Water Resources Research*, **23**, 181–190. (9)

Rodriguez-Iturbe, I., Entekhabi, D., Lee, J-S. and Bras, R. L. (1991) Non-linear dynamics of soil moisture at climate scales. 2. Chaotic analysis. *Water Resources Research*, **27**, 1907–1915. (269)

Russell, A. E. and Ewel, J. J. (1985) Leaching from a tropical Andept during big storms: a comparison of three methods. *Soil Sci.*, **139**, 181–189. (253)

Salter, S. H. (1989) World progress in wave energy 1988. *Int. J. Ambient Energy*, **10**, 3–24. (206)

SAS (1987) *SAS/STAT Guide for Personal Computers*, Version 6 edition. SAS Institute Inc., Cary, NC. (327)

SAS (1988) *SAS/ETS User's Guide*, Version 6, first edition. SAS Institute Inc., Cary, NC. (327)

Sas, H. (1989) *Lake Restoration by Reduction of Nutient Loading*. Academia, St. Augustin. (283)

Shaffer, K. A., Fritton, D. D. and Baker, D. E. (1979) Drainage water sampling in a wet dual-pore system. *J. Environ. Qual.*, **8**, 241–246. (253)

Shenker, N. (1985) Qualms about bootstrap confidence intervals. *J. Amer. Statist. Assoc.*, **80**, 360–361. (111)

Shennan, I. (1989) Holocene crustal movements and sea-level changes in Great Britain. *J. Quater. Sci.*, **4**, 77–89. (163, 166, 172, 178)

Shennan, I. and Woodworth, P. L. (1992) A comparison of late Holocene and twentieth century sea-level trends for the UK and North sea region. *Geophys. J. Int.*, **109**, 96–105. (166, 173)

Sichel, H. S. (1952) New methods in the statistical evaluation of mine sampling. *Trans. Inst. Min. Metall. (Lond.)*, **B61**, 261–288. (249, 255)

Slack, J. R., Wallis, J. R. and Matalas, N. C. (1975) On the value of information to flood frequency analysis. *Water Resources Research*, **11**, 629–647. (350)

Smith, A. F. M. (1975) A Bayesian approach to inference about a change-point in a sequence of random variables. *Biometrika*, **62**, 407–416. (176)

Smith, R. L. (1989) Extreme value analysis environmental time series: an application to trend detection in ground level ozone. *Statist. Sci.*, **4**, 367–393. (180, 222, 223)

Smith, R. L. (1992) The extremal index for a Markov chain. *J. Appl. Prob.*, **29**, 37–45. (72)

Smith, R. L. (1994) Spatial modelling of rainfall data. In Barnett, V. and Turkman, K. F. (eds) *Statistics for the Environment 2: Water Related Issues*, pp. 19–41. Wiley, Chichester. (43, 58)

Smith, R. L. (1994a) Multivariate threshold methods. Submitted to the *Proceedings of the NIST/Temple University Conference on Extreme Value Theory and its Applications*. (64)

Smith, R. L. (1994b) Regional estimation from spatially dependent data. Submitted. (75, 77)

Smith, R. L., Tawn, J. A. and Coles, S. G. (1994) Markov chain models for threshold exceedances. Submitted. (61, 64, 68)

Smith, R. V. (1977) Domestic and agricultural contributions to the inputs of phosphorus and nitrogen to Lough Neagh. *Water Research*, **11**, 453–459. (271, 273)

Smith, R. V. (1979) Sources of phosphorus in the Lough Neagh system and their reduction. *Prog. Water Technol.*, **11**, 209–217. (271)

Smith, R. V. (1986) The effects of reducing the urban load of phosphorus on Lough Neagh algal populations. In Solbe, J. F. and L. G. (eds) *Effects of Land Use on Freshwaters*, pp. 509–513. Ellis Harwood, Chichester. (272)

Smith, R. V. and Stewart, D. A. (1977) Statistical models of river loadings of nitrogen and phosphorus in the Lough Neagh system. *Water Research*, **11**, 631–636. (274)

Smith, R. V. and Stewart, D. A. (1989) A regression model for nitrate leaching in Northern Ireland. *Soil Use and Mgt*, **5**, 71–76. (274)

Sorooshian, S. and Dracup, J. A. (1980) Stochastic parameter estimation procedures for hydrologic rainfall-runoff models: correlated and heteroscedastic error cases. *Water Resources Research*, **16**, 430–442. (303)

Sorooshian, S. and Gupta, V. K. (1983) Automatic calibration of conceptual rainfall-runoff models: the question of parameter observability and uniqueness. *Water Resources Research*, **19**, 260–268. (303)

Sorooshian, S., Gupta, V. K. and Fulton, J. L. (1983) Evaluation of maximum likelihood parameter estimation techniques for conceptual rainfall-runoff models: influence of calibration data variability and length on model credibility. *Water Resources Research*, **19**, 251–259. (303)

Southgate, H. N. (1987) *Wave prediction in deep water and at the coastline.* Report SR 114, HR Wallingford, UK. (219)

Stein, A., Hoogerwerf, M. and Bouma, J. (1988) Use of soil-map delineations to improve (co-) kriging of point data on moisture deficits. *Geoderma,* **43**, 163–177. (267)

Stern, R. D. and Coe, R. (1984) A model fitting analysis of daily rainfall data (with discussion). *J. R. Statist. Soc.,* **A147**, 1–34. (3, 19, 20, 24, 26, 44)

Stewart, E. J. (1989) Areal reduction factors for design storm construction: joint use of raingauge and radar data. In *New Directions for Surface Water Modeling (Proceedings of Baltimore Symposium, May 1989),* pp. 31–40. IAHS Publication No. 181. (126)

Stewart, E. J. and Reynard, N. S. (1993) *Rainfall frequency estimation in England and Wales: survey.* Institute of Hydrology Report to National Rivers Authority (R & D Note 175), Wallingford. (126, 141)

Storey, W. C. (1990) Operational aspects of phosphorus removal from sewage by chemical treatment. In *Institute of Water and Environmental Management. Seminar on Eutrophication, Dundalk, Ireland.* (274)

Sumner, G. N. (1983) The use of correlation linkages in the assessment of daily rainfall patterns. *J. Hydrol.,* **66**, 169–182. (54)

Tawn, J. A. (1988) Bivariate extreme value theory—models and estimation. *Biometrika,* **75**, 397–415. (33)

Tawn, J. A. (1988) Bivariate extreme value theory: models and estimation. *Biometrika,* **75**, 245–253. (68)

Tawn, J. A. (1988) *Extreme value theory with oceanographic applications.* Ph.D. Thesis, Department of Mathematics, University of Surrey. (223, 224)

Tawn, J. A. (1992) Estimating probabilities of extreme sea levels. *Appl. Statist.,* **41**, 77–93. (221, 222, 223)

Tawn, J. A. (1993) Applications of multivariate extremes. Submitted to Proceedings of Conference on Extreme Value Theory and its Applications, Gaithersburg, MD, 1993. (79)

Tawn, J. A. (1993) Extreme sea-levels. In Barnett, V. and Turkman, K. F. (eds) *Statistics for the Environment,* pp. 243–263. Wiley, Chichester. (180, 220)

Tawn, J. A. (1994) Applications of multivariate extremes. In Galambos, J., Lechner, J. and Simm, E. (eds) *Extreme Value Theory,* pp. 249–268. Kluwer, Dordrecht. (222, 224)

Tawn, J. A. and Mitchell, W. M. (1993) A spatial analysis of Australian extreme sea-levels. Submitted. (180)

Tawn, J. A. and Vassie, J. M. (1989) Extreme sea levels: the joint probabilities method revisited and revised. *Proc. Instn Civ. Engrs, Part 2,* **87**, 429–442. (221)

Taylor, G. I. (1938) Statistical theory of turbulence. *Proc. R. Soc. Lond.,* **A164**, 476–490. (11)

Taylor, J. R. M. (1984) *Bending moments in long spines.* Edinburgh Wave Power Project Report No. 102. (210)

Thompson, K. R. (1980) An analysis of British monthly mean sea-levels. *Geophys. J. R. Astron. Soc.,* **63**, 57–73. (165)

Thompson, K. R. (1981) Monthly changes of sea-levels and the circulation of the North Atlantic. *Ocean Modelling,* **41**, 6–9. (162)

Thompson, K. R. (1986) North Atlantic sea-level and circulation. *Geophys. J. R. Astron. Soc.,* **87**, 15–32. (162)

Thorpe, T. W. (1993) *A review of wave energy.* UK Department of Trade and Industry, ETSU R 72. (216, 220)

Trupin, A. and Wahr, J. (1990) Spectroscopic analysis of global tide gauge sea level data. *Geophys. J. Int.*, **100**, 441–453. (163)

Tucker, M. J., Challenor, P. G. and Carter, D. J. T. (1984) Numerical simulation of a random sea: a common error and its effect upon wave group statistics. *Appl. Ocean Res.*, **6**, 118–122. (209)

Tukey, J. W., Brillinger, D. R. and Jones, L. V. (1978) *The Role of Statistics in Weather Resources Management. The Management of Weather Resources*, Vol 2. Department of Commerce, Washington DC. (110)

Tunney, H. (1990) A note on a balance sheet approach to estimating the phosphorus fertiliser needs of agriculture. *Ir. J. Agric. Res.*, **29**, 149–154. (283)

Urfer, W. (1993) Statistical analysis of climatological and ecological factors in forestry; spatial and temporal aspects. In Barnett, V. and Turkman, K. F. (eds) *Statistics for the Environment*, pp. 335–345. Wiley, Chichester. (323)

Urfer, W. and Pohlmann, H. (1992) Statistical analysis of spatially and serially correlated data. In Franke, J. and Roeder, A., (eds) *Mathematical Modelling of Forest Ecosystems*, pp. 160–167. J. D. Sauerländer's Verlag, Frankfurt/Main. (326)

Van Dantzig, D. (1956) Economic decision problems for flood prevention. *Econometrica*, **24**, 276–287. (195)

van der Ploeg, R.R. and Beese, F. (1977) Model calculations for the extraction of soil water by ceramic cups and plates. *Soil Sci. Soc. Amer. J.*, **41**, 466–470. (253)

van de Pol, R. M., Wierenga, P. J. and Nielsen, D. R. (1977) Solute movement in a field soil. *Soil Sci. Soc. Amer. J.*, **41**, 10–13. (263)

van Genuchten, M. Th. and Wierenga, P. J. (1976) Mass transfer studies in porous sorbing media. I. Analytical solutions. *Soil Sci. Soc. Amer. J.*, **40**, 473–480. (262)

van Straten, G. and Keesman, K. J. (1991) Uncertainty propagation and speculation in projective forecasts of environmental change: a lake eutrophication example. *J. Forecasting*, **10**, 163–190. (299, 316)

Vasseur, P., Ferard, J. F., Vial, J. and Larbaight, G. (1984) Comparaison des tests Microtox et Daphnie pour l'evaluation de la toxicité aigue d'effluents industriels. *Environ. Pollut.*, **A34**, 225–235. (286)

Veen, J. van (1948) *Dredge, Drain Reclaim: The Art of a Nation*, Trio, The Hague. (185)

Vonnegut, K., Jr (1969) *Slaughterhouse-Five*. Dell Publishing, New York (1990). (365)

Vries, M. de (1975) *A Morphological Time Scale for Rivers*. IAHR, Sao Paulo. (198)

Vrijling, J. K. (1992) What is acceptable risk? *Ann. Ponts Chaussees*, No. 64. (202)

Vrijling, J. K. (1994) Sea-level rise: a potential threat? In Barnett, V. and Turkman, K. F. (eds) *Statistics for the Environment 2: Water Related Issues*, pp. 183–204. Wiley, Chichester. (147)

Vrijling, J. K. and Beurden, J. van (1990) *Sea level rise; a probabilistic design problem*. ICCE, Delft. (201)

Wagenet, R. J. (1983) Principles of salt movement in soil. In Nelson, D. W. *et al.* (eds) *Chemical Mobility and Reactivity in Soil Systems*, pp. 123–140. Special Publication No. 11, Soil Science Society of America, Madison, WI. (261)

Wagenet, R. J. and Addiscott, T. M. (1987) Estimating the variability of unsaturated hydraulic conductivity using simple equations. *Soil Sci. Soc. Amer. J.*, **51**, 42–47. (262, 264)

Wagenet, R. J. and Rao, B. K. (1983) Description of nitrogen movement in the presence of spatially-variable soil hydraulic properties. *Agric. Water Manage.*, **6**, 227–242. (262, 264)

Wagner, G. H. (1962) Use of porous ceramic cups to sample soil water within the profile. *Soil Sci.*, **94**, 379–386. (252)

Walden, A. T. Prescott, P. and Webber, N. B. (1982) The examination of surge–tide interaction at two ports on the central south coast of England. *Coastal Engng.*, **6**, 59–70. (222)

Wallace, J. M. and Gutzler, D. S. (1981) Teleconnections in the 500 mb geopotential height field during the Northern hemisphere winter. *Mon. Weather Rev.*, **109**, 784–812. (51)

Wallis, J. R. (1980) Risk and uncertainties in the evaluation of flood events for the design of hydraulic structures. In: Guggino, E., Rossi, G. and Todini, E. (eds) *Piene e Siccità*, pp. 3–36. Fondazione Politecnica del Mediterraneo, Catania. (129)

WAMDI Group (1988) The WAM model—a third generation ocean wave prediction model. *J. Phys. Oceanog.*, **18**, 1775–1810.

Waymire, E., Gupta, V. and Rodriguez-Iturbe, I. (1984) A spectral theory of rainfall intensity at the meso-β scale. *Water Resources Research*, **20**, 1453–1465. (9, 11)

Webster, R. and Addiscott, T. M. (1990) Spatial averaging of water and solute flows in soil. In Roth, K., Flühler, H., Jury, W. A. and Parker, J. C. (eds) *Field-Scale Water and Solute Flux in Soils*, pp. 165–173. Birkhäuser, Basel. (260, 262)

Webster, R. and Burgess, T. M. (1980) Optimal interpolation and isarithmic mapping of soil properties. III. Changing drift and universal kriging. *J. Soil Sci.*, **31**, 505–524. (260)

Webster, R. and Oliver, M. A. (1989) Optimal interpolation and isarithmic maping of soil properties. VI. Disjunctive kriging and mapping the conditional probability. *J. Soil Sci.*, **40**, 497–512. (267)

Webster, R. and Oliver, M. A. (1990) *Statistical Methods for Soil and Land Resource Survey*. Clarendon Press, Oxford. (264)

Webster, C. P., Shepherd, M. A., Goulding, K. W. T. and Lord, E. I. (1993) Comparisons of methods for measuring the leaching of mineral nitrogen from arable land. *J. Soil Sci.*, **44**, 49–62. (251, 252, 253, 257, 260)

Wei, W. W. S. (1990) *Time Series Analysis. Univariate and Multivariate Methods*. Addison-Wesley, Redwood City. (334)

Wemelsfeler, P. J. (1949) Wetmatigheden in het optreden van stormloeden. *De Ingenieur (Den Haag)*, nr 9. (194)

White, R. E. (1989) Prediction of nitrate leaching from a structured clay soil using transfer functions derived from externally applied or indigenous solute fluxes. *J. Hydrol.*, **107**, 31–42. (263)

White, R. E., Haigh, R. A. and Macduff, J. H. (1987) Frequency distributions and spatially-dependent variability of ammonium and nitrate concentrations in soil under grazed and ungrazed grassland. *Fert. Res.*, **11**, 193–208. (257, 260)

Whitmore, A. P., Addiscott, T. M., Webster, R. and Thomas, V. H. (1983) *Spatial variation in leaching of solutes*. Report of Rothamsted Experimental Station for 1982, Part 1, p. 274. (260)

Whittaker, T. J. T. *et al.* (1992) *The UK's shoreline and nearshore wave energy resource*. UK Department of Trade and Industry, ETSU WV 1683. (219)

Wigley, T. M. L. and Raper, S. C. B. (1992) Implications for climate and sea-level of revised IPCC scerarios. *Nature*, **357**, 293–300. (147, 162, 174)

Wilks, D. S. (1989) Conditioning stochastic daily precipitation models on total monthly precipitation. *Water Resources Research*, **25**, 1429–1439. (21)

Williams, G. R. (1961) Cyclical variations in world-wide hydrologic data. *J. Hydraul.Engrg*, **87** (HY6), 71–88. (360)

Wiltshire, S. E. (1986) Regional flood frequency analysis, II: Multivariate classification of drainage basins in Britain. *Hydrol. Sci. J.*, **31**, 335–346. (128)

Wiltshire, S. and Beran, M. (1987) A significance test for homogeneity of flood frequency regions. In: Singh, V. P. (ed.) *Regional Flood Frequency Analysis*, pp. 147–158. Reidel, Dordrecht. (135)

Wolf, J. (1978) Interaction of tide and surge in semi-infinite uniform channel, with an application to surge propagation down the east coast of Britain. *Appl. Math. Model.*, **2**, 245–253. (222, 230)

Woodworth, P. L. (1987) Trends in UK mean sea-level. *Mar. Geol.*, **11**, 57–87. (160, 165, 167, 173, 175, 176)

Woodworth, P. L. (1990) A search for accelerations in records of European mean sea-level. *Int. J. Clim.*, **10**, 129–143. (147, 160, 164, 167)

Woodworth, P. L. and Jarvis, J. (1990) *A feasibility study of the use of short historical and short modern tide gauge records to investigate long term sea-level changes in the British Isles.* Proudman Oceanographic Laboratory, Internal Document No. 23. (167)

Woodworth, P. L., Spencer, N. E. and Alcock, G. A. (1990) On the availability of the European mean sea-level data. *Int. Hydrol. Rev.*, **67**, 131–146. (160)

Woolhiser, D. A. (1992) Modeling daily precipitation—progress and problems. Chapter 5 of *Statistics in the Environmental and Earth Sciences*, eds. A. Walden and P. Guttorp, Halsted Press, New York. (20, 21, 22)

Woolhiser, D. A. and Roldán, J. (1982) Stochastic daily precipitation models 2. A comparison of distribution of amounts. *Water Resources Research*, **18**, 1461–1468. (21, 29)

Yevjevich, V. (1963) Fluctuations of wet and dry years, Part I. In *Hydrology Papers*, No. 1. Colorado State University, Fort Collins. (364)

Yevjevich, V. (1968) Misconceptions in hydrology and their consequences. *Water Resources Research*, **4**, 225–232. (352)

Zawadski, I. (1973) Statistical properties of precipitation patterns. *J. Appl. Meteor.*, **12**, 459–472. (11)

Zeger, S. L. and Diggle, P. J. (1993) Semi-parametric models for longitudinal data with application to CD4 cell numbers in HIV seroconverters. *Biometrics*, to appear. (172)

Zeger, S. L. and Liang, K.-Y. (1992) An overview of methods for the analysis of longitudinal data. *Statistics in Medicine*, **11**, 1825–1839. (323)

Zellner, A. (1962) Seemingly unrelated regression. *J. Amer. Statist. Assoc.*, **57**, 348–368. (169)

Zellner, A. (1963) Seemingly unrelated regression. *J. Amer. Statist. Assoc.*, **58**, 977–992. (169)

Zellner, A. (1972) Seemingly unrelated regression. *J. Amer. Statist. Assoc.*, **67**, 255. (169)

Index

Index compiled by Geoffrey C. Jones